In what way is the square root of -1 useful for electric circuits? How can the Mont Blanc cable car help you understand constrained maximisation problems? Do prices of second-hand cars provide insights concerning partial derivatives? Is it possible to help a being on a distant planet distinguish right from left? This book explains. It also aims to make you feel very much at home with the mathematics needed for applications to actual physical problems. Its purpose is to enable you really to understand what you are doing, rather than blindly applying memorised techniques; and to make you think about whether your answers are sensible.

The author has been teaching this subject to first year undergraduates for many years. This book is based on that experience. It incorporates those explanations and examples of practical applications that have been found to be the most successful in enabling students to achieve a real appreciation of the subject.

ALL YOU WANTED TO KNOW ABOUT MATHEMATICS BUT WERE AFRAID TO ASK

ALL YOU WANTED TO KNOW ABOUT MATHEMATICS BUT WERE AFRAID TO ASK

Mathematics for science students
Volume 1

LOUIS LYONS
Jesus College, Oxford

Published by the Press Syndicate of the University of Cambridge
The Pitt Building, Trumpington Street, Cambridge CB2 1RP
40 West 20th Street, New York, NY 10011-4211, USA
10 Stamford Road, Oakleigh, Melbourne 3166, Australia

First published 1995

Printed in Great Britain at the University Press, Cambridge

A catalogue record of this book is available from the British Library

Library of Congress cataloguing in publication data

Lyons, Louis.
All you wanted to know about mathematics but were afraid to ask /
Louis Lyons.
p. cm.
Includes index.
Contents: v. 1. Mathematics applied to science
ISBN 0 521 43465 3 (v. 1). – ISBN 0 521 43600 1 (pbk. : v. 1)
1. Mathematics. I. Title.
QA39.2.L97 1995 94-28451 CIP
510–dc20

ISBN 0 521 43465 3 hardback
ISBN 0 521 43600 1 paperback

TAG

Contents

Chapter titles for Volume 2

סבותי אני ולבי לדעת ולתור ובקש חכמה וחשבון

קהלת ז׳ כ״ה

I have applied my heart to know and to search and to seek out wisdom and mathematics

Ecc. vii 25

Preface

If you understand something in only one way, then you do not really understand it at all.

Marvin Minsky, *Society of the Mind* (1987)

A few years ago, I was one of the examiners for the first year mathematics examination for Oxford undergraduates studying physics or engineering. I wanted to set a question which started as follows: 'A friend of yours has just been to his first lecture on the vector operators grad, div and curl. What would you tell him to help him appreciate their significance?' One of the other examiners, a mathematician, commented that this was a silly question since the students merely had to write down the standard definitions, as these determined all their properties.

This exchange of views convinced me that mathematicians have a very different approach to mathematics from that of scientists. It also persuaded me that, in order to enable science students to feel fully at home with the mathematics they need, at first for their courses and later for applying it effectively to scientific problems, it was necessary to supplement the conventional mathematical presentations with one that stressed explanations which are more meaningful to scientists.

This book aims to do just that. It is based on tutorials that I have been giving in Oxford for several years to physics students as part of their first year mathematics course. For their weekly tutorials, students come either singly or in pairs to see their tutor. Among other things, we discuss how well the students are coping with their work, what specific difficulties they may have encountered with lectures or with text-books they are reading, and the problems they are trying to solve. This hands-on experience has made me aware of what specific explanations and descriptions of applications are most useful in helping students overcome these difficulties, and become confident and efficient in applying the concepts. It is these ideas that I have tried to convey.

A consequence of this is that I do not set out to reproduce the proof of every formula that a student should know. The emphasis is more on conveying the reasons why a particular method is useful; what the results mean; how to check whether they are reasonable; and so on. Thus the ratio of words to equations is much higher than for conventional mathematics texts. Just as it sometimes requires more than one explanation to convince someone of a particular point, the book will from time to time repeat itself in order to provide a more comprehensive understanding. Thus the book will sometimes more or less repeat itself so as to give clearer insights.

Another specific feature of this book is that, as an experimental physicist, I often mention scientific applications of the relevant mathematical techniques. For example, the chapter on complex numbers contains a longish discussion of the phenomenon of resonance, which occurs in a very wide variety of physical situations. I make no apology for these digressions.

Although I hope that this book is readable and enjoyable, it is not a novel. I thus do not recommend anyone to start at the beginning, and to try to read steadily through to the end. Instead the relevant section should be used as a guide for that aspect of mathematics currently being studied. Furthermore, the arrangement of the material, while not completely random, is not such that you have to read and master all previous chapters before attempting the present one. Where some prior understanding is desirable, cross-references are provided. There are important and beautiful relationships among different branches of mathematics, with examples from one area providing additional insight onto another. This results in some topics being dealt with in more than one chapter. For example, the axes of an ellipsoid are mentioned in connection with simultaneous equations, saddle points, Lagrange multipliers and eigenvalues. By studying the same subject from more than one viewpoint, your understanding will be significantly strengthened, but this may require reading some sections out of order. In such cases you will clearly benefit from returning to the earlier section after the later material has been absorbed.

This book aims to cover material that is contained in a first year course of mathematics for students of the physical sciences. However, because I believe I can explain some areas of the subject better than others, the coverage of the various topics is not uniform. This, together with the omission of the standard proofs mentioned above, means that from time to time the reader should consult other texts in order to fill in the more mundane details of the course.

An additional feature of a face-to-face tutorial is that I can ask students to think about some particular problem for a couple of minutes before we continue. I am a great believer in the fact that, if you discover something for yourself, you will remember it far better than if you had simply been told the answer by someone else. At various places throughout the book, the reader is asked to try to work out some point. You are strongly recommended to do so before reading on.

There is a story about a famous theoretical physicist who wanted to learn to swim. After reading a book about the subject, he jumped into a pool and actually succeeded in swimming. This story is noteworthy because it is quite exceptional to be able to acquire a skill such as swimming, without considerable practice. Solving mathematical problems similarly requires much practical effort. It is thus essential not only to listen to lectures and to read books, but also to attempt to solve problems. It is otherwise all too easy to delude yourself into thinking that you understand what you have read and/or heard, only to be rudely awakened by the first problem you come across in an examination. The problems at the end of each chapter form an integral part of the book; it is essential to work through them. They are few in number, in order to encourage you to attempt most of them, and are designed to provide a deeper appreciation of the topics discussed in the relevant chapter.

By working your way through this book, thinking about the explanations provided, inserting your own comments, and solving the problems, you should make yourself feel very much at home with the subject matter. It is like discovering the streets of a city by wandering through them every day for several weeks, rather than trying to learn by heart their names and locations from a map. My aim is to help you understand each topic thoroughly, rather than just remember how to go through the mechanics of a proof or calculation. At best the latter may result in your storing away the information in that part of your brain which retains data required for examinations, and erases itself soon afterwards. More realistically, if you come across a problem which differs even slightly from the standard one you have seen, it is more than likely that you will lack the confidence and expertise to discover how to proceed.

Thus the message is that you should:

(i) attend lectures, reread the lecture notes, and correct mistakes in them;

(ii) read the relevant parts of this book, and others, and correlate the material here with your lecture notes;

(iii) solve problems from lecture hand-outs, books and copies of past examinations; and

(iv) discuss the subject with your friends and colleagues. (If you can convincingly explain a topic to them, there is a good chance you have really understood it yourself.)

Finally when you find a better explanation of your own than appears here, I would be only too happy to hear of it; it may even appear in the next edition!

The explanations in this book evolved through the tutorials I have given. I am first and foremost grateful to my students for their questions, and for signs of enlightenment after we had discussed the various topics. I also wish to thank numerous colleagues in the Nuclear Physics Laboratory and in Jesus College who patiently elucidated to me some of the more subtle points of the subject. I am particularly indebted to David Acheson, Ian Aitchison, Peter Clifford, Moshe Kugler and John Roe.

The first draft of this book was written while I was on sabbatical at the Nuclear Physics Department of the Weizmann Institute in Israel. The friendly atmosphere and congenial surroundings played a large part in giving me the impetus for writing. I am grateful both to my host, Prof. Yehuda Eisenberg, for inviting me, and to the Royal Society for a Visiting Professorship which made my stay possible.

Finally I am extremely grateful to Sue Geddes, who performed remarkably in producing a beautifully typed version of the text; and to Paula Collins and Irmgard Smith for so elegantly interpreting my version of the diagrams.

Oxford Louis Lyons
1994

1

Simultaneous equations

There is no such thing as simultaneity.
(Albert Einstein, 1905)

In this chapter we are going to consider the very familiar topic of the solution of a set of n linear equations for n unknowns. The mathematical manipulations involved will be no more complicated than multiplication, division and simple properties of determinants. However, interesting results will emerge.

A further motivation for this topic is that simultancous cquations arise naturally in the solution of problems in many branches of physics and mathematics. These include: the intersection of geometrical shapes; a whole variety of coupled motions (e.g. two pendula on a common support, nearby electrical circuits interacting via their mutual inductance, etc.); the solution of simultaneous differential equations; finding stationary values of a function in two or more dimensions, with or without constraints; determining the axes of an ellipsoid; eigenvalue problems; etc., etc. In simple cases, the simultaneous equations will be linear in the variables, and it is on this type that we concentrate for the remainder of this chapter.

1.1 Two equations for two unknowns

This is the simplest non-trivial case. We can write the equations in their most general form as

$$\left.\begin{aligned} ax + by &= c \\ dx + ey &= f \end{aligned}\right\} \qquad (1.1)$$

The traditional method of using these equations to eliminate one of the variables yields

$$\left.\begin{aligned} x &= \frac{ce - bf}{ae - bd} \\ y &= \frac{cd - af}{bd - ae} \end{aligned}\right\} \tag{1.2}$$

For later developments in this chapter, it is useful to write this solution in determinant form:

$$\left.\begin{aligned} x &= \frac{\begin{vmatrix} c & b \\ f & e \end{vmatrix}}{\begin{vmatrix} a & b \\ d & e \end{vmatrix}} \\ y &= \frac{\begin{vmatrix} a & c \\ d & f \end{vmatrix}}{\begin{vmatrix} a & b \\ d & e \end{vmatrix}} \end{aligned}\right\} \tag{1.3}$$

where the y equation is related to that of (1.2) by having overall minus signs in both the numerator and denominator.

The form of the solution (1.3) is very easy to remember, since it consists of the ratios of two determinants. The ones in the denominator are made up of the coefficients of x and y in eqns (1.1), in the same format as they appear there, i.e. $\begin{vmatrix} a & b \\ d & e \end{vmatrix}$. The numerator for x consists of the denominator, but with the column of the x coefficients replaced by the constants from the right hand sides of eqns (1.1), i.e. $\begin{vmatrix} c & b \\ f & e \end{vmatrix}$; and correspondingly for y.

There exist other recipes for writing down the determinants, which differ from the determinants we have described by at most an overall sign. The advantage of the prescription described here is that, in the solutions (1.3) for the variables, the signs in front of the determinant ratios are always positive. The other methods require some extra minus signs.

We can check that our solution remains correct if the two equations in (1.1) are written in the opposite order (i.e. the second one first); or with the y terms preceding the x ones in both equations. In either case, all the determinants would change sign, and their ratios remain unaltered.

It is a valid point that the solution of the eqns (1.1) is so simple that it is really not worth bothering about determinants. However, they are so useful for solving linear simultaneous equations in a larger number

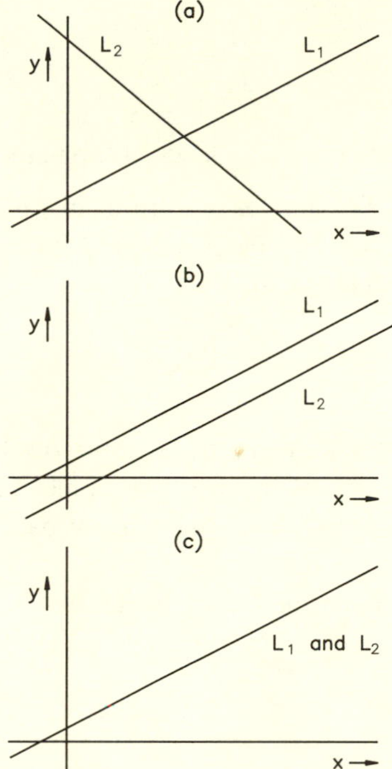

Fig. 1.1 (a) Two lines L_1 and L_2 intersect at a point. (b) Two lines L_1 and L_2 are parallel and do not intersect. (c) Two lines L_1 and L_2 are coincident; any point on the lines is a solution of the simultaneous equations.

of variables that it is a good idea to become accustomed to using them even in the simple case of two unknowns.

We now consider the types of solution that we can obtain to the simultaneous equations (1.1) from a geometrical point of view. Each equation is a straight line, and so usually we would expect the solution to be the point at which the two lines intersect (see fig. 1.1(a)). However, there are two other interesting possibilities. The first is that the two lines are parallel and hence do not intersect and so there is no solution (fig. 1.1(b)). Alternatively, our two lines could be coincident, and then any point on the lines satisfies the two equations (fig. 1.1(c)).

How do our solutions (1.3) cope with these possibilities? The regular situation of fig. 1.1(a) is straightforward, in that we simply calculate the values of the determinants, find their ratios, and that gives us the intersection point. For the cases where the two lines are parallel or identical, their common gradient is

$$-\frac{a}{b} = -\frac{d}{e} \tag{1.4}$$

This makes the determinant $\begin{vmatrix} a & b \\ d & e \end{vmatrix}$, which appears in the two denominators of (1.3), vanish. We thus have a problem with our solutions.

What distinguishes the two cases is the following. If the lines are parallel but not coincident, then

$$\frac{a}{d} = \frac{b}{e} \neq \frac{c}{f} \tag{1.5}$$

and hence the determinants in the numerators of (1.3) are usually non-zero. Thus the ratios are of the form of some finite number divided by zero. This is consistent with the lines being parallel, and 'meeting at infinity'.

On the other hand, when the lines coincide

$$\frac{a}{d} = \frac{b}{e} = \frac{c}{f} \tag{1.6}$$

and all the determinants in (1.3) vanish. Thus the solution is of the form zero divided by zero, and is indeterminate. This shows that this approach to the solution is not productive in this case. In fact the 'two' equations are now completely equivalent, and the second adds no new information. We thus have effectively only one equation, from which of course we cannot determine both x and y.

An example of this last case is

$$\left.\begin{aligned} x - 2y &= 1 \\ 2x - 4y &= 2 \end{aligned}\right\} \tag{1.7}$$

For the two parallel lines, we could have

$$\left.\begin{aligned} x - 2y &= 1 \\ x - 2y &= 2 \end{aligned}\right\} \tag{1.8}$$

while two lines intersecting in a point are

$$\left.\begin{aligned} x - 2y &= 1 \\ x + 2y &= 3 \end{aligned}\right\} \tag{1.9}$$

1.2 Three equations for three unknowns

We now go over to the situation of three linear equations for three unknowns x, y and z. We will not devote the rest of this book to examining further cases where the number of variables is increased by one at each step, but will assume that the intelligent student will be able to generalise the results of this and the preceding section to any such cases as necessary.

We write the equations for this case as

$$\left.\begin{aligned} a_{11}x + a_{12}y + a_{13}z &= c_1 \\ a_{21}x + a_{22}y + a_{23}z &= c_2 \\ a_{31}x + a_{32}y + a_{33}z &= c_3 \end{aligned}\right\} \tag{1.10}$$

where the coefficients a_{ij} on the left-hand side have two subscripts, referring to the equation number and to the variable respectively, while the constants c_k on the right-hand side have only an equation subscript.

If we apply the prescription of the previous section to the three variable case, we arrive at the tentative solution

$$\left.\begin{aligned} x &= \frac{\begin{vmatrix} c_1 & a_{12} & a_{13} \\ c_2 & a_{22} & a_{23} \\ c_3 & a_{32} & a_{33} \end{vmatrix}}{D} \\ y &= \frac{\begin{vmatrix} a_{11} & c_1 & a_{13} \\ a_{21} & c_2 & a_{23} \\ a_{31} & c_3 & a_{33} \end{vmatrix}}{D} \\ z &= \frac{\begin{vmatrix} a_{11} & a_{12} & c_1 \\ a_{21} & a_{22} & c_2 \\ a_{31} & a_{32} & c_3 \end{vmatrix}}{D} \end{aligned}\right\} \tag{1.11}$$

where

$$D = \begin{vmatrix} a_{11} & a_{12} & a_{13} \\ a_{21} & a_{22} & a_{23} \\ a_{31} & a_{32} & a_{33} \end{vmatrix} \tag{1.12}$$

A way of checking whether this is plausible is by choosing a particularly simple set of equations, and seeing if we obtain the correct solution. Thus if our 'simultaneous' equations are

$$\left.\begin{aligned} x &= 1 \\ y &= 2 \\ z &= 3 \end{aligned}\right\} \qquad (1.13)$$

the solution is obvious, and indeed (1.11) gives the right answer. In fact (1.11) is correct in all cases, provided that the determinant D of eqn (1.12) is non-zero.

I like this type of check to determine whether I have written the determinant ratios correctly, because I tend to forget whether it applies with the constants c_k appearing on the right-hand sides of eqns (1.10), or on their left (in which case the signs of all the answers would change).

As before, the simplest situation is when the determinant D is non-zero, and we obtain a unique solution for x, y and z. Geometrically, eqns (1.10) represent planes in three-dimensional space (see Section 2.4), and the standard situation is that three planes meet in a point, which of course corresponds to the solution of the three simultaneous equations. All other cases have $D = 0$.

The various geometrical possibilities are listed in Table 1.1. As we see, it is only case (i) with $D \neq 0$ which yields a point solution. For the others, the fact that $D = 0$ implies that the left-hand sides of eqns (1.10) are not independent, but in the examples given, that of the third equation can be expressed as a linear combination of the left-hand sides of the first two. Then if the coefficients on the right-hand sides satisfy the same linear relationship (examples (iv), (vi) and (vii)), one of the equations is redundant and can be thrown away. We are then left with two equations for three unknowns. In general this will give us a line of solutions (examples (vi) and (vii)); for case (iv), even the two remaining equations are identical, and so any point in the plane defined by the first equation is a satisfactory solution.

Again for the case where $D = 0$, if the constants c_k do not satisfy the same linear relationship as do the left-hand sides of the equations, then the equations are inconsistent, and there is no solution. Thus in example (viii) the left-hand side of the last equation is simply twice that of the first equation minus that of the second, i.e.

$$[x + 2y + 3z] = 2[x + y + z] - [x - z]$$

Table 1.1. *Solutions of three simultaneous equations.*

	D	Configuration of planes	Sketch	Solution	Example	Solution
(i)	$\neq 0$	E.g. floor, two adjacent walls	*	Point	$x+y=1;\ x+y+z=-1;\ 2x-z=4$	$x=1,\ y=0,\ z=-2$
(ii)	0	Three parallel planes		None	$x+y+z=1;\ x+y+z+2;\ x+y+z=3$	
(iii)	0	Two planes coincident, 1 parallel		None	$x+y+z=1;\ 2x+2y+2z=2;$ $x+y+z=2$	
(iv)	0	Three coincident		Plane	$x+y+z=1;\ 2x+2y+2z=2;$ $3x+3y+3z=3$	$x+y+z=1$
(v)	0	Two planes parallel. 1 inclined		None	Ceiling; floor; side wall	
(vi)	0	Two planes coincident, one inclined		Line	$x-z=1;\ x+y+z=1;\ 2x+2y+2z=2$	$\frac{x}{1}=\frac{y-2}{-2}=\frac{z+1}{1}$
(vii)	0	Pages of book		Line	$x+y+z=1;\ x-z=1;\ x+2y+3z=1$	As above
(viii)	0	Toblerone		None	$x+y+z=1;\ x-z=1;\ x+2y+3z=2$	

* Requires full three-dimensional diagram; in the other entries in this column, planes are viewed end-on.

The constants on the right, however, are such that

$$[2] \neq 2[1] - [1],$$

and so there is no common point that lies on all three planes.

There is another interesting interpretation of the condition $D = 0$. As explained in Section 2.4, the coefficients (a_{i1}, a_{i2}, a_{i3}) are the components of a vector normal to the ith plane. The condition $D = 0$ implies that the normals to the three planes are themselves coplanar (see eqns (3.36) and (3.30)). For any three planes this will not in general be so, and indeed the normals of three planes which meet in a point will not be coplanar.†

What do we do about solving the equations for the case $D = 0$, when there is a line solution? Our determinant solution (1.11) fails since in this case we find that x, y and z are all given as zero divided by zero.

The first step is to remember that one of the equations is a linear combination of the other two, and to throw it away. Then we are left with two equations, which clearly cannot be solved completely for x, y and z. Now two intersecting planes define a line, and so the two remaining equations specify the required line. It is more usual, however, to write the line's equation in terms of its direction, and any point on it (see eqn (2.32)). A simple way to do this is to choose two separate values of z, and to determine the corresponding values of x and y in each case. Thus for the example (vii) of Table 1.1, we find that for $z = 1$, the first two equations yield $x = 2$, $y = -2$; while for the choice $z = -1$, $x = 0$ and $y = 2$. (It is advisable to check that these points do indeed lie on the third plane.) Thus the direction of the line is

$$(\Delta x, \Delta y, \Delta z) = (2, -4, 2)$$

and so we can write its equation as

$$\frac{x}{1} = \frac{y-2}{-2} = \frac{z+1}{1}, \tag{1.14}$$

where we have taken the arbitrary point on the line as the second of our two choices.

This method needs modifying slightly if it turns out that the line of intersection is perpendicular to the z axis. Then the z value along the line is uniquely defined, and choosing some arbitrary value of z will in general make the two equations inconsistent. In that case, we need to

† Another minor feature is that it is easier to draw the $D = 0$ situations, since we can look end-on to all three planes (along the normal to the plane in which the three normals lie); for $D \neq 0$, we cannot do this, and so we need a full three-dimensional diagram (see Table 1.1).

choose two arbitrary values of x (or of y), and to solve for the other variables in each case. An example of this type is provided by the two planes

$$\left.\begin{aligned} 2x + 4y + 3z &= 4 \\ x + 2y + z &= 0 \end{aligned}\right\} \tag{1.15}$$

whose intersection has $z = 4$. Two points on the line are $(0, -2, 4)$ and $(2, -3, 4)$, and so its equation can be written as

$$\frac{x}{2} = \frac{y+2}{-1}, \quad z = 4 \tag{1.16}$$

An arbitrary choice like $z = 0$ would have yielded the inconsistent equations

$$\left.\begin{aligned} 2x + 4y &= 4 \\ x + 2y &= 0 \end{aligned}\right\} \tag{1.17}$$

A related way to find the line is to set one of the variables, say x, to some value λ. Then, for example (vii), solving the first two equations for y and z yields

$$\left.\begin{aligned} y &= 2 - 2\lambda \\ z &= \lambda - 1 \end{aligned}\right\} \tag{1.18}$$

which we can rewrite as

$$x = \frac{y-2}{-2} = z + 1 = \lambda, \tag{1.14$'$}$$

which is identical to our earlier result (1.14).

If instead of setting $x = \lambda$ we had chosen $z = \nu$, we would have obtained in an analogous manner

$$x - 1 = \frac{y}{-2} = z = \nu \tag{1.19}$$

We leave it to the reader to work out why this and eqn (1.14) represent the same line, even though they look different.

This method works for all cases, including the one where the line of intersection of the planes is perpendicular to one of the axes.

A third way of finding the direction of the common line is to note that it is perpendicular to the normals to the two planes. The vector defining the normal to a plane is given by the coefficients of the variables (see Section 2.4), and the vector product (see Section 3.3) of the two normals is then the required direction. Thus for (1.15), we calculate

$$(2, 4, 3) \wedge (1, 2, 1) = (-2, 1, 0)$$

This approach too works in all cases. To define the line, however, we still have to find one arbitrary point on it, and that is why I prefer the second method.

1.3 Special case : $c_k = 0$

A special case of eqns (1.10), which is worthy of separate consideration, arises when the constants c_k on the right-hand sides of the equations are all zero. Thus the equations read

$$\left.\begin{aligned} a_{11}x + a_{12}y + a_{13}z &= 0 \\ a_{21}x + a_{22}y + a_{23}z &= 0 \\ a_{31}x + a_{32}y + a_{33}z &= 0 \end{aligned}\right\} \tag{1.20}$$

It is fairly obvious that a solution is provided by

$$x = y = z = 0 \tag{1.21}$$

This is hardly surprising since eqns (1.20) each describe a plane passing through the origin, and hence the origin is a common point. Alternatively the three determinants in (1.11) are all zero (since each of them has a column of zeroes), and so x, y and z are similarly zero (since in general D is non-zero).

Now $x = y = z = 0$ is not a very exciting solution, so we can ask if there are any more interesting possibilities. The answer is yes, provided that the determinant D of eqn (1.12) is also zero. As in Section 1.2, this implies that the normals to the three planes are coplanar, that the equations of the planes are not independent, and that (at least) one of them can be discarded. We are then left with two planes, which define a line passing through the origin (compare examples (vi) or (vii) of Table 1.1); or in the extreme case where the three planes are identical, any point in this plane is a solution (as in example (iv)). When the coefficients c_k are all zero, we cannot have a situation where the equations are inconsistent (contrast examples (ii), (iii), (v) and (viii) of Table 1.1).

Finding the common line when $D = 0$ is simple. We first discard one of the redundant equations, and then divide through both of the remaining equations by one of the variables, say z.

We then have

$$\left.\begin{aligned} a_{11}(x/z) + a_{12}(y/z) &= -a_{13} \\ a_{21}(x/z) + a_{22}(y/z) &= -a_{23} \end{aligned}\right\} \tag{1.22}$$

These are two equations for two unknowns (x/z) and (y/z), of the type discussed in Section 1.1. Their solution then defines the direction of the line through the origin which is common to the three planes of (1.20).

Thus for example, if the equations are

$$\left.\begin{aligned} x/z + y/z &= -1 \\ 2x/z - y/z &= 4 \end{aligned}\right\} \tag{1.23}$$

then $x/z = 1$ and $y/z = -2$, so that the required line is

$$x = -y/2 = z \tag{1.24}$$

Of course, when we discussed the 'two equations for two unknowns' problem in Section 1.1, we could have considered the special case where the constants are zero there as well. However, this situation is so trivial that it is not worth separate examination.

Equations of the type (1.20) arise in the so-called eigenvalue problem to which we will return later (see Chapter 16 of Volume 2). For the three variable case, the equations are

$$\left.\begin{aligned} a_{11}x + a_{12}y + a_{13}z &= \lambda x \\ a_{21}x + a_{22}y + a_{23}z &= \lambda y \\ a_{31}x + a_{32}y + a_{33}z &= \lambda z \end{aligned}\right\} \tag{1.25}$$

where the same constant λ occurs on the right-hand side of each equation. These can be rewritten as

$$\left.\begin{aligned} (a_{11} - \lambda)x + a_{12}y + a_{13}z &= 0 \\ a_{21}x + (a_{22} - \lambda)y + a_{23}z &= 0 \\ a_{31}x + a_{32}y + (a_{33} - \lambda)z &= 0 \end{aligned}\right\} \tag{1.26}$$

Then to have a more interesting solution than just $x = y = z = 0$, the determinant of coefficients in (1.26) must be zero, i.e.

$$\begin{vmatrix} a_{11} - \lambda & a_{12} & a_{13} \\ a_{21} & a_{22} - \lambda & a_{23} \\ a_{31} & a_{32} & a_{33} - \lambda \end{vmatrix} = 0$$

When the determinant is expanded, this gives a cubic equation for λ, and hence there can be three different solutions for λ. For each of these values of λ, we can then find the corresponding direction in three-dimensional $x - y - z$ space for which this type of solution for eqns (1.25) exists.

One easily-visualised situation in which eqns (1.25) arise is in finding the principal axes of an ellipsoid. (If you don't know what an ellipsoid looks like, think of a rugby ball or an American football, which has been squashed by someone sitting on it.) Then it turns out that the three directions corresponding to the three different values of λ are the required principal axes, and are mutually perpendicular; and the values of λ are related to the lengths of these axes. For the corresponding two-dimensional case of determining the axes of an ellipse, the derivation of equations equivalent to (1.25) can be found in Section 8.3.3. (See also Problem 8.2; and Section 16.4.3 of Volume 2.)

Finally we clarify a point which is sometimes confusing to newcomers to the subject. Why do we sometimes want the determinant D of eqn (1.12) to be zero, and at other times non-zero? The answer is that $D \neq 0$ will always give us a unique point solution, while $D = 0$ results in something else. Now when the constants c_k are non-zero, we are usually interested in obtaining this point solution, while when all the c_k are zero, generally we want to find something in addition to the solution at the origin; these different situations respectively require $D \neq 0$ and $D = 0$.

Problems

1.1 Solve the simultaneous equations

$$\left.\begin{aligned} x + y + z &= 6 \\ 2x + y - z &= 1 \\ x + 2y - 3z &= -4 \end{aligned}\right\}$$

(i) by eliminating variables; and (ii) by using determinant ratios.

1.2 Find the solution (if it exists) for the equations

$$\left.\begin{aligned} x + y + z &= 1 \\ x - y + 2z &= 0 \\ 2x + 4y + z &= c \end{aligned}\right\}$$

(i) for the case $c = 2$; (ii) for $c = 3$

Find the angles between the normals to the planes. What do you notice about these angles?

For the case where there is a solution, find what linear combination of the first two equations gives the third one.

1.3 Three planes

$$a_i x + b_i y + c_i z = d_i \quad (i = 1, 2, 3)$$

meet in a common point. A fourth plane

$$a_4 x + b_4 y + c_4 z = d_4$$

passes through the same point. Show that

$$\begin{vmatrix} a_1 & b_1 & c_1 & d_1 \\ a_2 & b_2 & c_2 & d_2 \\ a_3 & b_3 & c_3 & d_3 \\ a_4 & b_4 & c_4 & d_4 \end{vmatrix} = 0$$

1.4 An ellipsoid centred on the origin is given by the equation

$$4x^2 + 4y^2 + 3z^2 + 4xy - 2xz + 2yz = 1$$

The axes of this ellipsoid are determined as follows. First we set up the eigenvalue equations

$$\begin{aligned} 4x + 2y - z &= \lambda x \\ 2x + 4y + z &= \lambda y \\ -x + y + 3z &= \lambda z \end{aligned}$$

Now find the three values of λ for which these equations have a solution, other than just $x = y = z = 0$. For each value of λ, find the direction in space of the line passing through the origin which is the solution of the above simultaneous equations. These are the directions of the axes of the ellipsoid. The lengths of the axes are then determined by finding the points on each axis that satisfies the equation of the ellipsoid.

Find the lengths of the axes, and the angles between them.

1.5 Solve the simultaneous equations

$$\begin{aligned} 2w + 2x + y + 2z &= 8 \\ w - 3y &= 7 \\ w + x - 2y + z &= 9 \\ x - y + z &= 6 \end{aligned}$$

for w, x, y and z.

2

Three-dimensional geometry

A man comes out of a small hut, and walks a mile due south. He then walks another mile due east, at which point he sees a bear. The hut he started from is now exactly a mile away. What colour was the bear?

(Well-known conundrum)

2.1 Geometry is relevant

What have the following situations in common?

Situation (i) Two aeroplanes are flying in the neighbourhood of an airport on defined paths. Will they collide?

Situation (ii) A satellite is on a trajectory towards Jupiter. Measurements made near the end of its journey indicate that the relative positions of Jupiter and its moons are not quite as expected. How far off course is the satellite?

Situation (iii) A telescope is designed so as to be able to investigate a particular star over a period of several months. Given that the earth rotates on its axis, and also revolves around the sun, how does the required orientation of the telescope vary with time?

Situation (iv) An experiment requires a large and complicated apparatus, the various components of which are being constructed in several different laboratories. When it is to be finally assembled, it is crucial that the separate parts which are supposed to fit together do so; and also that the different components and their associated electrical leads, vacuum pipework, etc. do not lay claim to the same region of space.

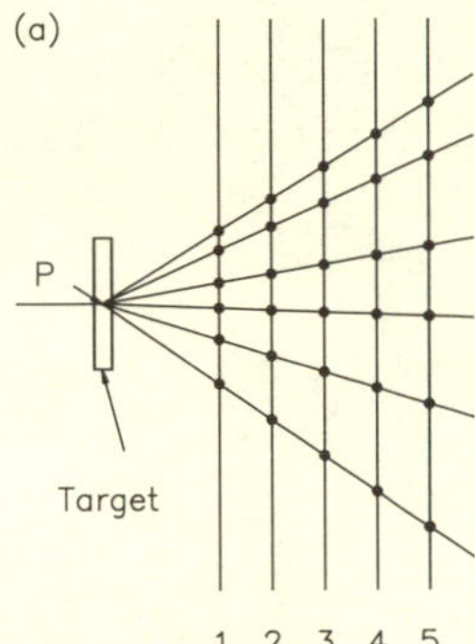

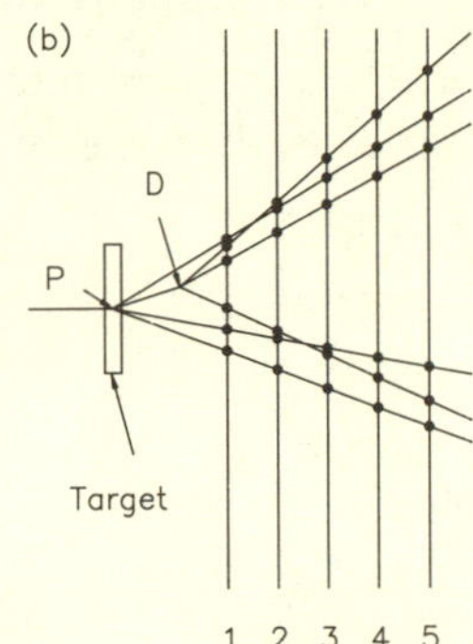

Fig. 2.1 Schematic diagram showing a detector for elementary particles. The locations of the particles are recorded at each of the five planes of the detector (labelled 1–5), enabling the trajectories of the particles to be reconstructed. For simplicity, these paths are assumed in the diagram to be straight lines. (a) The incident beam interacts in the target at P, and produces six outgoing tracks which point straight back to a common point P in the target. (b) One of the four particles emerging from P decays at D. Thus three of the tracks found in the detector point straight back at P; three others do not, but are consistent with coming from a decay at the downstream point D.

Situation (v) Detectors for high energy elementary particle experiments often consist of counters in the form of planes or cylinders which record the positions of charged particles as they pass through. If the apparatus is in a uniform magnetic field of constant strength, charged particles travel along approximately helical paths. By fitting such a helix to each set of space points, we can determine whether all the observed particles originate from some common point, or whether some of them were produced from a downstream point, corresponding to the decay position of a short-lived particle (see fig. 2.1).

The feature that is common to the above physical situations is that they all require a geometrical description of what is happening. Geometry itself may not be the most exciting of subjects, but in a wide variety of applications it is vital. This is our motivation for becoming familiar with lines, planes, spheres, cylinders, etc., as well as with translations and rotations.

2.2 Resumé of two-dimensional geometry

Before launching ourselves into three dimensions, it is a good idea to remind ourselves of some of the simple results of two-dimensional geometry.

2.2.1 The straight line

The equation of a line is

$$y = mx + c \tag{2.1}$$

where m is the gradient and c is the intercept. In order to extend this to three dimensions (see Section 2.5), it is useful to recall alternative ways of writing the line, *viz*:

$$ax + by = d \tag{2.1$'$}$$

or

$$a(x - x_0) + b(y - y_0) = 0 \tag{2.1$''$}$$

Here a and b determine the direction of the line, in that the gradient m is given by

$$m = -a/b \tag{2.2}$$

In the language of Chapter 3, the vector (a, b) is perpendicular to that defining the direction of the line.

The constant d determines how close to the origin the line passes. The shortest distance from the origin to the line is

$$p = \frac{d}{\sqrt{a^2 + b^2}} \tag{2.3}$$

where, for a given a and b, the sign of p determines whether the origin is above or below the line. Finally x_0 and y_0 in eqn (2.1$''$) are the coordinates of any point on the line (see fig. 2.2).

Clearly there is some arbitrariness in writing a given line in the forms (2.1$'$) or (2.1$''$)). Thus while the relative values of a, b and d are important, their overall scale has no physical significance (e.g. if a line is defined by $2x + 3y = 1$, then $4x + 6y = 2$ is equally valid); and similarly (x_0, y_0) are the coordinates of any point on the line.

A useful way to specify a point on a line is to use its parametric form. This means that we have a single parameter, say λ, and whatever its

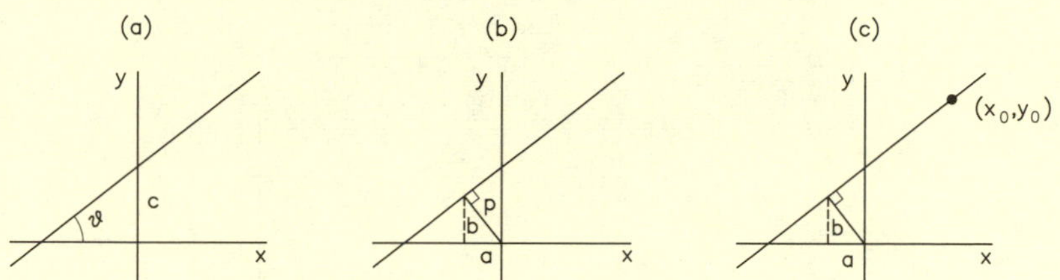

Fig. 2.2 Straight lines in two dimensions. (a) $y = mx + c$. The gradient $m = \tan\theta$. (b) $ax + by = d$. The ratio of a and b is determined by the normal from the origin to the line. The constant d is given in terms of the shortest distance p from the origin via eqn (2.3). (c) $a(x - x_0) + b(y - y_0) = 0$. The constants a and b are determined as in (b), while (x_0, y_0) is any point on the line.

magnitude, the resulting point will be on the required line. We can, for example, rewrite eqn (2.1) as

$$y = mx + c = \lambda,$$

so that in parametric form

$$\left.\begin{aligned} x &= (\lambda - c)/m \\ y &= \lambda \end{aligned}\right\} \tag{2.4}$$

Alternatively, from (2.1″),

$$a(x - x_0) = -b(y - y_0) = \lambda$$

to give

$$\left.\begin{aligned} x &= x_0 + \lambda/a \\ y &= y_0 - \lambda/b \end{aligned}\right\} \tag{2.4'}$$

By writing a point on the line in terms of one parameter, we emphasise that there is one single variable (λ) which specifies where we are along the line.

2.2.2 The circle

A circle of radius R centred on the origin is defined by Pythagoras' Theorem as

$$x^2 + y^2 = R^2 \tag{2.5}$$

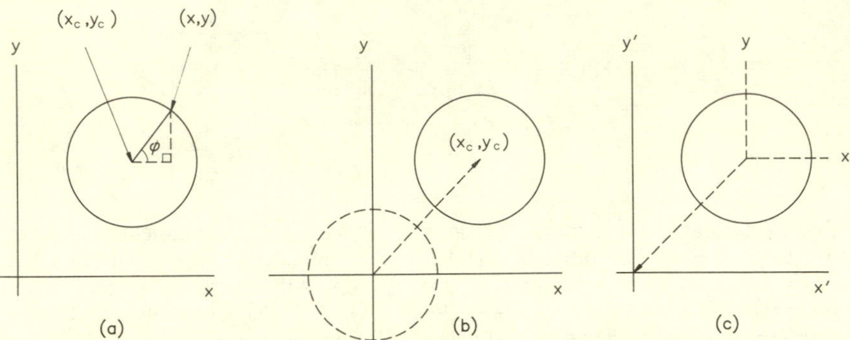

Fig. 2.3 (a) Circle of radius R centred at (x_c, y_c). Eqn (2.6) follows from the fact that the sides of the right angle triangle are of lengths $x - x_c$, $y - y_c$ and R. The parameter ϕ of (2.8) is the angle between the horizontal and the direction from the centre of the circle to the point on the circumference. (b) The circle of (a) can be obtained by shifting one centred on the origin by a distance x_c in x, and by y_c in y. The new equation is obtained by performing the replacements (2.7) in the old one (eqn (2.5)). (c) We can alternatively achieve a displaced circle by moving the axes so that the origin of the $x' - y'$ system is at $x = -x_c$, $y = -y_c$.

2.2.2.1 *Translations*

If our circle is centred on (x_c, y_c) rather than on the origin, we can again use Pythagoras' Theorem (see fig. 2.3(a)) to give

$$(x - x_c)^2 + (y - y_c)^2 = R^2 \tag{2.6}$$

Alternatively we could have derived this equation by starting from (2.5), and using

$$\left.\begin{aligned} x &\to x - x_c \\ y &\to y - y_c \end{aligned}\right\} \tag{2.7}$$

as a general recipe for how x and y transform when whatever is being plotted is moved (without rotation or stretching) so that what was at the origin – in our case, the centre of the original circle – now appears at (x_c, y_c).

Instead of moving the circle so that its centre is now at (x_c, y_c), we could equivalently have imagined moving the axes so that the origin is taken to $(-x_c, -y_c)$ (see fig. 2.3(c)). This produces the same transformation (2.7), and hence the same equation (2.6) for the circle with respect to the new axes.

Because the transformation (2.7) applies to translations for any curve or point, it is important to remember it. The only possible source of confusion is whether (2.7) transforms an object so that it moves by $(+x_c, +y_c)$ or $(-x_c, -y_c)$; or in which sense the origin moves if we consider that it is the axes which are translated. The easiest way of deciding is to consider what happens to a point $(x = x_0, y = y_0)$ which we wish to move along by, say, +2 units in x. Then our new x is $x_0 + 2$, which is equivalent to writing $x - 2 = x_0$.

Thus we see that when we translate our point by +2, we need to subtract 2 from x to change the old equation into the new one. This is as stated in (2.7).

A useful parametric description of the circle of eqn (2.6) is

$$\left.\begin{aligned} x &= x_c + R\cos\phi \\ y &= y_c + R\sin\phi \end{aligned}\right\} \tag{2.8}$$

where ϕ is the parameter which specifies the angle between the x axis and the direction from the centre of the circle to our given point (see fig. 2.3(a)).

2.2.3 The ellipse

An ellipse centred on the origin and with its major and minor axes parallel to the coordinate axes can be written

$$ax^2 + by^2 = 1 \tag{2.9}$$

The lengths of the semi-axes are $1/\sqrt{a}$ and $1/\sqrt{b}$ (see Fig. 2.4(a)), so a and b must both be positive.

If the ellipse is centred at (x_c, y_c), we can use (2.7) to write its equation as

$$a(x - x_c)^2 + b(y - y_c)^2 = 1 \tag{2.9$'$}$$

A parametric description of this ellipse is

$$\left.\begin{aligned} x &= x_c + \cos\phi/\sqrt{a} \\ y &= y_c + \sin\phi/\sqrt{b}, \end{aligned}\right\} \tag{2.10}$$

2.2.3.1 Rotations

We now want to consider an ellipse as shown in fig. 2.4(c). We will approach this as a special case of the general problem of what happens when we rotate a point by an angle ϑ with respect to the origin in an

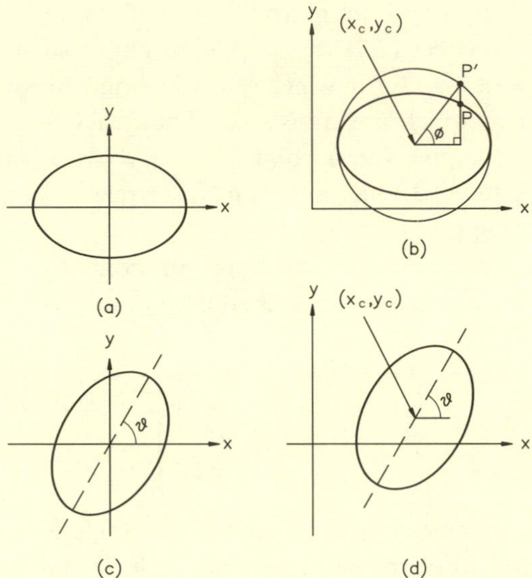

Fig. 2.4 (a) The ellipse of eqn (2.9). (b) The displaced ellipse of eqn (2.9′), whose centre is at (x_c, y_c). The parameter ϕ for the point P is the angular position of the point P′ on the circle of radius $1/\sqrt{2}$, where P and P′ have the same x coordinate. (c) The ellipse of (a) has been rotated anticlockwise by an angle ϑ. Its equation is given by (2.14). (d) The ellipse of (a) has been rotated, and then translated so that its centre is at (x_c, y_c). Its equation is now of the form (2.15).

anticlockwise direction (see fig. 2.5(a)). Since in polar coordinates, the angle is conventionally measured in an anticlockwise sense starting from $\vartheta = 0$ at the x axis, our choice for the direction of rotation is sensible.

After a little bit of trigonometry, we find that the new coordinates (x', y') are given in terms of the old ones (x, y) by

$$\left.\begin{aligned} x' &= x\cos\vartheta - y\sin\vartheta \\ y' &= x\sin\vartheta + y\cos\vartheta \end{aligned}\right\} \tag{2.11}$$

This in fact is easily remembered. The relationships necessarily involve the angle ϑ, and in the limit where ϑ tends to zero, we want

$$\left.\begin{aligned} x' &\sim x \\ y' &\sim y \end{aligned}\right\} \tag{2.12}$$

which is achieved with the $\cos\vartheta$ and $\sin\vartheta$ factors as shown in eqn (2.11). The only remaining question is whether the minus sign occurs in front of the $y\sin\vartheta$ term in the x' equation, or the $x\sin\vartheta$ term in that for y'.

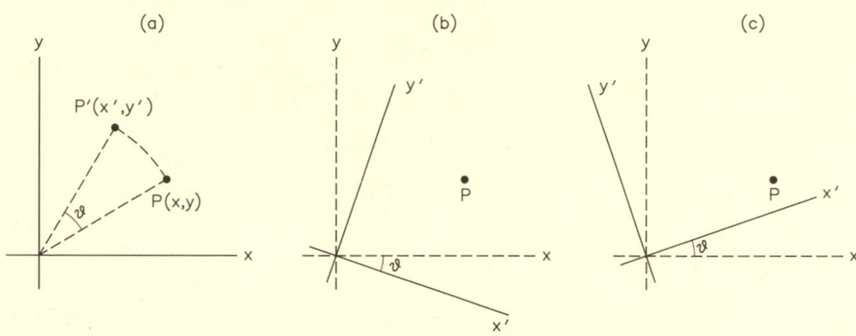

Fig. 2.5 (a) The point P′ is obtained by rotating P about the origin by an angle θ in an anticlockwise direction. The relationship between the new and old coordinates is given by eqn (2.11). (b) The same result could have been achieved by leaving P alone, but instead rotating the coordinate axes x, y clockwise by an angle ϑ, to give new axes x', y'. (c) If alternatively we had used an anticlockwise rotation of the axes, equations (2.11′) would apply. For the situation shown, the point P has y' smaller than its original y.

The effect of a small rotation on a point in the first quadrant is to make x' smaller than x (see fig. 2.5(a)); since x, y and $\sin\vartheta$ are all positive, we need the minus sign in the equation for x'.

Of course we could have achieved the identical result by rotating the coordinate axes by an angle ϑ in a clockwise sense (see fig. 2.5(b)). However, if we adopt the more conventional choice of rotating the axes in an anticlockwise direction, the transformation becomes

$$\left.\begin{aligned} x' &= x\cos\vartheta + y\sin\vartheta \\ y' &= -x\sin\vartheta + y\cos\vartheta \end{aligned}\right\} \tag{2.11$'$}$$

This can be regarded as the same as eqn (2.11), but with the effective ϑ defined in the opposite sense, and hence the opposite sign (so that $\sin\vartheta$ changes sign, but $\cos\vartheta$ does not). Another way of realising that the sign choices in (2.11′) are sensible is to see that, with x and y both positive and ϑ as a small anticlockwise rotation of the axes, y' will be smaller than y (see fig. 2.5(c)).

The conclusion is that the general structure of eqns (2.11) or (2.11′) is easily remembered, and that the choice of where to put the minus sign is readily determined by considering a simple example.

Now we return to our ellipse of fig. 2.4(c). The transformation relations between the new and old coordinates are given by (2.11). In order to replace x and y by x' and y' in eqn (2.9), we need to regard (2.11) as simultaneous equations for x and y; on solving them, we obtain

$$\left.\begin{aligned} x &= x' \cos\vartheta + y' \sin\vartheta \\ y &= -x' \sin\vartheta + y' \cos\vartheta \end{aligned}\right\} \tag{2.13}$$

(See Problem 2.1.) Substituting into (2.9) then yields

$$(a\cos^2\vartheta + b\sin^2\vartheta)x'^2 + (a\sin^2\vartheta + b\cos^2\vartheta)y'^2 + 2(a-b)\sin\vartheta\cos\vartheta x'y' = 1 \tag{2.14}$$

We can rewrite this more neatly as

$$Ax^2 + By^2 + 2Cxy = 1 \tag{2.14$'$}$$

where we have now dropped the primed symbols off x and y, and A, B and C are defined in terms of a, b and ϑ.

The important feature of eqn (2.14′) for the rotated ellipse is that it contains a term in xy, which was absent in the equations from the nicely orientated ellipses of fig. 2.4(a) and (b).

If our ellipse is both rotated and shifted (see fig. 2.4(d)), its equation will be of the form

$$A(x - x_c)^2 + B(y - y_c)^2 + 2C(x - x_c)(y - y_c) = 1 \tag{2.15}$$

2.2.4 General considerations

The straight lines, circles and ellipses we have considered are all examples of curves. Each of them can be represented by a single equation relating x and y (eqn (2.1) for the line, (2.6) for the circle, and (2.15) for the ellipse). That a curve in two dimensions can be defined by one equation is very sensible. We can think of the single equation in general as defining the y value(s) allowed at any given x coordinate. (For the straight line, there will be one such y value, while for the circle or ellipse there will be two or less, depending on the particular x value.) When we join up all these points, we obtain some sort of curve. For the curve to be a straight line, our equation must be linear in x and y (i.e. there are no terms like x^2, y^2, xy or higher powers).

We can similarly regard our single equation as imposing one constraint between the previously independent variables x and y of our two-dimensional space. This leaves one of them free, so that we effectively have one free parameter to describe where we are along our curve. (See eqn (2.4) and (2.8), and Problem 2.3, for the line, circle and ellipse respectively).

If instead we had two equations, then our solutions would have been in the form of points. (Whether it is one point, or several, or none depends on the exact form of the two equations.) Thus the two lines

$$\left.\begin{aligned} 2x + 3y &= 1 \\ x + y &= 0 \end{aligned}\right\}$$

define the point $(-1, 1)$. An even simpler example could be

$$\left.\begin{aligned} x &= 3 \\ y &= 4 \end{aligned}\right\}$$

which again specifies a point.

However, a single equation, like

$$x = 3, \tag{2.16}$$

defines a line. In this case it is parallel to the y axis, since y does not appear in the equation, and hence any value of y will satisfy our eqn (2.16), provided of course that $x = 3$.

The above sections clearly do not cover the whole of analytic geometry in two dimensions, but they are sufficient to enable us now to advance to the three spatial dimensions of the real physical world.

2.3 Three dimensions : Basics

The remainder of this chapter deals with geometry in three dimensions. In fact a lot of this material will occur again (and in a much neater way) in Chapter 3, which is about vectors. In this section, we deal with some preliminaries.

2.3.1 Axes

The first decision we have to make is how to construct axes in three dimensions. It is very useful to use a right-handed system of axes, since then we are likely to be using the same convention as other people. (This is even more important when we come to vectors.) As an extension of the way we label axes in two dimensions, it is sensible to draw x and y axes as shown in fig. 2.6. Then it is necessary to draw the z axis coming *out* of the page. (If a right-handed screw is rotated from x to y, it advances along the positive z direction). If you choose it going into the page, that would be a left-handed system.

As an aside, the question of whether it matters at a basic physics level

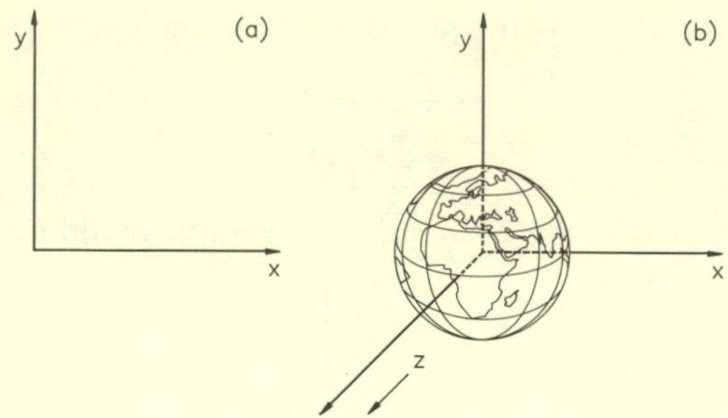

Fig. 2.6 Coordinate axes. In two dimensions these are usually drawn as in (a). For three dimensions, we add the z axis coming *out* of the paper, in order for the set of axes to be right-handed.

if we use left- or right-handed axes is a very fundamental one. This subject is all to do with 'parity', and it has been known since 1957 that the physical world can distinguish between these systems when β-decay interactions are involved.

Be that as it may, it is important to know which axes you are using when specifying mundane things like where parts of your apparatus are located in space. So you are highly recommended to keep to right-handed systems, and have your z axes emerging out of the paper.

2.3.2 Directions in space

We now consider how directions in three dimensions are specified. We can do this by imagining an appropriate line drawn from the origin, and going to a point (x_0, y_0, z_0); then these coordinates determine the direction of the line. For a given direction, only two of these are independent, since we could equally have specified another point on the line, given by (kx_0, ky_0, kz_0), where k is any arbitrary non-zero constant.

Alternatively we could quote the angles ϑ_x ϑ_y and ϑ_z that the given direction makes with the three coordinates axes (see fig. 2.7). Again only two of these are independent, since once two of the angles are specified, the third is determined (up to a sign ambiguity) from Pythagoras' Theorem by the relation

$$\cos^2 \vartheta_x + \cos^2 \vartheta_y + \cos^2 \vartheta_z = 1 \tag{2.17}$$

These three cosines are called the 'direction cosines' of the line, and are

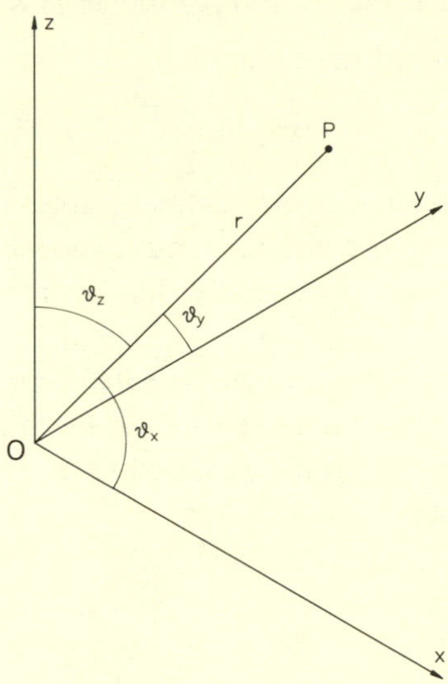

Fig. 2.7 The point P has coordinates (x_0, y_0, z_0), and is a distance r from the origin. The direction cosines of the line from the origin to P are $\cos\vartheta_x$, $\cos\vartheta_y$ and $\cos\vartheta_z$, where $\cos\vartheta_x = x_0/r$, etc. Then (x_0, y_0, z_0) are direction cosine ratios for the line OP.

determined by

$$\cos\vartheta_x = \frac{x_0}{\sqrt{x_0^2 + y_0^2 + z_0^2}}, \tag{2.18}$$

and similarly for y and z. Then the ratios

$$\cos\vartheta_x : \cos\vartheta_y : \cos\vartheta_z = x_0 : y_0 : z_0$$

Thus the coordinates (x_0, y_0, z_0) are known as 'direction cosine ratios' of the line. Because they are merely ratios, the direction cosine ratios $\alpha x_0 : \alpha y_0 : \alpha z_0$ (where α is an arbitrary constant) would specify the same direction.

Finally the angle β between two directions with direction cosine ratios $(a_1,\ b_1,\ c_1)$ and $(a_2,\ b_2,\ c_2)$ is given by

$$\cos\beta = \frac{a_1a_2 + b_1b_2 + c_1c_2}{\sqrt{(a_1^2 + b_1^2 + c_1^2)(a_2^2 + b_2^2 + c_2^2)}} \tag{2.19}$$

(See Problem 2.5.) Had the $(a_i,\ b_i,\ c_i)$ been direction cosines, rather than merely direction cosine ratios, the denominator of (2.19) would have been unity, and could have been omitted.

2.4 Planes in three dimensions

We have already discussed the equation

$$ax + by = d \tag{2.1'}$$

which in two-dimensional geometry defines a line. What does this equation represent when we move into three dimensions? The obvious answer is that it is again a line; unfortunately this is wrong. There are several ways of seeing this.

First let us consider how we might specify a simple line, such as the x axis. This is completely determined by the fact that, for any point on this line, the x value is arbitrary, but both y and z are zero. Thus to represent the line, we can write

$$\left.\begin{aligned} y &= 0 \\ \text{and } z &= 0 \end{aligned}\right\} \tag{2.20}$$

Thus we need two equations, and it turns out that this does not depend on our specific choice of line. Since (2.1′) is only one equation, it must represent something different.

A second approach is to realise that in three dimensions we start off with three independent coordinates. As discussed in Section 2.2.4, a point on a line involves one degree of freedom (e.g. the parameter specifying how far along the line we are), and this is independent of whether we are considering space to be two-, three- or multidimensional. To reduce the x, y and z in our three-dimensional case down to one degree of freedom, we need two relationships among the coordinates; these are the two equations required to specify a line in three dimensions.

As our final method, we can attempt to draw (2.1′) in order to find out what it represents. This presents a problem because we have only two-dimensional paper, so we must draw some sort of projection, for example at $z = 0$. This certainly results in a line, but we have to consider what happens when we vary z. Then we imagine making a three-dimensional model, with the appropriate 'curve' in the x–y plane at each value of z. For our case of eqn (2.1′), this is particularly easy, since z does not appear in the equation at all. Thus whatever combinations of values of x and y satisfy the equation when $z = 0$ will also do so for any other value of z (e.g. $z = 613$). Thus our model for eqn (2.1′) would look like a whole series of lines, but all stacked up behind and in front of the one for $z = 0$. Since they are all in the same orientation, the net result is a plane.

Thus a single relationship among the x, y and z coordinates in three dimensions represents a surface; when the relationship is linear,† the surface is a plane. The various ways of specifying a line are discussed in Section 2.5.

The plane of eqn (2.1′) was somewhat special in that the equation did not involve z. We can write a plane in a completely general orientation as

$$ax + by + cz = d \tag{2.21}$$

We can in fact guess most of the properties of the constants occuring in this equation, by analogy with those of the line in two dimensions (see Section 2.2.1).

We first note that only three out of a, b, c and d are independent because eqn (2.21) can be arbitrarily scaled up or down by any constant factor (except zero) without affecting the equality.

That we need three constants to specify a plane is very sensible. For example we could use the three positions where the plane cuts the coordinate axes. Since the x axis is defined by $y = z = 0$, the plane cuts it at a point with coordinates $(x_c, 0, 0)$, where

$$x_c = d/a$$

Similarly the intercepts on the other axes are

$$y_c = d/b$$

$$\text{and } z_c = d/c$$

With an arbitrary choice of one of the constants (e.g. $d = 1$), these three equations then specify eqn (2.21) completely.

Another property of the constants a, b and c is that they define the direction of the normal to the plane. This should not be surprising because of the analogy with the two-dimensional line (see the sentence below eqn (2.2)). We can demonstrate that this is so by showing that the direction defined by (a, b, c) is perpendicular to that joining any two points (x_1, y_1, z_1) and (x_2, y_2, z_2) that lie within the plane. This we now do.

Since the two points lie in the plane, their coordinates each satisfy eqn (2.21). On subtraction, we obtain

$$a(x_1 - x_2) + b(y_1 - y_2) + c(z_1 - z_2) = 0 \tag{2.22}$$

† 'Linear' means that there are no powers of x, y and z higher than 1; and there are no terms involving products of any of the variables x, y and z (e.g. yz or xyz).

Now $((x_1 - x_2), (y_1 - y_2), (z_1 - z_2))$ defines the direction of the line joining our two points in the plane. Then, using eqn (2.19) for the angle β between the two directions, we have

$$\cos\beta = \frac{a(x_1 - x_2) + b(y_1 - y_2) + c(z_1 - z_2)}{\sqrt{a^2 + b^2 + c^2}\sqrt{(x_1 - x_2)^2 + (y_1 - y_2)^2 + (z_1 - z_2)^2}} = 0 \quad (2.23)$$

where the last equality follows from eqn (2.22). This demonstrates that (a, b, c) is normal to any direction in the plane. Incidentally this also shows that eqn (2.21) does indeed represent a plane, since all directions within it are perpendicular to (a, b, c); previously we had merely argued by analogy that it was plausible that (2.21) was a plane.

Finally we come to the significance of d. Again in analogy with the two-dimensional line, d determines the shortest distance p from the origin to the plane. Not surprisingly, the direction from the origin to the point in the plane which is closest to the origin is perpendicular to the plane. As shown in Problem 2.6

$$p = \frac{d}{\sqrt{a^2 + b^2 + c^2}} \quad (2.24)$$

(compare eqn (2.3)). As with the line in two dimensions, there is some arbitrariness in how we define the sign of p, but for a set of planes with a, b and c identical, the sign of d determines on which side of the plane the origin lies.

If the values of a, b and c have been normalised so that

$$a^2 + b^2 + c^2 = 1, \quad (2.25)$$

then the constant d directly gives the distance p. The condition (2.25) corresponds to (a, b, c) being the direction cosines of the normal to the plane, rather than merely direction cosine ratios.

An alternative way of writing the plane is

$$a(x - x_0) + b(y - y_0) + c(z - z_0) = 0 \quad (2.26)$$

Here (x_0, y_0, z_0) are the coordinates of a specific point in the plane, and as before (a, b, c) specifies the direction of the plane's normal. It appears at first sight that we have six parameters defining our plane. However, they are not all independent, since there are two degrees of freedom in choosing the coordinates (x_0, y_0, z_0) of a point within the plane; and there is also the arbitrary overall scaling factor of the direction cosine ratios (a, b, c) of the normal to the plane. Thus once again we really require only three independent constants to specify a plane.

We can write an arbitrary point in the plane (2.21) in parametric form.

For any surface in three dimensions, we have two degrees of freedom, and hence can freely choose two of the coordinates. Thus our parametric form could be

$$(x, y, (d - ax - by)/c) \tag{2.27}$$

where x and y are now our parameters. Alternatively, for the plane represented by eqn (2.26), we could choose

$$(x_0 + \lambda/a,\ y_0 + \mu/b,\ z_0 - (\lambda + \mu)/c) \tag{2.28}$$

with λ and μ as parameters. A third more symmetric possibility is

$$(x_0 + \lambda v_1 + \mu w_1,\ y_0 + \lambda v_2 + \mu w_2,\ z_0 + \lambda v_3 + \mu w_3), \tag{2.29}$$

where the plane passes through (x_0, y_0, z_0), and (v_1, v_2, v_3) and (w_1, w_2, w_3) are direction cosine ratios of two directions within the plane. (They do not specify the directions from the origin to points in the plane, but rather from (x_0, y_0, z_0) to points in the plane.)

2.5 Lines in three dimensions

2.5.1 Equation of a line

Unless they happen to be parallel to each other, two planes intersect in a straight line. This means that if we specify the two planes exactly, this uniquely defines a particular line. Thus if the planes are

$$\left.\begin{aligned} a_1x + b_1y + c_1z &= d_1 \\ \text{and } a_2x + b_2y + c_2z &= d_2, \end{aligned}\right\} \tag{2.30}$$

then the two equations of (2.30) are the equations of the line.

We might be interested in knowing some points through which our line passes. Now we certainly cannot solve these two equations for three unknowns x, y and z. Indeed we do not expect to be able to do so uniquely, since (2.30) are the equations of a line, and so there is not one unique point on it; there is in fact one degree of freedom in how we choose an arbitrary point along our line. One way of proceeding is to choose some specific value for z (e.g. z_0, or -5.7, or 0), and then to solve the two simultaneous equations for x and y. If for $z = z_0$, the solution is $x = x_0$ and $y = y_0$, then (x_0, y_0, z_0) is one of the points on our line.†

† This method breaks down in the case where the line is perpendicular to the z axis. Then substitution of an arbitrary value for z could result in inconsistent equations for x and y. In that case, we simply start again choosing an arbitrary value for one of the other variables x or y.

Another feature of interest concerning the line is the direction in which it is pointing. We assume this has direction cosine ratios (l, m, n) which we want to determine. Since the line is defined as the intersection of the two planes specified by the two separate equations of (2.30), it follows that the line lies in each of the planes. A further consequence is that our line is perpendicular to the normals of each of the planes; these have directions cosine ratios (a_1, b_1, c_1) and (a_2, b_2, c_2). Then from eqn (2.19)

$$\left.\begin{aligned} a_1 l + b_1 m + c_1 n = 0 \\ \text{and } a_2 l + b_2 m + c_2 n = 0 \end{aligned}\right\} \tag{2.31}$$

since each of the angles is 90°. These two equations enable us to determine l, m and n (up to an arbitrary normalisation factor, which is unimportant since l, m and n are only direction cosine ratios).

From eqns (2.30) we can thus obtain one arbitrary point on our line, and its direction cosine ratios. This then enables us to rewrite the line in an alternative form as

$$\frac{x - x_0}{l} = \frac{y - y_0}{m} = \frac{z - z_0}{n} \tag{2.32}$$

If we cover up the z term, this looks just like the equation of a line in two dimensions, with gradient m/l. Thus we should not be surprised that eqns (2.32) provide the suitable extension when we have three dimensions.

Another way of looking at eqns (2.32) is to rewrite them as

$$\left.\begin{aligned} \frac{x - x_0}{l} = \frac{y - y_0}{m} \\ \text{and } \frac{y - y_0}{m} = \frac{z - z_0}{n} \end{aligned}\right\} \tag{2.32$'$}$$

Each of these is the equation of a plane (see Section 2.4), and hence we immediately have the equations of the line in previous form (2.30).

Eqns (2.32) are also useful for obtaining a parametric representation of a point on our line. We first set each of the terms equal to λ, which determines how far along the line any given point (x, y, z) is from our chosen one (x_0, y_0, z_0). Then we solve the equations for x to obtain

$$x = x_0 + \lambda l$$

and similarly y and z. This gives us the parametric form as

$$(x_0 + \lambda l, y_0 + \lambda m, z_0 + \lambda n) \tag{2.33}$$

As expected, we have one free parameter λ in order to specify an arbitrary point along a line.

Before we leave the equation of a line, it is worth considering how many constants are needed to specify a given line. We could do this by giving the coordinates of two points on the line. This involves six pieces of information (x, y and z for each of the points), but for each there is one degree of freedom in how we choose this point along the line. Thus there are actually only four independent numbers. (For example, we could choose $x_1 = 0$ and $y_2 = 0$, so that the constants specifying the line are y_1, z_1, x_2 and z_2.)

Now we can look back at eqn (2.30) and see that there appear to be eight constants (a_i b_i c_i and d_i for $i = 1$ or 2). Why is this more than the required four?

The answer is in two steps. First we can reduce the eight to six by remembering that in each equation we can multiply by any factor and the equation remains satisfied. Thus, for example, we could adopt the convention that d_i is always chosen to be unity, or alternatively that a_i b_i and c_i satisfy (2.25).

What decreases the number of constants from six to four is the fact that, in order to define a line, each of the two planes of eqn (2.30) can be chosen arbitrarily. For example, if you hold a booklet so that the pages are open in a fan-like fashion, the line of the binding can be specified as the intersection of any page with any other page. There is thus one further degree of freedom in choosing a_i, b_i, c_i and d_i for each plane, and so once again we really do need only four independent constants.

2.5.2 *Worked example : Shortest distance from a point to a line*

We want to find the shortest distance from the point P with coordinates (x_p, y_p, z_p) to the line given by eqn (2.32). For example, we may have observed a nuclear particle coming more or less from the region in which a point-like source is known to be. If we make measurements to determine the straight line trajectory of the nuclear particle, we may be interested to know whether it is consistent with coming directly from the source (i.e. what is the distance of the point source from the extrapolated line, and whether it is consistent with zero).

To do this, we first consider a general point Q on the line, described by the parameter λ, and find its distance s from P. We then minimise s with respect to λ.

The general point Q is given by eqn (2.33), and so s satisfies

$$s^2 = (x_0 - x_p + \lambda l)^2 + (y_0 - y_p + \lambda m)^2 + (z_0 - z_p + \lambda n)^2 \qquad (2.34)$$

The normal way to find the minimum of s involves setting $ds/d\lambda = 0$. However the expression for s involves a square root sign, which will result in a messy expression when we differentiate. It is much better instead to set $ds^2/d\lambda = 0$; this relies on the fact that a stationary value for s will also be one for s^2.†

Then

$$\frac{1}{2}\frac{d(s^2)}{d\lambda} = (x_0 - x_p + \lambda l)l + (y_0 - y_p + \lambda m)m + (z_0 - z_p + \lambda n)n = 0 \quad (2.35)$$

This is readily solved for λ to give

$$\lambda = -\frac{(x_0 - x_p)l + (y_0 - y_p)m + (z_0 - z_p)n}{l^2 + m^2 + n^2} \quad (2.36)$$

where the denominator will be unity if (l, m, n) are direction cosines. Finally we substitute this value of λ into eqn (2.34) to obtain the required distance, and into (2.33) if we also want the coordinates of the closest point on the line.

Eqn (2.35) tells us something else that is interesting. The form of this equation is such that the directions specified by

$$\left.\begin{array}{l} (l, m, n) \\ \text{and } (x_0 + \lambda l - x_p, y_0 + \lambda m - y_p, z_0 + \lambda n - z_p) \end{array}\right\} \quad (2.37)$$

are perpendicular (see eqn (2.19)). The first of these specifies the direction of the line, while the second is the direction from P to that point on the line which is closest to P. This is thus a proof of the not very surprising fact that, even in three dimensions, the direction of the shortest path from a point to a line is perpendicular to the line.

The corresponding situation of the shortest distance from a point to a plane is considered in Problem 2.6.

2.5.3 Skew lines

2.5.3.1 What they are

Although we regarded the worked example of Section 2.5.2 as a problem in three dimensions, we could in fact have reduced it to a two-dimensional one by choosing our $x - y$ plane as containing the point and the line. Now, however, we deal with a situation which really does involve three dimensions in an unavoidable way.

† We perhaps ought to be a little careful of what happens if $s = 0$. However, this does not cause any difficulties in our example.

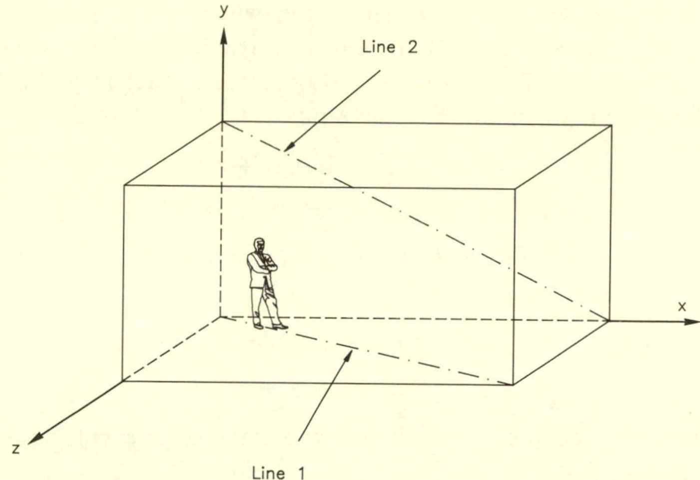

Fig. 2.8 A room of size 2, 1 and 2 units in x, y and z. The first line goes diagonally across the floor, and the second diagonally on a wall. These lines are not parallel, but never meet; they are called 'skew'.

We consider two lines in arbitrary orientations. In two dimensions, there are only two possibilities : either the lines are parallel or they intersect. Now we have a new feature. In three dimensions, lines which are not parallel need not meet; they can simply miss each other.

As a simple example of this, stand in the corner of an ordinary room where the floor is rectangular. Now imagine two lines, the first of which is drawn diagonally across the floor to the opposite corner. The second line passes through the corner of the ceiling directly above you, and goes diagonally down the wall (see fig. 2.8). Clearly these lines are not parallel, but equally they never meet even when extrapolated. Such lines are referred to as 'skew'.

2.5.3.2 Distance of closest approach

One interesting quantity for skew lines is their closest distance of approach. That is, if we are free to choose any points on each of the lines, what is the minimum distance we can have between these points, for the given lines? We will calculate this for the specific example of fig. 2.8 soon, but first we look at the general problem.

We assume that our two lines are given by

$$\left.\begin{aligned} \frac{x-x_1}{l_1} &= \frac{y-y_1}{m_1} = \frac{z-z_1}{n_1} = \lambda \\ \text{and } \frac{x-x_2}{l_2} &= \frac{y-y_2}{m_2} = \frac{z-z_2}{n_2} = \mu, \end{aligned}\right\} \qquad (2.38)$$

where λ and μ are parameters specifying which point we choose on each line (compare eqn (2.32). If you have trouble recalling why these equations represent lines, go back and read Section 2.5.1 again). Then the parametric representation of the points is

$$\left.\begin{aligned} &(x_1 + \lambda l_1, y_1 + \lambda m_1, z_1 + \lambda n_1) \\ \text{and } &(x_2 + \mu l_2, y_2 + \mu m_2, z_2 + \mu n_2) \end{aligned}\right\} \tag{2.39}$$

The distance D between these two arbitrary points is now readily expressed as

$$\begin{aligned} D^2 = [(x_1 + \lambda l_1) - (x_2 + \mu l_2)]^2 + [(y_1 + \lambda m_1) - (y_2 + \mu m_2)]^2 \\ + [(z_1 + \lambda n_1) - (z_2 + \mu n_2)]^2 \end{aligned} \tag{2.40}$$

We now need to choose λ and μ in order to make D^2 as small as possible. This requires us to set

$$\frac{\partial D^2}{\partial \lambda} = \frac{\partial D^2}{\partial \mu} = 0 \tag{2.41}$$

(see Chapter 6 and Section 7.6), so we need to evaluate the two partial derivatives:

$$\left.\begin{aligned} \frac{1}{2}\frac{\partial D^2}{\partial \lambda} &= [(x_1 + \lambda l_1) - (x_2 + \mu l_2)]l_1 + [(y_1 + \lambda m_1) - (y_2 + \mu m_2)]m_1 \\ &\quad + [(z_1 + \lambda n_1) - (z_2 + \mu n_2)]n_1 = 0 \\ -\frac{1}{2}\frac{\partial D^2}{\partial \mu} &= [(x_1 + \lambda l_1) - (x_2 + \mu l_2)]l_2 + [(y_1 + \lambda m_1) - (y_2 + \mu m_2)]m_2 \\ &\quad + [(z_1 + \lambda n_1) - (z_2 + \mu n_2)]n_2 = 0 \end{aligned}\right\} \tag{2.42}$$

Eqns (2.42) are two simultaneous equations for λ and μ. They look cumbersome, but we can simplify them by noting that the terms in square brackets are just the direction cosine ratios of the line joining the point on the first line to that on the second. When λ and μ are the solutions of eqn (2.42), these will correspond to the points of closest approach, and we will denote the direction cosine ratios of the line joining them by $(s,\ t,\ u)$ i.e.

$$(s, t, u) = ((x_1+\lambda l_1)-(x_2+\mu l_2), (y_1+\lambda m_1)-(y_2+\mu m_2), (z_1+\lambda n_1)-(z_2+\mu n_2)) \tag{2.43}$$

with λ and μ chosen so as to give the shortest distance. Then eqns. (2.42) become

$$\left.\begin{aligned} sl_1 + tm_1 + un_1 = 0 \\ \text{and } sl_2 + tm_2 + un_2 = 0 \end{aligned}\right\} \tag{2.42$'$}$$

This shows that the direction specified by (s, t, u) is perpendicular to both (l_1, m_1, n_1) and (l_2, m_2, n_2) (see eqn (2.19)). That is, the direction between the points of closest approach on two skew lines is perpendicular to both of the lines. This is a very plausible generalisation of the result in Section 2.5.2 about the shortest distance from a point to a line.

To determine D, we need to find the two closest points. This we do by solving eqn (2.42) for λ and μ. (These equations look complicated, but are merely two linear simultaneous equations for λ and μ, and so are readily solved – see Section 2.5.3.3 below). Then we substitute the values of λ and μ into the parametric forms (2.39), which gives us the closest points. Finally we determine D by Pythagoras' Theorem (eqn (2.40)).

This approach needs modifying slightly for the simpler problem of two parallel lines. Then the two equations (2.42) become identical, and we cannot use them to find both λ and μ. This is consistent with the fact that for parallel lines there are no unique points corresponding to the distance of closest approach.

However, the solution of the problem is straight-forward. In this case, eqn (2.40) becomes a function of $\lambda - \mu$ only, rather than λ and μ separately. Thus we can choose λ arbitrarily (say, $\lambda = 0$) and then solve eqn (2.42) for μ. This will then give us two points whose separation is the distance between the parallel lines.

2.5.3.3 *Specific example*

Having solved the problem for the general case, we now do so for the specific lines shown in fig. 2.8. We assume that, in suitable units, the dimensions of the sides of the room are 2, 1 and 2 in x, y and z respectively. Then the equations of the lines are

$$\left.\begin{aligned} x = z, \quad & y = 0 \text{ for line 1} \\ x + 2y = 2, \quad & z = 0 \text{ for line 2} \end{aligned}\right\} \tag{2.44}$$

We can thus express the lines in parametric form as $(\lambda, 0, \lambda)$ and $(2 - 2v, v, 0)$ respectively.

The distance between these points is given by

$$D^2 = (\lambda - 2 + 2v)^2 + v^2 + \lambda^2 \tag{2.45}$$

For this to be a minimum

$$\left.\begin{aligned} \frac{1}{2}\frac{\partial D^2}{\partial \lambda} &= (\lambda - 2 + 2v) + \lambda = 0 \\ \text{and } \frac{1}{2}\frac{\partial D^2}{\partial v} &= 2(\lambda - 2 + 2v) + v = 0 \end{aligned}\right\} \tag{2.46}$$

The solution of these two linear simultaneous equations is

$$\left.\begin{aligned} \lambda &= \frac{1}{3} \\ \nu &= \frac{2}{3} \end{aligned}\right\} \tag{2.47}$$

so that the points of closest approach are

$$\left.\begin{aligned} &\left(\frac{1}{3}, 0, \frac{1}{3}\right) \text{ for line 1} \\ &\left(\frac{2}{3}, \frac{2}{3}, 0\right) \text{ for line 2} \end{aligned}\right\} \tag{2.48}$$

The required shortest distance is simply the distance D between these points, and is $\sqrt{\frac{2}{3}}$. One simple check we can make is that, as is necessary, D is not larger than the vertical distance between the lines at the point where you are standing (which is 1 unit).

Finally we can check that the direction of the line joining the closest points is perpendicular to each of the two lines. This has direction cosine ratios $(\frac{1}{3}, \frac{2}{3}, -\frac{1}{3})$, as compared with $(1, 0, 1)$ and $(2, -1, 0)$ for the lines 1 and 2 respectively. Since

$$\left.\begin{aligned} \frac{1}{3} \times 1 + \frac{2}{3} \times 0 - \frac{1}{3} \times 1 = 0 \\ \text{and } \frac{1}{3} \times 2 + \frac{2}{3} \times (-1) - \frac{1}{3} \times 0 = 0 \end{aligned}\right\} \tag{2.49}$$

they are indeed perpendicular. If this had not turned out to be so, we would have made a mistake and would have to go back and check what had gone wrong.

2.5.3.4 Intersecting lines

As a final topic concerning lines in three dimensions, we will find the condition for the two lines of (2.38) in fact to intersect. If this is so, then the point of intersection is common to the two lines. That is, for a particular choice of λ and μ,

$$\left.\begin{aligned} x_1 + \lambda l_1 &= x_2 + \mu l_2 \\ y_1 + \lambda m_1 &= y_2 + \mu m_2 \\ \text{and } z_1 + \lambda n_1 &= z_2 + \mu n_2 \end{aligned}\right\} \tag{2.50}$$

We can rewrite these as

$$\left.\begin{aligned} \lambda l_1 - \mu l_2 &= x_2 - x_1 \\ \lambda m_1 - \mu m_2 &= y_2 - y_1 \\ \lambda n_1 - \mu n_2 &= z_2 - z_1 \end{aligned}\right\} \qquad (2.50')$$

We thus have three equations for our two unknowns λ and μ. In general these equations will be incompatible with each other, and the lines will be skew (i.e. there is no point of intersection). If we insist that they do meet, we have to ensure that the eqns (2.50′) are consistent. The condition for this is that

$$\begin{vmatrix} l_1 & l_2 & x_2 - x_1 \\ m_1 & m_2 & y_2 - y_1 \\ n_1 & n_2 & z_2 - z_1 \end{vmatrix} = 0 \qquad (2.51)$$

This then is a requirement on the coefficients of the lines in order for them to meet.

The significance of this condition will be investigated further in Section 3.5.1, where we deal with this subject using vectors.

2.6 Spheres, etc.

In Sections 2.2.2 and 2.2.3 on two-dimensional geometry we dealt with circles, and with ellipses in various orientations. Now we consider the corresponding three dimensional objects.

Given eqn (2.5) for a circle in two dimensions, it does not take too much imagination to guess that the equation of a sphere of radius R centred on the origin is

$$x^2 + y^2 + z^2 = R^2 \qquad (2.52)$$

If we remain dubious, we can convince ourselves by remembering that a sphere is such that any point on its surface is a distance R from the origin; and the square of this distance can be obtained as $x^2 + y^2 + z^2$ by applying Pythagoras' Theorem twice, first for example in the x–z plane, and then incorporating y as well (see fig. 2.9).

Often spheres are centred not at the origin, but at some other point. If this has coordinates (x_c, y_c, z_c), then the equation of our sphere becomes

$$(x - x_c)^2 + (y - y_c)^2 + (z - z_c)^2 = R^2 \qquad (2.53)$$

This follows from applying Pythagoras' Theorem to distances measured with respect to (x_c, y_c, z_c). Alternatively we can use the obvious extension to three dimensions of the rule (2.7) for translations.

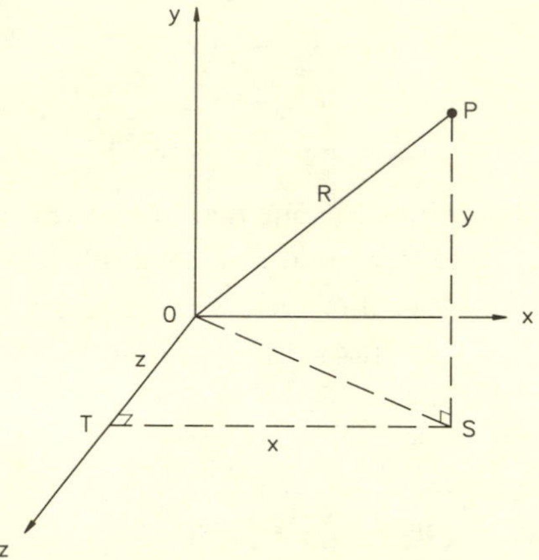

Fig. 2.9 The distance of the point P from the origin is R. The point S is the projection of P onto the x–z plane, and T is the projection of S onto the z axis. Thus OTS and OSP are right angle triangles, and TS, SP and OT have lengths equal to the x, y and z coordinates of P. Eqn (2.52) follows immediately.

The equivalent of an ellipse in three dimensions (called an ellipsoid) with axes of lengths $2a$, $2b$ and $2c$ aligned along the coordinate axes is described by the equation

$$\frac{x^2}{a^2} + \frac{y^2}{b^2} + \frac{z^2}{c^2} = 1 \tag{2.54}$$

If this is reorientated by a rotation about the z axis, the x and y coordinates transform according to eqns (2.11), while the z coordinate remains unchanged. Had the rotation instead been by an angle ϕ about the x axis, by analogy we can write down the transformation as

$$\left.\begin{aligned} x' &= x \\ y' &= y\cos\phi - z\sin\phi \\ z' &= y\sin\phi + z\cos\phi \end{aligned}\right\} \tag{2.55}$$

Finally a rotation χ about the y axis is given by

$$\left.\begin{aligned} x' &= x\cos\chi + z\sin\chi \\ y' &= y \\ z' &= -x\sin\chi + z\cos\chi \end{aligned}\right\} \tag{2.56}$$

More complicated rotations about a general axis (i.e. not one of the coordinate axes) are beyond the scope of this chapter, but can be built up out of these simpler rotations.

Translations of the ellipsoid are dealt with in the same way as for any equation (compare eqns (2.52) and (2.53) for the sphere).

We started this section with a circle in two dimensions. If we actually want to write down a circle in three dimensions, we remember that a circle is given by the intersection, for example, of a plane with a sphere, or by two spheres. Thus in analogy with a three-dimensional line being defined by the equations of two planes,

$$\left.\begin{aligned}(x-x_c)^2+(y-y_c)^2+(z-z_c)^2=R^2\\ \text{and } ax+by+cz=d\end{aligned}\right\} \tag{2.57}$$

is a circle, provided that the distance of the point (x_c, y_c, z_c) to the plane is less than R.

Similarly

$$\left.\begin{aligned}(x-x_1)^2+(y-y_1)^2+(z-z_1)^2=R_1^2\\ \text{and } (x-x_2)^2+(y-y_2)^2+(z-z_2)^2=R_2^2\end{aligned}\right\} \tag{2.58}$$

is another circle, provided that the distance between the points (x_1, y_1, z_1) and (x_2, y_2, z_2) is in the range $|R_1 - R_2|$ to $R_1 + R_2$ – otherwise the spheres do not intersect at all.

Some other objects also have relatively simple equations. Thus

$$(x-x_c)^2+(y-y_c)^2=R^2 \tag{2.6}$$

regarded as an equation in three dimensions, describes a right circular cylinder of radius R and arranged with its axis parallel to the z axis (compare the discussion in Section 2.4 of $ax + by = d$ as a plane in three dimensions).

Finally we consider a cone (see fig. 2.10), which is used in geometry to define circles, ellipses, parabolae and hyperbolae, by the intersection of appropriately orientated planes with it. For the cone shown in the figure, slices at constant z are circles, whose radii are proportional to the absolute magnitude of the z coordinate. Thus

$$x^2+y^2=(kz)^2 \tag{2.59}$$

is the required equation.

2.7 Conclusion

Looking back over this chapter, we discover that although the ideas are interesting, some of the algebra is tedious and repetitive, especially in the way the x, y and z coordinates occur in a very similar manner. Thus for

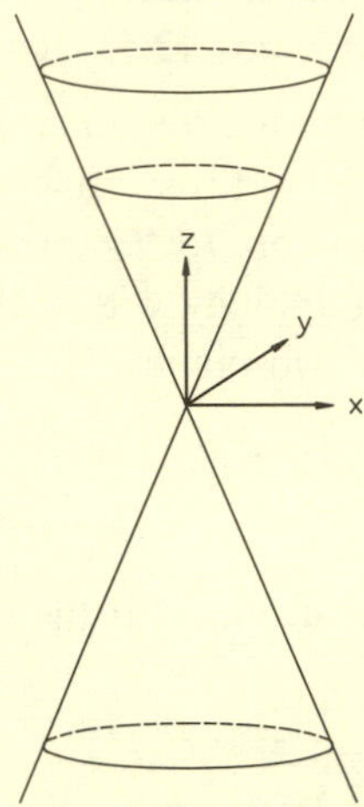

Fig. 2.10 A cone, whose axis is parallel to the z axis, is such that slices at constant z are circles, whose radii are proportional to $|z|$. Its equation is thus (2.59), where the semi-angle of the cone is $\tan^{-1} k$.

spheres, x, y and z occur in a completely symmetric manner, while for planes (e.g. eqn (2.21)), ax, by and cz are equivalent to each other in any formulae. Similarly the problem of Section 2.5.3.2 of the shortest distance between two skew lines was such that after we had written down the x terms (for example, in eqns (2.40) or (2.42) or (2.43), the y and z ones follow in a completely obvious manner.

We can thus look forward to the next chapter on vectors, where several of the above topics will be considered again. The concise notation there not only leads to neater formulae, but also is helpful in enabling us to see to the heart of a problem, rather than being bogged down in long equations.

Problems

2.1 (i) Derive the transformation (2.11) for a rotation.

(ii) Show that if we solve eqns (2.11) for x and y in terms of x' and y', we obtain eqns (2.13), which are of the same form as (2.11), but with the sign of ϑ reversed. (This is as it should be, since if P′ is obtained by rotating P through an angle ϑ, then P can be obtained from P′ by a rotation of $-\vartheta$.)

2.2 Show that the equation of a circle centred on the origin is unchanged if we apply the transformation (2.11) for a rotation by an angle ϑ.

2.3 Devise a parametric representation for a point on the ellipse of eqn (2.15).

2.4 The area of an ellipse is π times the product of the lengths of its semi-minor and semi-major axes. Find the area enclosed by the ellipse

$$5x^2 + 5y^2 - 6xy + 22x - 26y + 35 = 0$$

2.5 The cosine formula states that in a triangle with sides of length l, m and n, the angle α between the sides of length l and m is given by

$$n^2 = l^2 + m^2 - 2lm\cos\alpha$$

Use this to derive eqn (2.19) for the angle between two directions specified by direction cosine ratios (a_1, b_1, c_1) and (a_2, b_2, c_2).

Assuming the earth is a sphere of radius 6370 kilometres, what is the shortest distance on the surface of the earth between Jerusalem (lattitude 31°45′ north, longitude 35°13′ east) and Sydney (33°50′ south, 151°15′ east)?

2.6 Find the shortest distance from the point (x_0, y_0, z_0) to the plane $ax + by + cz = d$. Show that the point in the plane closest to (x_0, y_0, z_0) is such that the direction between the two points is normal to the plane.

2.7 Check whether or not the following lines in three dimensions are identical:

(i) $(x-1)/2 = y+2 = -z/3$.
(ii) $x + y + z + 1 = 0$ and $2x + 5y + 3z + 8 = 0$.

2.8 Find the shortest distance from the origin to the line defined in Problem 2.7(ii).

2.9 Determine the equation of the line which passes through the point $(1, -2, 0)$, and is perpendicular to both lines defined below:

Line 1 : $x - 3 = (y-3)/2 = -z$.
Line 2 : $3x - 3y + z = 9$ and $-x + y + 3z = 2$.

2.10 Find the equation of the line with the following properties

(i) It is perpendicular to line (i) of Problem 2.7.
(ii) It passes through the origin.
(iii) It intersects line (i) of Problem 2.9.

2.11 A regular tetrahedron has edges of length l. Find the shortest distance s between opposite edges (i.e. edges which do not meet at a vertex) as follows:

(i) Imagine the base ABC is situated in the x–y plane. Choose a convenient orientation and location, and write down the coordinates of each corner of the base.
(ii) Find the coordinates of the apex D of the tetrahedron, by ensuring that its distance from each of the vertices A, B and C is l.
(iii) Write down the equations of the lines corresponding to, for example, sides AB and CD. Express a general point on each of these lines in parametric form.
(iv) Express the distance s in terms of these parameters, and then minimise s^2 with respect to them.
(v) Use the values of the parameters you have just obtained in order to find the points of closest approach, and hence s. If this is larger than l, go back and find your mistake.

For such a simple situation, you can also use a symmetry argument to find the points of closest approach. Check that this agrees with the answer you have just obtained.

2.12 Three planes are given by the following equations:

plane 1: $x + 2y + 3z = 9$
plane 2: $x - y + z = 4$
plane 3: $x + y - z = 1$

A line is defined by the intersection of planes 1 and 2, and a second line by the intersection of planes 2 and 3. Find the shortest distance between these lines.

2.13 (i) What are the radii of the circles defined by eqn (2.57) and by eqns (2.58)?
(ii) Draw sketches to show what happens when the conditions below eqn (2.58) are not satisfied.

2.14 Assume that the earth is aproximately an ellipsoid with polar axis of length $2R_p$, and equatorial axes of length $2R_e$; in this approximation, the equator is a circle. The axis of the earth, which initially was along the z axis, is tipped by rotating the earth about its x axis by an angle γ, in a sense such that the North Pole which was at positive z acquires a positive y component. At some particular instant, the centre of the earth has coordinates $(x_c, y_c, 0)$ with respect to axes centred on the sun. What is the equation of the earth's surface (i.e. ellipsoid) with respect to these axes?

3
Vectors

Question: What happens if you cross a tsetse fly with a rock-climber?
Answer: You cannot cross them, because a rock-climber is a scaler.

(Anonymous riddle)

3.1 Preliminaries

Today my particle accelerator in Geneva is not operating, and I have decided to have the day off. I take a bus ride to a nearby village, and get off at the bus-stop in the main square. The village is very pretty, and I wander around enjoying the chalets and the views. I also venture into the surrounding hillside, with its criss-crossing paths. After a couple of hours, I begin to think it is time to return to the village, to catch the bus back to Geneva. If I know that I have walked a total of 6 miles, how far am I from the bus-stop?

The answer of course is that we cannot tell – it could be anything from zero distance to 6 miles. It all depends on how straight or crooked my walk had been up till then. This is because the problem is basically one involving vectors, in which not only a magnitude is involved, but also a direction. The distance that I am from the bus-stop is given by the length of the vector formed by adding all the vectors corresponding to the various straight bits of path along which I had walked.

Vectors are thus very useful in any problem in which directions are implied. Such quantities include: positions, velocities and accelerations; forces and fields (e.g. electric or magnetic); momentum; etc. Examples of scalar quantities which do not have a direction associated with them include temperature, mass, energy (e.g. potential or kinetic), time, etc.

Even in problems where directions are involved, it is not essential to use vectors explicitly, because we can instead describe what is happening to the various components along each of the coordinate axes. This, however, is more cumbersome, and equations written in vector notation look very much neater. This has the consequence that, once we are familiar with vectors, we can often see how to solve a problem much more readily, than if it had been written out long-hand in component form. We will come across an extreme example of this in Chapter 10 of Volume 2 on vector operators. Maxwell's electromagnetic equations can be written so neatly in vector notation that it is a simple operation to use them to derive the existence of electromagnetic waves. In component form, it needed someone of the genius of Maxwell to see his way through the mass of algebra required.

Throughout this book, we shall denote vectors as bold-type characters, e.g. **v**. This is to be taken to represent the magnitude and the direction of whatever **v** represents. It is conventional, however, not to regard **v** as being attached to any specific point in space. Thus a set of parallel lines of the same length would be represented by the same vector direction. This implies that in three dimensions, a vector needs three scalar numbers to specify it completely. These could be its components along the coordinate axes; or its length and two direction cosine ratios (see Section 2.3.2); etc. If it is important to locate, for example, the space point at which a force **F** is applied, we would have to do this separately from the definition of **F**; we would simply state that it acts at a point **r** with respect to the origin.

However, when we write a vector **v** in terms of its components† (v_x, v_y, v_z), we are refering to the coordinates of the end-point of the vector, when its starting point coincides with the origin. The other situation where we also imply that a vector starts at the origin is when we specify a position vector, such as **r** in the previous paragraph.

If we wish to specify just a direction, we often use unit vectors, i.e. vectors whose length is 1. A convention we adopt throughout this book is to represent unit vectors along the x, y and z coordinates by **i**, **j** and **k** respectively. Another example is $\hat{\mathbf{r}}$, a unit vector along the radial direction.

Some vectors are very specific, and apply to one particular quantity. For example, **a** could be the acceleration of the sun with respect to

† Throughout this book, we use the convention that, unless otherwise explicitly stated, the components of a vector **F** with respect to cartesian axes x, y and z are denoted by F_x, F_y and F_z. Of course when we consider components, it may well be that there are better systems to use than rectangular axes. Thus spherical or cylindrical polars may be preferable in some cases.

a system of axes at the centre of the Milky Way; or it could be the frictional force at the bottom of a ladder when someone is half-way up it. In contrast other vectors apply over a whole region of space. An example of this is the gravitational field due to the earth at any point in its vicinity; or the wind velocity anywhere in the atmosphere. In these last two examples, the vector will vary as we move around in space. We describe this as a vector field $\mathbf{F}(\mathbf{r})$, where our vector $\mathbf{F}$ depends on which space point $\mathbf{r}$ we are at. The unit radial vector $\hat{\mathbf{r}}$ is another example of a vector which varies with position.†

Of course vector quantities, be they specific vectors or vector fields, can also depend on time. Thus the position of the sun varies with time; and the strength and direction of the winds in the atmosphere do so too. In this case our vector $\mathbf{F}(t)$ is also a function of a scalar variable t.

A vector equation, such as

$$\mathbf{a} = \mathbf{b} \tag{3.1}$$

is more powerful than a scalar equation. The equality of two vectors means that if they are considered as starting at the same point, then their ends will coincide. This requires that not only are their lengths equal, but so are their orientations. Thus in three dimensions, the vector equation (3.1) is equivalent to three scalar equations; each of the corresponding components of $\mathbf{a}$ and $\mathbf{b}$ must be equal.

Another point about vector equations is that we should ensure that we never write anything like $\mathbf{v} = 3$.‡ This has a vector equal to a scalar, which is impossible. (It is possible that the length of the vector $\mathbf{v}$ is 3 units, but we would write this differently – see Section 3.3.4). This injunction seems to be so obvious as to be hardly worth mentioning. However, by the time we use vector and scalar products (see Sections 3.3 and 3.4) and then vector operators (Chapter 10 of Volume 2), the scope for mistakes of this kind is increased. It is then important to check that each term in an equation is indeed either a scalar or a vector as required.

A tactical point about solving vector equations is that this is usually most readily achieved by using the various vector relations which we

† This feature of quantities varying with spatial position is not confined to vectors. Thus although scalars contain examples of specific quantities (e.g. the cost of a kilogram of aubergines), we could consider the temperature T at any point in the atmosphere – this is an example of a scalar field. We return to vector and scalar fields in Chapter 10 of Volume 2.

‡ However, an equation like $\mathbf{v} = 0$ is valid, and means that all components of $\mathbf{v}$ are zero. This is not really an exception to our general rule, since the zero can be regarded as a vector of zero length.

discuss later in this chapter, and in Chapter 10 of Volume 2. Trying to write out the equation in components is generally not the best approach. The main exception to this rule is that the proofs of some basic vector identities may require the use of components (see Problem 3.2).

Throughout most of this book, we shall be dealing with vectors in three dimensions. However, many of the results readily carry over into two (or indeed four or more) dimensions. Indeed a lot of the discussion on complex numbers in Chapter 4 bears a striking similarity to two-dimensional vectors. In relativity, time takes on a role which in many ways is analogous to the spatial coordinates, and hence it is convenient to regard them together as being the components of a four-dimensional vector.

3.2 Multiplication by a scalar and vector addition

We now introduce three simple operations to do with vectors. The first is multiplication of a vector by a scalar k. Thus the equation

$$\mathbf{a} = k\mathbf{b} \tag{3.2}$$

means that the vector $\mathbf{a}$ is k times as long as the vector $\mathbf{b}$, but they are in the same direction. If k is negative, it is probably simpler to think of $\mathbf{a}$ as $|k|$ times as long as $\mathbf{b}$, but in the opposite direction. In terms of components along the coordinate axes, (3.2) implies

$$\left.\begin{aligned} a_x &= kb_x \\ a_y &= kb_y \\ \text{and } a_z &= kb_z \end{aligned}\right\} \tag{3.3}$$

Two vectors can be added together, e.g.

$$\mathbf{c} = \mathbf{a} + \mathbf{b} \tag{3.4}$$

Again in terms of components, this implies that

$$\left.\begin{aligned} c_x &= a_x + b_x \\ c_y &= a_y + b_y \\ \text{and } c_z &= a_z + b_z \end{aligned}\right\} \tag{3.5}$$

Diagramatically, we add $\mathbf{b}$ to $\mathbf{a}$ by moving it parallel to itself so that its starting point coincides with where $\mathbf{a}$ ends. Then $\mathbf{c}$ is the vector whose starting point coincides with that of $\mathbf{a}$, and which ends where $\mathbf{b}$ does

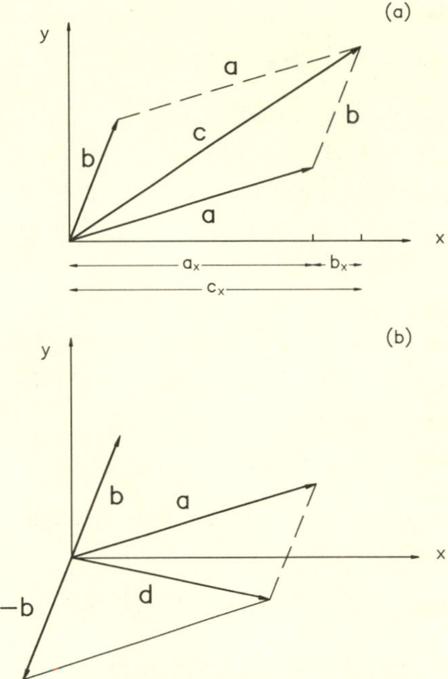

Fig. 3.1 (a) Diagram illustrating vector addition $\mathbf{c} = \mathbf{a}+\mathbf{b}$. The dashed vectors are parallel to the original vectors **a** and **b**, and hence have been labelled identically. Then **c** lies along the diagonal of the parallelogram so formed. This diagram is only in two dimensions, but a corresponding procedure is used for addition in three (or more) dimensions. (b) Subtraction of **b** from **a** is equivalent to adding **a** and $-\mathbf{b}$.

(See fig. 3.1(a)). Alternatively, we could have achieved the same result by interchanging **a** and **b** in this sequence of operations.

Subtraction of vectors involves nothing new, since we can regard the equation

$$\mathbf{d} = \mathbf{a} - \mathbf{b} \tag{3.6}$$

as simply involving the addition of $-\mathbf{b}$ (see fig. 3.1(b)). The vector $-\mathbf{b}$ is of course the same length as **b**, but opposite in direction (see eqn (3.2)). Again in terms of components

$$d_i = a_i - b_i, \tag{3.7}$$

where i is taken to mean x, y or z.

Because of the way vector addition is defined, we can write any vector **F** as

$$\mathbf{F} = F_x\mathbf{i} + F_y\mathbf{j} + F_z\mathbf{k} \tag{3.8}$$

where **i**, **j** and **k** are unit vectors along the x, y and z axes respectively. Thus $F_x\mathbf{i}$ is a vector along the x axis of magnitude equal to the component of **F** in that direction. It is important to distinguish between the vector $F_x\mathbf{i}$ and F_x, which is a scalar quantity. It is thus impossible to write $\mathbf{F} = F_x + F_y + F_z$, because this would involve equating a vector to a scalar.

A special case of (3.8) occurs when the vector is the position vector **r**. Since the components of **r** are simply the x, y and z coordinates of the point,

$$\mathbf{r} = x\mathbf{i} + y\mathbf{j} + z\mathbf{k} \tag{3.9}$$

If we want to specify the unit vector $\hat{\mathbf{r}}$, we need to divide the above equation by the magnitude of **r**. Then

$$\hat{\mathbf{r}} = \cos\vartheta_x\mathbf{i} + \cos\vartheta_y\mathbf{j} + \cos\vartheta_2\mathbf{k} \tag{3.10}$$

since $x/r = \cos\vartheta_x$, where ϑ_x is the angle that the direction of **r** makes with the x axis. Thus $\cos\vartheta_x$ is just the relevant direction cosine.

Another consequence of the definitions (3.2) and (3.4) is that repeated addition of the same vector is equivalent to scalar multiplication by the appropriate number, e.g.

$$\mathbf{a} + \mathbf{a} + \mathbf{a} = 3\mathbf{a} \tag{3.11}$$

Equations of the form (3.2) occur frequently in physics, and include examples where the scalar k has physical units, rather than being a dimensionless constant. For example, the relation

$$\mathbf{F} = m\mathbf{a}$$

relates the acceleration **a** produced on a body of mass m by a force **F**. This implies that the acceleration is always in the same direction as the force, and that the ratio of their magnitudes is m.

Another example is

$$\mathbf{E} = \frac{1}{4\pi\epsilon_0}\frac{q}{r^2}\hat{\mathbf{r}} \tag{3.12}$$

where **E** is the electric field produced at a distance r from the origin, where there is a point charge q. The constant ϵ_0 is known as the permittivity of free space, and $\hat{\mathbf{r}}$ is a unit vector pointing away from the

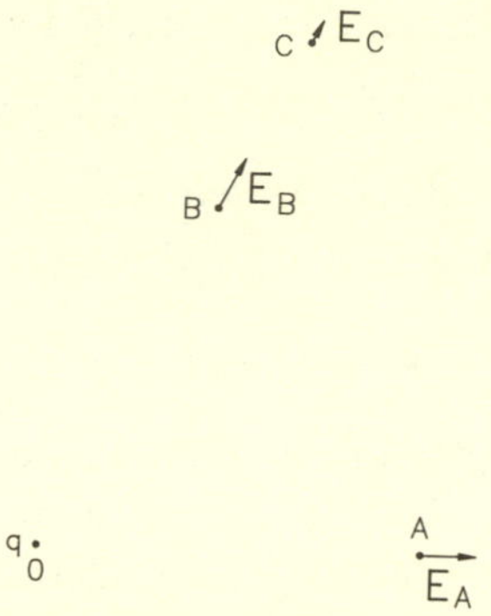

Fig. 3.2 A point charge q is situated at the origin O. At each point in space, the electric field **E** is parallel to the radial vector $\hat{\mathbf{r}}$. At B and C, the radial vectors are parallel, but at A it is in a different direction. The field at C is smaller than that at B because of the r^2 factor in the denominator of eqn (3.12). Even though A and B are equidistant from the origin, the electric field vectors at A and B are not equal, because they point in different directions.

origin. An interesting feature of this equation is that **E** varies not only in magnitude as r varies, but also in direction as we consider different points at constant r, i.e. **E** is a vector field. This is built into eqn (3.12), because $\hat{\mathbf{r}}$, being radial, is not always in the same direction (see fig. 3.2).

It is important in eqn (3.12) to distinguish between the unit vector $\hat{\mathbf{r}}$ and the scalar distance r.

Vector addition is used, for example, in deciding the net force **F** acting on an object which is subject to several different forces $\mathbf{F}_1, \mathbf{F}_2, \mathbf{F}_3 \ldots$ simultaneously. Thus a small charged ball moving in an electric and a magnetic field would experience gravitational, electrostatic and magnetic forces, the sum of them giving the total force, which determines the ball's instantaneous acceleration.

A physical example of vector subtraction is provided by relative velocities. If two cars are travelling with velocities $\mathbf{v}_1$ and $\mathbf{v}_2$, it is their relative velocity $\mathbf{v}_1 - \mathbf{v}_2$ which is relevant for determining the damage that would occur in a collision, and the direction from which the impact occurs.

3.3 Scalar and vector products of vectors

We have already dealt with the product of a vector with a scalar (see eqns (3.2) and (3.3)). Now we consider the product of two vectors. Since this is a new concept, there is freedom in choosing what is meant by this.

In fact two different products are defined, called the scalar and the vector products, because the answers are respectively a scalar and a vector.

3.3.1 Definitions

The scalar product † is given by

$$\mathbf{a} \cdot \mathbf{b} = ab \cos \vartheta \tag{3.13}$$

where a and b are the lengths of the vectors **a** and **b**, and ϑ is the angle between them. Thus the scalar product can be thought of as the length of **a**, times the component of **b** along **a** (or alternatively, as the length of **b**, multiplied by the component of **a** along it.)

For the vector product,

$$\mathbf{a} \wedge \mathbf{b} = ab \sin \vartheta \hat{\mathbf{n}} \tag{3.14}$$

where $\hat{\mathbf{n}}$ is a unit vector perpendicular to the plane containing **a** and **b**. Of course there are two antiparallel directions normal to any plane. By convention, the one chosen here is such that **a**, **b** and $\hat{\mathbf{n}}$ make a right-handed set of axes. That is, if a right-handed screw is rotated from **a** to **b**, it advances along $\hat{\mathbf{n}}$. (Thus the axes defined as follows are right-handed: x in the direction of lines of lattitude, pointing east; y parallel to lines of longitude, and pointing north; and z vertically upwards.)

The scalar product is always written with a dot between the two vectors being multiplied, while the vector product has a $\wedge$ or a $\times$ between the vector factors. These products are thus also referred to as dot and cross products. A combination of two vectors written side by side with no symbol between them (e.g. **ab**) is meaningless.

3.3.2 Why are they defined like this?

On looking back at eqns (3.13) and (3.14), we may think that it is sensible to define separate scalar and vector products, and that it is reasonable that the products of the lengths of the vectors appear in the answers, but we may wonder why the particular choices were made for the angular factors and for $\hat{\mathbf{n}}$. The answer is that it turns out that these combinations of factors tend to be useful in applications of vectors, either within mathematics or to physical situations (see Section 3.3.6).

† It is important to distinguish between the 'scalar product' discussed here, which is the product of two vectors and is a scalar, and 'multiplication by a scalar' (see Section 3.2), where a vector is multiplied by a scalar, and the result is a vector.

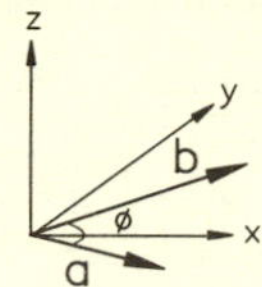

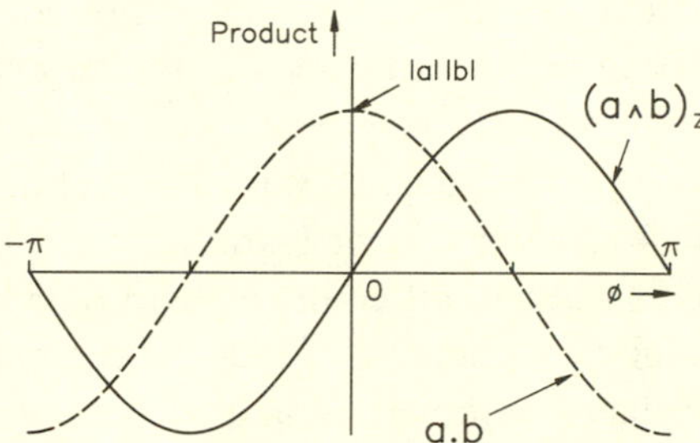

Fig. 3.3 Scalar and vector products for two vectors in the x–y plane. The vector product lies along the z axis; it vanishes when **a** and **b** are parallel. When ϕ is negative, the angle θ between the vectors is $|\phi|$. For ϕ between zero and $-\pi$, the z component of $\mathbf{a} \wedge \mathbf{b}$ is negative because the definition of $\hat{\mathbf{n}}$ in eqn (3.14) is such that it has flipped over, as compared with its orientation for ϕ in the range zero to π. The scalar product (dashed line) vanishes when **a** and **b** are perpendicular.

Also as far as the vector product (3.14) is concerned, the two vectors **a** and **b** define a plane (provided they are not parallel), and the unique direction as far as a plane is concerned is its normal. Thus if we want a special direction for the vector product of **a** and **b**, the choice of the normal to both of them is natural.

When **a** and **b** are parallel, their normal is no longer uniquely defined, but this is not a problem as the $\sin\vartheta$ factor is zero. Thus the length of the vector answer is zero, and its direction is unimportant.

In contrast the scalar product is zero whenever the two vectors are perpendicular to each other.

Fig. 3.3 shows the way the scalar product and the magnitude of the vector product vary for the case where **a** is a fixed vector in the x–y plane, and **b** is a vector of fixed length also in the x–y plane, but making a variable angle ϕ with **a**. In this case, the vector product lies along the z axis; its magnitude is shown as positive or negative depending on whether it lies along the positive or negative z direction.

3.3.3 Non-commutation of vector product

The vector product contains a new feature in that

$$\mathbf{a} \wedge \mathbf{b} = -\mathbf{b} \wedge \mathbf{a} \tag{3.15}$$

i.e. $\mathbf{a} \wedge \mathbf{b}$ is not equal to $\mathbf{b} \wedge \mathbf{a}$. This is because of our definition of $\hat{\mathbf{n}}$ for the direction of the answer being given by a right-handed screw rotation from the first vector to the second. If we interchange the order of the two vectors, the direction of $\hat{\mathbf{n}}$ reverses. This is not compensated by any possible change in sign in the other factors of the definition (3.14) for the vector product: a and b are lengths of vectors and hence necessarily positive, while ϑ is simply the angle between the vectors, and hence is the same whether we consider $\mathbf{a}$ and $\mathbf{b}$, or $\mathbf{b}$ and $\mathbf{a}$.

Eqn (3.15) is exciting because it contrasts with the behaviour of products of ordinary numbers. There the order of the factors is irrelevant; multiplication of ordinary numbers is said to be commutative. In fact this is also true for vector addition and for scalar products (see eqns (3.5) and (3.13)) where the order of $\mathbf{a}$ and $\mathbf{b}$ is irrelevant. For the vector product, this is decidedly not the case.

This does not mean that the definition of the vector product is wrong. It is just that the process is non-commutative. We must thus be careful about the order in which the factors occur. If we interchange them, we must remember to include the minus sign of eqn (3.15). We shall find other examples of non-commutation when we come to matrices in Chapter 15 of Volume 2.

3.3.4 Special cases

Some special cases of scalar products merit attention. The scalar product of any vector with itself is just the square of its length:

$$\mathbf{a} \cdot \mathbf{a} = a^2 \tag{3.16}$$

Thus if we want an equation expressing the fact that a vector $\mathbf{v}$ is of length 3, we can write $\mathbf{v} \cdot \mathbf{v} = 9$.

For unit vectors along the axes, the scalar products are thus

$$\left.\begin{aligned} \mathbf{i} \cdot \mathbf{i} = \mathbf{j} \cdot \mathbf{j} = \mathbf{k} \cdot \mathbf{k} = 1 \\ \text{but } \mathbf{i} \cdot \mathbf{j} = \mathbf{j} \cdot \mathbf{k} = \mathbf{k} \cdot \mathbf{i} = 0 \end{aligned}\right\} \tag{3.17}$$

The first line of these contains just special cases of (3.16), while the second line follows immediately from the orthogonality of the axes.

If two vectors are given in component form, we can use eqns (3.17) to obtain

$$\begin{aligned}\mathbf{a}\cdot\mathbf{b} &= (a_x\mathbf{i}+a_y\mathbf{j}+a_z\mathbf{k})\cdot(b_x\mathbf{i}+b_y\mathbf{j}+b_z\mathbf{k})\\ &= a_xb_x+a_yb_y+a_zb_z\end{aligned} \tag{3.18}$$

Now we investigate some simple examples of vector products. For any vector multiplied by itself, we have

$$\mathbf{v}\wedge\mathbf{v}=0 \tag{3.19}$$

since the relevant angle ϑ is 0.

For the unit vectors along the coordinate axes,

$$\left.\begin{aligned}\mathbf{i}\wedge\mathbf{i}=\mathbf{j}\wedge\mathbf{j}=\mathbf{k}\wedge\mathbf{k}=0\\ \mathbf{i}\wedge\mathbf{j}=\mathbf{k},\ \mathbf{j}\wedge\mathbf{k}=\mathbf{i},\ \mathbf{k}\wedge\mathbf{i}=\mathbf{j}\\ \text{and } \mathbf{i}\wedge\mathbf{k}=-\mathbf{j},\ \mathbf{j}\wedge\mathbf{i}=-\mathbf{k},\ \mathbf{k}\wedge\mathbf{j}=-\mathbf{i}\end{aligned}\right\} \tag{3.20}$$

where the top line is a special case of (3.19).

The first equality in the second line follows from eqn (3.14), and from the facts that the length of $\mathbf{i}\wedge\mathbf{j}$ is unity, because $\mathbf{i}$ and $\mathbf{j}$ are each of unit length, and they are at right angles to each other; and the direction of $\mathbf{i}\wedge\mathbf{j}$ is normal to both of them (i.e. $\pm\mathbf{k}$, with the positive sign being selected by the right-hand screw rule). The second and third equalities on the same line can be obtained from the first by a cyclic increase in the index of the unit vector (i.e. $\mathbf{i}\to\mathbf{j}\to\mathbf{k}\to\mathbf{i}$), and similarly for the third line. In the second and third lines, the equation contains a + sign if $\mathbf{i}$, $\mathbf{j}$ and $\mathbf{k}$ occur in cyclic order, and a − sign otherwise.

We can now write out the vector product of two vectors given in component form:

$$\begin{aligned}\mathbf{a}\wedge\mathbf{b} &= (a_x\mathbf{i}+a_y\mathbf{j}+a_z\mathbf{k})\wedge(b_x\mathbf{i}+b_y\mathbf{j}+b_z\mathbf{k})\\ &= (a_yb_z-a_zb_y)\mathbf{i}+\dots\end{aligned} \tag{3.21}$$

where the dots indicate that there are two more terms obtained by cyclic increases of the components and the unit vectors. The answer is identical to the following determinant:

$$\mathbf{a}\wedge\mathbf{b}=\begin{vmatrix}\mathbf{i} & \mathbf{j} & \mathbf{k}\\ a_x & a_y & a_z\\ b_x & b_y & b_z\end{vmatrix} \tag{3.22}$$

Determinants we have met so far probably contained only real numbers, and the value of the determinant would then have been a number. Here the top row consists of three vectors. If we just multiply out the various elements according to the usual rules for determinants, we obtain a vector for the answer; this turns out to be identical to (3.21).

It is much easier to remember (3.22) than (3.21). We might worry that it is not obvious whether **i**, **j** and **k** should appear as the top line or the bottom; or alternatively why **i**, **j**, **k** and the components of **a** and of **b** each occur as a row rather than as a column. In fact, because of the properties of determinants, these changes make no difference. Of course, if we interchange the row of a_x, a_y and a_z with that of b_x, b_y and b_z, the answer will change sign; this corresponds to the fact that the vector product depends on the order of the factors. Thus it is important to have the rows (or alternatively, the columns) of the components in the same order as the corresponding vectors in the vector product. Other plausible changes are unimportant.

The determinant (3.22) contains one row of vectors, and two rows of components. It could not have contained two or three rows of vectors, because then we would have had two or three factors of vectors sitting next to each other in the answer, without a dot or a cross between them. As explained in the last paragraph of Section 3.3.1, this is forbidden.

3.3.5 Equations with vector and scalar products

We sometimes have equations like

$$\mathbf{a} \cdot \mathbf{b} = \mathbf{a} \cdot \mathbf{c}$$

It is tempting, but in general incorrect, to deduce from this that $\mathbf{b} = \mathbf{c}$. The reason this is wrong is that the equation relates only the components of **b** and of **c** along the **a** direction; it says absolutely nothing about any components perpendicular to **a**. To make this point obvious, we can extend the above equation as

$$\mathbf{a} \cdot \mathbf{b} = \mathbf{a} \cdot \mathbf{c} = \mathbf{a} \cdot (\mathbf{c} + \lambda \mathbf{d})$$

where **d** is perpendicular to **a** (and hence $\mathbf{a} \cdot \mathbf{d}$ is zero), and λ is an arbitrary parameter. Then **b** cannot be equal both to **c** and to $\mathbf{c} + \lambda \mathbf{d}$.

The only situation where we can deduce from $\mathbf{a} \cdot \mathbf{b} = \mathbf{a} \cdot \mathbf{c}$ that $\mathbf{b} = \mathbf{c}$ is if the scalar products are equal for all possible vectors **a**, in any arbitrary directions. For various choices of **a**, we can ensure that each component of **b** is equal to the corresponding component of **c**, and hence the vectors themselves are equal.

Similar remarks apply to the equation

$$\mathbf{a} \wedge \mathbf{b} = \mathbf{a} \wedge \mathbf{c}$$

In this case, we can add to $\mathbf{b}$ or $\mathbf{c}$ any vector parallel to $\mathbf{a}$, without affecting the equality. Thus again we cannot conclude that $\mathbf{b} = \mathbf{c}$, unless the relationship above is true for all possible choices of $\mathbf{a}$.

3.3.6 Applications

Here we give a few random examples of the use of these products, in order to provide a feeling for their widespread applicability.

3.3.6.1 Scalar products

The definition of the scalar product is such that $\mathbf{v} \cdot \hat{\mathbf{a}}$ gives the component of $\mathbf{v}$ along $\hat{\mathbf{a}}$, where $\hat{\mathbf{a}}$ is a unit vector. Thus we can express $\mathbf{v}$ in terms of components along three orthogonal axes parallel to the vectors $\mathbf{a}, \mathbf{b}$ and $\mathbf{c}$, as

$$\mathbf{v} = \frac{(\mathbf{v} \cdot \mathbf{a})\mathbf{a}}{\mathbf{a} \cdot \mathbf{a}} + \frac{(\mathbf{v} \cdot \mathbf{b})\mathbf{b}}{\mathbf{b} \cdot \mathbf{b}} + \frac{(\mathbf{v} \cdot \mathbf{c})\mathbf{c}}{\mathbf{c} \cdot \mathbf{c}} \tag{3.23}$$

where the scalar products in the denominator are required in order to convert $\mathbf{a}$ into $\hat{\mathbf{a}}$, etc.

The angle ϑ between two directions specified by unit vectors $\hat{\mathbf{u}}$ and $\hat{\mathbf{v}}$ is given by

$$\cos\vartheta = \hat{\mathbf{u}} \cdot \hat{\mathbf{v}} \tag{3.24}$$

This again follows directly from the definition of the scalar product.

A plane through the origin is such that the vector $\mathbf{r}$ from the origin to any point in the plane is perpendicular to the plane's normal $\mathbf{n}$. Then, from eqn (3.24), the equation of the plane is

$$\mathbf{r} \cdot \mathbf{n} = 0 \tag{3.25}$$

A geometry problem that can be solved elegantly with the use of vectors is the proof that the diagonals of a rhombus are perpendicular. If two adjacent sides of the rhombus are denoted by $\mathbf{a}$ and $\mathbf{b}$, the diagonals are given by $\mathbf{a} + \mathbf{b}$ and $\mathbf{a} - \mathbf{b}$ (see fig. 3.4). The angle between them is determined from the scalar product

$$(\mathbf{a} + \mathbf{b}) \cdot (\mathbf{a} - \mathbf{b}) = \mathbf{a} \cdot \mathbf{a} - \mathbf{b} \cdot \mathbf{b}$$

But $\mathbf{a} \cdot \mathbf{a}$ and $\mathbf{b} \cdot \mathbf{b}$ are just the squares of the lengths of the sides, which for a rhombus are equal. Thus the dot product of the vectors describing the diagonals is zero, and hence the diagonals are perpendicular.

We consider other aspects of vector geometry in Section 3.5.

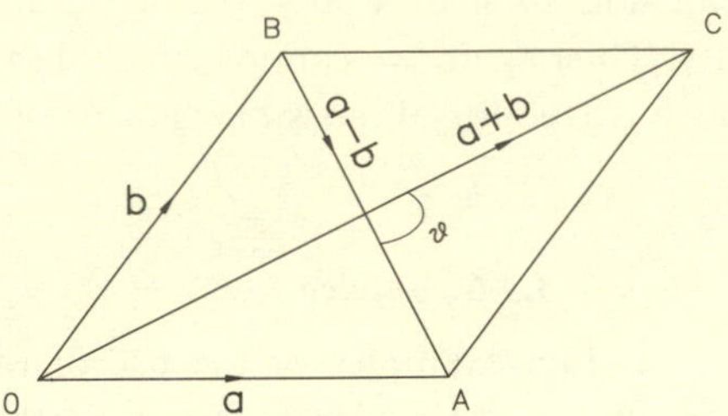

Fig. 3.4 OACB is a rhombus, with sides OA and OB being given by the vectors **a** and **b**. The diagonals OC and BA are then $\mathbf{a}+\mathbf{b}$ and $\mathbf{a}-\mathbf{b}$ respectively. The cosine of the angle ϑ between the diagonals is determined in terms of the scalar product of $\mathbf{a}+\mathbf{b}$ and $\mathbf{a}-\mathbf{b}$; as this turns out to be zero, ϑ must be 90°.

If a force **F** acts on a body which moves through a distance **s**, the work done is

$$W = \mathbf{F} \cdot \mathbf{s} \tag{3.26}$$

The $\cos\vartheta$ factor in the scalar product allows for the fact that it is only the component of **F** along the direction of motion that does work.

3.3.6.2 Vector products

The area of a parallelogram with sides **a** and **b** is given by $\mathbf{a} \wedge \mathbf{b}$. This is a vector, normal to the plane of the parallelogram. Usually we think of areas as scalars; if we want to keep to this convention, we should write it as $|\mathbf{a} \wedge \mathbf{b}|$. Sometimes, however, it is useful to consider areas as vectors (see Section 3.4.2); however, we should beware of the fact that the area then depends (up to a sign) on the order in which we consider the sides.†

If a body has a force **F** acting on it at a point whose position with respect to some origin is **r**, this force produces a torque $\mathbf{r} \wedge \mathbf{F}$ about the origin. The vector product takes care of the angle between **F** and **r** as required for the calculation of the torque. The result is again of course a vector, whose direction specifies the direction of the axis about which the body acquires angular momentum (with respect to the origin). The sense of the torque along this axis determines the sense of the angular acceleration; if we imagine holding the axis with our right hand so that our outstretched thumb points along the direction of the torque's vector, our curled-up fingers show the direction of motion. For the situation

† Another problem is that for a curved surface, the vectors corresponding to the different surface elements are not all parallel, and hence when they are added, the magnitude of the resultant vector area is less than the conventional surface area. Thus the total vector area of a closed surface is zero.

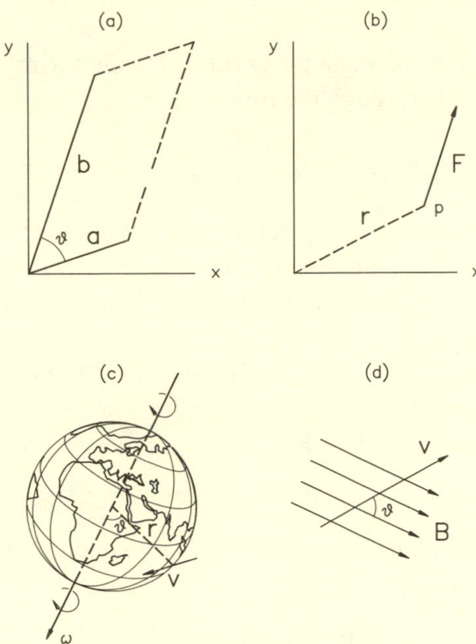

Fig. 3.5 Examples of vector products: (a) The parallelogram with sides of length a and b has area $ab \sin \vartheta$. This is just the magnitude of the vector $\mathbf{a} \wedge \mathbf{b}$ (see eqn (3.14)). The vector representing the area is normal to the plane of the parallelogram. (b) The force $\mathbf{F}$ acting at a point P exerts a torque about the origin. Its magnitude is $Fr \sin \vartheta$, where r is the distance of P from the origin. This is equal to the magnitude of $\mathbf{r} \wedge \mathbf{F}$. (c) The earth is rotating about its axis as shown. Then $\boldsymbol{\omega}$ gives the magnitude of the angular velocity, and the direction of the axis. At a point on the surface, defined by a vector $\mathbf{r}$ from a point on the axis of rotation, the speed due to the rotation and the direction of motion are given by the velocity vector $\mathbf{v} = \boldsymbol{\omega} \wedge \mathbf{r}$. (d) A particle of electric charge q moves with velocity $\mathbf{v}$ in a region of magnetic induction $\mathbf{B}$. It experiences a force $\mathbf{F}$ whose magnitude is $qvB \sin \vartheta$, and whose direction is perpendicular to both $\mathbf{v}$ and $\mathbf{B}$. Then $\mathbf{F} = q\mathbf{v} \wedge \mathbf{B}$ gives the correct magnitude and direction for the force. The effect of the force is to deflect the particle.

shown in fig. 3.5(b), the torque $\mathbf{r} \wedge \mathbf{F}$ acts out of the page, and the angular motion with respect to the origin is anticlockwise.

Angular momentum is another example from mechanics involving vector products; a particle of momentum $\mathbf{p}$ at a position $\mathbf{r}$ with respect to the origin has angular momentum

$$\mathbf{L} = \mathbf{r} \wedge \mathbf{p}$$

about the origin. Its angular velocity $\boldsymbol{\omega}$ is $\mathbf{r} \wedge \mathbf{v}/r^2$, where $\mathbf{v}$ is the particle's velocity.

If a solid body (for example the earth) is rotating with angular velocity $\boldsymbol{\omega}$ about a given axis through the origin, then a point with position $\mathbf{r}$ has velocity

$$\mathbf{v} = \boldsymbol{\omega} \wedge \mathbf{r}$$

where $\boldsymbol{\omega}$ is (fairly obviously) a vector whose magnitude is ω, and whose direction specifies the rotation axis. The cross product results in a vector which is in the correct direction (and of the right magnitude) to describe the motion at the specified point.

Electromagnetism abounds in situations where vector products occur. Thus the force $\mathbf{F}$ on a charge q moving with velocity $\mathbf{v}$ through a region of magnetic field of induction $\mathbf{B}$ is

$$\mathbf{F} = q\mathbf{v} \wedge \mathbf{B} \tag{3.27}$$

The Biot–Savart law for the magnetic induction due to a current I flowing through a short wire is

$$\mathbf{B} = \frac{\mu_0}{4\pi} I\,d\mathbf{l} \wedge \mathbf{r}/\mathbf{r} \cdot \mathbf{r} \tag{3.28}$$

where $\mathbf{r}$ is the vector from the current element to the point where we want to know $\mathbf{B}$, $d\mathbf{l}$ gives the length of the wire and its direction, and μ_0 is the permeability of free space. Yet another example is the force $\mathbf{F}$ on a short length of wire carrying a current, when in a region of magnetic induction:

$$\mathbf{F} = I\,d\mathbf{l} \wedge \mathbf{B} \tag{3.29}$$

In all these cases, the $\sin\vartheta$ factor in the definition of the vector product is exactly what is required to give the correct magnitude of the effect, and the direction of the vector product specifies its direction (i.e. that of the force in (3.27) and (3.29), and of the induction $\mathbf{B}$ in (3.28)).

3.4 Triple scalar and vector products

3.4.1 Definitions and simple properties

Because $\mathbf{a} \wedge \mathbf{b}$ is a vector, we can use it in either of the definitions of Section 3.3 for the product of two vectors. Thus $(\mathbf{a} \wedge \mathbf{b}) \cdot \mathbf{c}$ is termed the triple scalar product, because the answer is a scalar (even though it involves three vectors, and a vector product); and $(\mathbf{a} \wedge \mathbf{b}) \wedge \mathbf{c}$ is called the triple vector product (guess why!).

For **a**, **b** and **c** in component form, the triple scalar product is readily written down using eqns (3.22) and (3.18) as

$$(\mathbf{a} \wedge \mathbf{b}) \cdot \mathbf{c} = \begin{vmatrix} c_x & c_y & c_z \\ a_x & a_y & a_z \\ b_x & b_y & b_z \end{vmatrix} \tag{3.30}$$

This is because taking the scalar product of any two vectors involves writing one of them in the form of eqn (3.8), and then replacing the **i**, **j** and **k** by the relevant components of the other vector. With $\mathbf{a} \wedge \mathbf{b}$ as in eqn (3.22) and substituting the components of **c** for the unit vectors, we immediately obtain (3.30).

Because the order of the factors in a scalar product is unimportant, $\mathbf{c} \cdot (\mathbf{a} \wedge \mathbf{b})$ is equal to the previous triple scalar product. What is somewhat more surprising is that these are also equal to $\mathbf{a} \cdot (\mathbf{b} \wedge \mathbf{c})$. This is because, from eqn (3.30), we have

$$\mathbf{a} \cdot (\mathbf{b} \wedge \mathbf{c}) = (\mathbf{b} \wedge \mathbf{c}) \cdot \mathbf{a} = \begin{vmatrix} a_x & a_y & a_z \\ b_x & b_y & b_z \\ c_x & c_y & c_z \end{vmatrix} \tag{3.30$'$}$$

and as the determinants in (3.30) and (3.30′) differ merely in a cyclic reordering of the rows, they are equal. Thus the positions of the dot and the cross in the triple scalar product are irrelevant, and

$$\mathbf{a} \cdot (\mathbf{b} \wedge \mathbf{c}) = (\mathbf{a} \wedge \mathbf{b}) \cdot \mathbf{c} \tag{3.31}$$

Similarly cyclic permutations of **a**, **b** and **c** leave the triple scalar product unaltered.

There is a convention that the triple scalar product is sometimes written as $\{\mathbf{a}\,\mathbf{b}\,\mathbf{c}\}$, it being understood that a dot and a cross are required somewhere. The triple scalar product is completely symmetric in the three vectors, even though we introduced it as the vector product of the first two, dotted with the third.

Since $(\mathbf{b} \wedge \mathbf{c}) = -(\mathbf{c} \wedge \mathbf{b})$,

$$\mathbf{a} \cdot (\mathbf{b} \wedge \mathbf{c}) = -\mathbf{a} \cdot (\mathbf{c} \wedge \mathbf{b}) \tag{3.32}$$

Thus if the cyclic order is broken, the triple scalar product changes sign, i.e.

$$\{\mathbf{a}\,\mathbf{b}\,\mathbf{c}\} = -\{\mathbf{a}\,\mathbf{c}\,\mathbf{b}\} \tag{3.32$'$}$$

with it again being irrelevant where we insert the dot and the cross on each side of the equation, or what is the ordering of the three vectors on each side, provided they are cyclic.

One other point to note is that we placed brackets around $\mathbf{a} \wedge \mathbf{b}$ in eqn (3.30). This was because we first take this vector product, before performing the scalar product with $\mathbf{c}$. The opposite order of operations (i.e. $\mathbf{a} \wedge (\mathbf{b} \cdot \mathbf{c})$) is impossible; $\mathbf{b} \cdot \mathbf{c}$ is a scalar product, and there is no way we can take the vector product of this with $\mathbf{a}$. Thus if we write the triple scalar product as in eqn (3.32′) without any internal brackets, it is very definitely to be understood that we first perform the vector product, and then the scalar one. Thus although, as we said earlier, we are free to put the dot and the cross where we want, their positions define which two of the vectors are multiplied first. Eqn (3.31) of course satisfies this requirement.

The triple vector product $(\mathbf{a} \wedge \mathbf{b}) \wedge \mathbf{c}$ is a vector. Now $\mathbf{a} \wedge \mathbf{b}$ is perpendicular to $\mathbf{a}$ and to $\mathbf{b}$ (i.e. it is in the direction of the normal to the plane containing $\mathbf{a}$ and $\mathbf{b}$), and $(\mathbf{a} \wedge \mathbf{b}) \wedge \mathbf{c}$ is perpendicular to $(\mathbf{a} \wedge \mathbf{b})$. It is thus in the plane containing $\mathbf{a}$ and $\mathbf{b}$. This implies that it can be written as

$$(\mathbf{a} \wedge \mathbf{b}) \wedge \mathbf{c} = \lambda\mathbf{a} + \mu\mathbf{b} \tag{3.33}$$

where λ and μ are scalar constants. It turns out that

$$(\mathbf{a} \wedge \mathbf{b}) \wedge \mathbf{c} = (\mathbf{a} \cdot \mathbf{c})\mathbf{b} - (\mathbf{b} \cdot \mathbf{c})\mathbf{a} \tag{3.34}$$

as can be verified by writing out each side in component form† (see Problem 3.2). In fact it takes a fair amount of tedious work in order to deduce relations like (3.34). Thus if we can make use of them in problems involving vectors, they will save us a lot of algebra that would otherwise have been required.

Eqn (3.34) is indeed of the form of eqn (3.33), and the fact that the two terms on the right-hand side each involve $\mathbf{a}$, $\mathbf{b}$ and $\mathbf{c}$ has an air of plausibility. The factors multiplying $\mathbf{a}$ and $\mathbf{b}$ are each scalars (they are in fact scalar products) so the right-hand side is a vector as is required.

My only problem with eqn (3.34) is that I cannot remember which term has the minus sign, partly because it depends on whether we are writing $(\mathbf{a} \wedge \mathbf{b}) \wedge \mathbf{c}$ or $\mathbf{c} \wedge (\mathbf{a} \wedge \mathbf{b})$. In the absence of a list of formulae, the following procedure can be used:

† Both here and in eqn (3.30) earlier, we seem to be contradicting our earlier maxim that wherever possible we should avoid using components. The point is that components can provide a boring but useful way of proving basic vector identities. From then on, however, we should be able to utilise such identities without resorting to components again.

(i) The argument above eqn (3.33) gives the correct form of the answer.
(ii) Remember that each term involves all three vectors and has a scalar product, and that one term has a minus sign.
(iii) Guess where the minus sign goes, and then check it with a simple specific example.

Thus eqn (3.34) gives

$$\begin{aligned}(\mathbf{i}\wedge\mathbf{j})\wedge\mathbf{i} &= (\mathbf{i}\cdot\mathbf{i})\mathbf{j}-(\mathbf{j}\cdot\mathbf{i})\mathbf{i}\\ &= \mathbf{j}\end{aligned} \tag{3.35}$$

where we have used eqns (3.17) for the scalar products of the unit vectors. Alternatively, by direct vector multiplication and using eqns (3.20), we obtain

$$(\mathbf{i}\wedge\mathbf{j})\wedge\mathbf{i}=\mathbf{k}\wedge\mathbf{i}=\mathbf{j} \tag{3.35$'$}$$

thereby confirming that the sign choice in eqn (3.34) is correct.

With triple vector products, it is most important to include brackets round the pair of vectors to be multiplied first. This is because $\mathbf{a}\wedge(\mathbf{b}\wedge\mathbf{c})$ is in general different from $(\mathbf{a}\wedge\mathbf{b})\wedge\mathbf{c}$; the former is in the plane of $\mathbf{b}$ and $\mathbf{c}$, while the latter is in the plane of $\mathbf{a}$ and $\mathbf{b}$, and so usually they will not even be parallel. This contrasts with the situation for triple scalar products, where the brackets can be omitted (because we know it is essential to perform the vector product first).

Another difference is that the triple scalar product vanishes if two of the vectors are parallel, but this is not so for the triple vector product, provided they are not in the first bracket.

Of course it is possible to construct products involving even more vectors (for example, $\mathbf{a}\wedge[\mathbf{b}\wedge(\mathbf{c}\wedge\mathbf{d})]$, $(\mathbf{a}\wedge\mathbf{b})\cdot(\mathbf{c}\wedge\mathbf{d})$, $(\mathbf{a}\wedge\mathbf{b})\wedge(\mathbf{c}\wedge\mathbf{d})$, etc.). However, once you have achieved familiarity with triple products, the properties of these more complex products follow in a straightforward manner.

3.4.2 Applications

The extension of a parallelogram to three dimensions is called a parallelepiped. Its volume is given by its base area times its perpendicular height. If three sides meeting at the origin are described by vectors $\mathbf{a}$, $\mathbf{b}$ and $\mathbf{c}$, this volume is $|\,(\mathbf{a}\wedge\mathbf{b})\cdot\mathbf{c}\,|$ (The modulus sign is required because the triple scalar product can be negative, while a volume is conventionally positive.) The vector area $\mathbf{a}\wedge\mathbf{b}$ points in just the right direction so

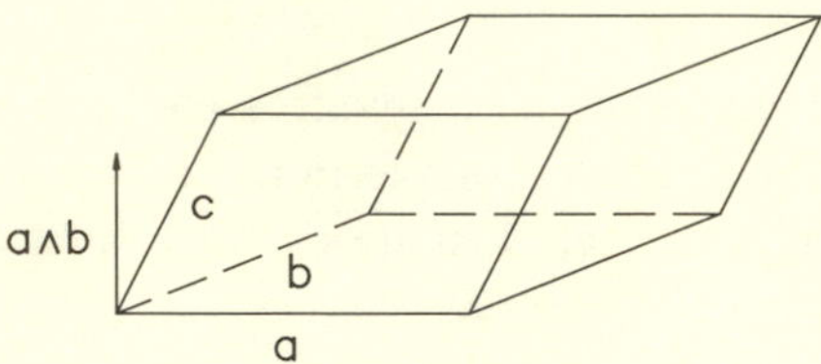

Fig. 3.6 A parallelepiped, whose sides meeting at the origin are described by vectors **a**, **b** and **c**. The area of the base is $\mathbf{a} \wedge \mathbf{b}$, which is a vector perpendicular to the base. Then $(\mathbf{a} \wedge \mathbf{b}) \cdot \mathbf{c}$ is this area multiplied by the projection of **c** along this normal, i.e. it is the base area times the perpendicular height, which is just the volume of the parallelepiped.

that when it is dotted with **c**, the triple scalar product is the volume (see fig. 3.6). While areas are given by vector products and thus turn out to be vectors, volumes are triple scalar products, and hence scalars.

The volume is symmetric with respect to the three sides of the parallelepiped; this is consistent with the modulus of the triple scalar product being independent of the way the individual vectors are considered.

Another use of the triple scalar product is for deciding whether three vectors **p**, **q** and **r** are coplanar. If they are, then

$$\{\mathbf{pqr}\} = 0 \tag{3.36}$$

There are several ways of seeing that this is so. First, the volume of the parallelepiped made out of these vectors would be zero if they were coplanar. Alternatively, $\mathbf{p} \wedge \mathbf{q}$ is perpendicular to the plane of **p** and **q**; if **r** also lies in this plane, it is perpendicular to $\mathbf{p} \wedge \mathbf{q}$, and so $(\mathbf{p} \wedge \mathbf{q}) \cdot \mathbf{r}$ is zero. Similarly, if **r** is in the plane of **p** and **q**, it can be expressed as $\lambda\mathbf{p} + \mu\mathbf{q}$, and then

$$\begin{aligned}(\mathbf{p} \wedge \mathbf{q}) \cdot \mathbf{r} &= (\mathbf{p} \wedge \mathbf{q}) \cdot (\lambda\mathbf{p} + \mu\mathbf{q}) \\ &= \lambda\{\mathbf{pqp}\} + \mu\{\mathbf{pqq}\} \\ &= 0\end{aligned} \tag{3.37}$$

Other examples of a triple scalar product occur in considering the shortest distance between two skew lines (see Section 3.5.1), and in the definition of reciprocal vector sets (Section 3.6).

As an example of the use of the triple vector product, we consider the force **F** on a short current-carrying wire, due to the magnetic field produced by a second short wire with another current flowing in it. The force is given by eqn (3.29) as

$$\mathbf{F} = I_1 d\mathbf{l}_1 \wedge \mathbf{B}, \tag{3.29$'$}$$

where I_1 and $d\mathbf{l}_1$ refer to the first wire, and $\mathbf{B}$ is the magnetic flux produced there by the second current. But from (3.28),

$$\mathbf{B} = \frac{\mu_0}{4\pi} I_2(d\mathbf{l}_2 \wedge \mathbf{r})/\mathbf{r}\cdot\mathbf{r} \tag{3.28$'$}$$

where I_2 and $d\mathbf{l}_2$ are for the second wire, and $\mathbf{r}$ is the vector from the second to the first wire. Substituting (3.28′) into (3.29′) yields

$$\mathbf{F} = \frac{\mu_0}{4\pi} I_1 I_2 d\mathbf{l}_1 \wedge (\mathbf{dl}_2 \wedge \mathbf{r})/\mathbf{r}\cdot\mathbf{r} \tag{3.38}$$

Thus the force is proportional to a triple vector product.

Another example is when we wish to construct three mutually orthogonal vectors in space, making use of two physical vectors $\mathbf{a}$ and $\mathbf{b}$ in a problem. For example, we may have an axis defining the rotation of a body (e.g. the earth), and another vector giving the position of a given point in this body; to describe the new position of this point after a specific rotation requires us to employ three axes (compare Section 3.5.2).

If we choose $\mathbf{a}$ as one of our three vectors, then $\mathbf{a}\wedge\mathbf{b}$ is perpendicular to it. A vector perpendicular to both of them is $\mathbf{a}\wedge(\mathbf{a}\wedge\mathbf{b})$. Thus $\mathbf{a}$, $\mathbf{a}\wedge\mathbf{b}$ and $\mathbf{a}\wedge(\mathbf{a}\wedge\mathbf{b})$ are a mutually orthogonal set, whatever the angle between $\mathbf{a}$ and $\mathbf{b}$ (provided they are not parallel).

3.5 Vector geometry

3.5.1 Simple shapes

In Chapter 2, the equations of various three-dimensional objects were written in terms of the coordinates x, y and z. We now briefly repeat the process, but using the much neater notation of vectors.

The plane

$$ax + by + cz = d \tag{2.21}$$

becomes

$$\mathbf{n}\cdot\mathbf{r} = d \tag{3.39}$$

where $\mathbf{r}$ as usual is the vector (x, y, z) and $\mathbf{n}$ is in the direction of the normal to the plane, and has components (a, b, c). If $\mathbf{n}$ is a unit vector, the shortest distance from the origin to the plane is d (up to an ambiguity of sign, caused by the two possible choices of the direction for the normal); otherwise it is $d/|\mathbf{n}|$ (compare eqn (2.24)).

A plane through the point (x_0, y_0, z_0) can be written as

$$a(x - x_0) + b(y - y_0) + c(z - z_0) = 0 \tag{2.26}$$

In vector notation, this is

$$\mathbf{n} \cdot (\mathbf{r} - \mathbf{r}_0) = 0 \tag{3.40}$$

This really is of the same form as (3.39), since we can take the scalar product $-\mathbf{n} \cdot \mathbf{r}_0$ to the right-hand side of the equation, and call it d.

Finally, the parametric representation of a point in a plane

$$(x_0 + \lambda v_1 + \mu w_1, y_0 + \lambda v_2 + \mu w_2, z_0 + \lambda v_3 + \mu w_3) \tag{2.29}$$

gives rise to the vector equation

$$\mathbf{r} = \mathbf{r}_0 + \lambda \mathbf{v} + \mu \mathbf{w}, \tag{3.41}$$

where $\mathbf{v}$ and $\mathbf{w}$ are any two non-parallel vector directions within the plane, and λ and μ are two scalar parameters; as they vary, $\mathbf{r}$ moves over the whole plane.

A line in three dimensions can be specified as the intersection of two planes. In terms of coordinates, this yields eqns (2.30); for vectors, we could have, for example, two equations of the form (3.39). More useful is the analogy with the alternative way of writing the line as

$$\frac{x - x_0}{l} = \frac{y - y_0}{m} = \frac{z - z_0}{n} = \lambda, \tag{2.32}$$

which becomes

$$\mathbf{r} = \mathbf{r}_0 + \lambda \mathbf{v} \tag{3.42}$$

or equivalently

$$(\mathbf{r} - \mathbf{r}_0) \wedge \mathbf{v} = 0 \tag{3.42$'$}$$

Here $\mathbf{r}_0$ is a specific point on the line, with coordinates (x_0, y_0, z_0); $\mathbf{v}$ is a vector in the direction of the line, and has components (l, m, n); and λ is the one parameter specifying how far along the line the point $\mathbf{r}$ is – see fig. 3.7. Since there is one degree of freedom for a line compared with two for a plane, we have one parameter here, but there were two in eqn (3.41).

As mentioned in Section 2.5.3, two straight lines

$$\left.\begin{aligned} \mathbf{r}_1 &= \mathbf{s} + \lambda \mathbf{v} \\ \text{and } \mathbf{r}_2 &= \mathbf{t} + \mu \mathbf{w} \end{aligned}\right\} \tag{3.43}$$

in three dimensions can fail to meet, even if they are not parallel. The derivation of the shortest distance d between them requires the result that the vector joining the closest points is perpendicular to both of the lines. Either we accept this from our previous analysis in terms of coordinates

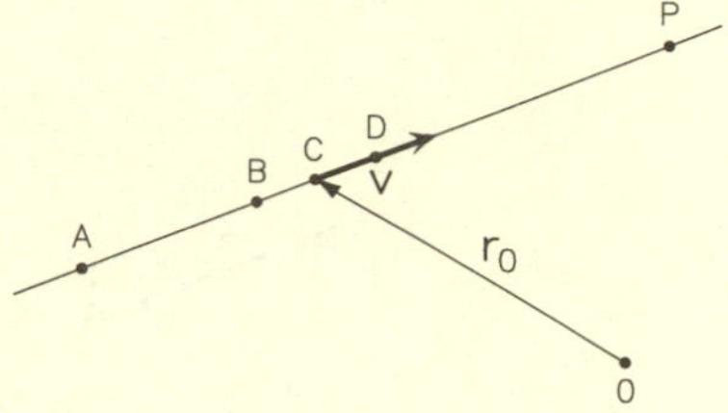

Fig. 3.7 The vector $\mathbf{r}$ describes the position of any point P on the straight line, whose equation is $\mathbf{r} = \mathbf{r}_0 + \lambda\mathbf{v}$; $\mathbf{r}_0$ is any specific point on the line, and $\mathbf{v}$ is a vector in the direction of the line. The various points A, B, C, D and P on the line correspond to $\lambda = -2, -\frac{1}{2}, 0, \frac{1}{2}$ and 3 respectively.

(see eqn (2.42′) and the discussion following it), or we prove it by a vector method as follows.

The vector between any two points on the lines (3.43) is $\mathbf{u} = \mathbf{r}_1 - \mathbf{r}_2$. If we change λ or μ slightly, $\mathbf{u}$ will also change a bit and become $\mathbf{u} + \delta\mathbf{u}$, where

$$\delta\mathbf{u} = \delta\lambda\,\mathbf{v} \text{ or } -\delta\mu\,\mathbf{w}$$

If we have chosen λ and μ so that the corresponding points are the closest to each other, $\mathbf{u}\cdot\mathbf{u}$ is a minimum and so $\delta(\mathbf{u}\cdot\mathbf{u})$ is zero. But

$$\begin{aligned}\delta(\mathbf{u}\cdot\mathbf{u}) &= \mathbf{u}\cdot\delta\mathbf{u} + \delta\mathbf{u}\cdot\mathbf{u}\\ &= 2\mathbf{u}\cdot\delta\mathbf{u}\end{aligned}$$

Thus $\mathbf{u}\cdot\delta\mathbf{u}$ is zero, and hence $\delta\mathbf{u}$ is perpendicular to $\mathbf{u}$. But $\delta\mathbf{u}$ is in the direction of the vector $\mathbf{v}$ or $\mathbf{w}$. We have thus demonstrated that the vector $\mathbf{u}$ joining the points of closest approach is indeed perpendicular to the directions of both lines.

The distance between the two points of closest approach is then given by resolving the vector $\mathbf{r}_1 - \mathbf{r}_2$ joining any two points on the lines along the mutually perpendicular direction $\mathbf{v}\wedge\mathbf{w}$. Thus we need to take the scalar product of $\mathbf{r}_1 - \mathbf{r}_2$ with a unit vector in the direction of $\mathbf{v}\wedge\mathbf{w}$ i.e.

$$\begin{aligned}d &= \frac{(\mathbf{r}_1 - \mathbf{r}_2)\cdot(\mathbf{v}\wedge\mathbf{w})}{|\,\mathbf{v}\wedge\mathbf{w}\,|}\\ &= \frac{(\mathbf{s} - \mathbf{t})\cdot(\mathbf{v}\wedge\mathbf{w})}{|\,\mathbf{v}\wedge\mathbf{w}\,|}\end{aligned} \tag{3.44}$$

where the last line is obtained by using eqns (3.43) and the fact that the triple scalar products $\{\mathbf{vvw}\}$ and $\{\mathbf{wvw}\}$ are both zero. Alternatively, we could have written down (3.44) directly because we were considering any two points on the lines, and these could have been taken as $\mathbf{s}$ and $\mathbf{t}$.

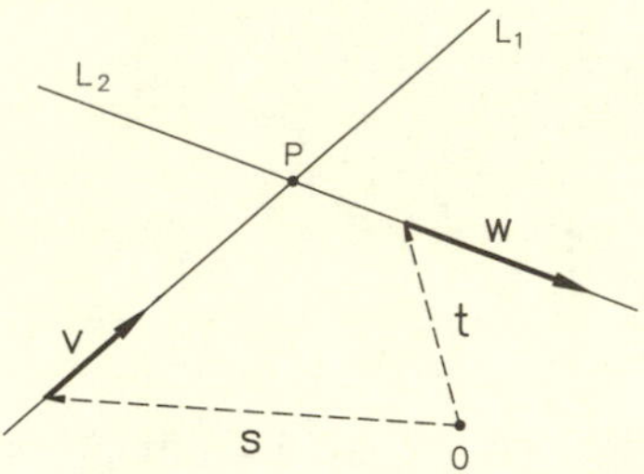

Fig. 3.8 Two lines which both lie in the plane of the page, and intersect at P. The line L_1 is given by $\mathbf{r} = \mathbf{s} + \lambda\mathbf{v}$, and L_2 by $\mathbf{r} = \mathbf{t} + \mu\mathbf{w}$. The shortest distance between two skew lines is proportional to the triple scalar product $(\mathbf{s} - \mathbf{t}) \cdot (\mathbf{v} \wedge \mathbf{w})$. For our coplanar lines, $\mathbf{v} \wedge \mathbf{w}$ is perpendicular to $\mathbf{s} - \mathbf{t}$, so the triple scalar product vanishes and the shortest distance is zero.

The only case in which this breaks down is when $\mathbf{v}$ and $\mathbf{w}$ are parallel. Then the normal to the two lines is not uniquely defined, $\mathbf{v} \wedge \mathbf{w}$ vanishes, and eqn (3.44) yields 0/0 for the answer. This, of course, mirrors the failure in this situation of the method of Section 2.5.3.2 using coordinates. It is, of course, simple to find the separation of two parallel lines, even in three dimensions, either by coordinate or vector methods. We can thus choose an arbitrary point (for example $\mathbf{s}$) on the first line, and then find the shortest distance from it to the second line – see Problem 3.5

If the two lines of eqns (3.43) actually do intersect, the shortest distance between them is zero. This requires that the triple scalar product $(\mathbf{s} - \mathbf{t}) \cdot (\mathbf{v} \wedge \mathbf{w})$ in the numerator of eqn (3.44) vanishes, while $|\mathbf{v} \wedge \mathbf{w}|$ does not. This in turn implies that the three vectors $\mathbf{s} - \mathbf{t}$, $\mathbf{v}$ and $\mathbf{w}$ are coplanar. Fig. 3.8 shows two intersecting lines, drawn so that both lie in the plane of the page; the vectors $\mathbf{s} - \mathbf{t}$, $\mathbf{v}$ and $\mathbf{w}$ are indeed coplanar.

We now turn to the sphere. Its equation is

$$(x - x_c)^2 + (y - y_c)^2 + (z - z_c)^2 = R^2 \tag{2.53}$$

which becomes

$$(\mathbf{r} - \mathbf{r}_c) \cdot (\mathbf{r} - \mathbf{r}_c) = R^2$$

This can be verified by writing the scalar product in terms of the components of $\mathbf{r} - \mathbf{r}_c$, when eqn (2.53) is obtained.

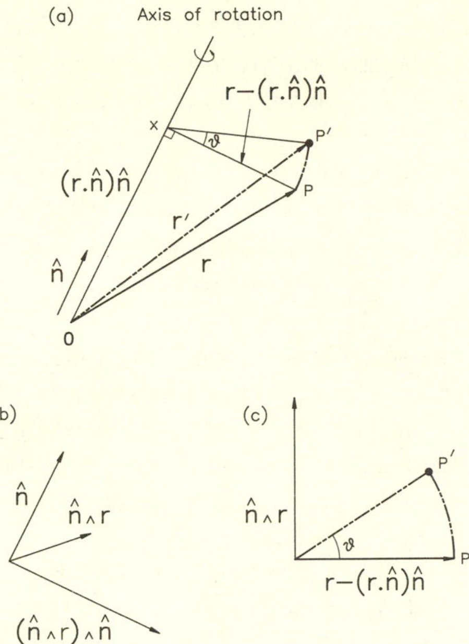

Fig. 3.9 (a) The effect of a rotation by an angle ϑ about an axis defined by a unit vector $\hat{\mathbf{n}}$. The point P at position $\mathbf{r}$ is moved to P′ at $\mathbf{r}'$. The point X is at the foot of the perpendicular from P onto the axis $\hat{\mathbf{n}}$. (b) The axes used to define the new position. $\hat{\mathbf{n}}$ is along the axis of rotation. $(\hat{\mathbf{n}} \wedge \mathbf{r}) \wedge \hat{\mathbf{n}}$ is perpendicular to $\hat{\mathbf{n}}$, and lies in the plane containing $\mathbf{r}$ and the axis of rotation. $\hat{\mathbf{n}} \wedge \mathbf{r}$ is perpendicular to the other two axes; $\mathbf{r}$ has no component along it, but $\mathbf{r}'$ does. (c) View looking down on the axis of rotation towards the origin O. The vector $\mathbf{r}$ has a component $|\mathbf{r} - (\mathbf{r} \cdot \hat{\mathbf{n}})\hat{\mathbf{n}}|$ along the $(\hat{\mathbf{n}} \wedge \mathbf{r}) \wedge \hat{\mathbf{n}}$ axis. Then P′ has this component reduced by a factor of $\cos\vartheta$, and a similar component reduced by $\sin\vartheta$ instead of $\cos\vartheta$ is developed along the $\hat{\mathbf{n}} \wedge \mathbf{r}$ direction.

3.5.2 Translations and rotations

Translations of axes are achieved by replacing $\mathbf{r}$ by $\mathbf{r} + \mathbf{r}_0$, where $\mathbf{r}_0$ is the vector specifying the new origin with respect to the original axes. Alternatively, if we move the object we are interested in so that the point that was at the origin is now at $\mathbf{r}_c$, we need to replace $\mathbf{r}$ by $\mathbf{r} - \mathbf{r}_c$ in the equation defining it (compare Section 2.2.2.1).

Now we turn to rotations. We discussed the effect of rotations about one of the coordinate axes in Section 2.6. Here we consider the effect on a vector $\mathbf{r}$ of rotating it by an angle ϑ around a general axis specified by a unit vector $\hat{\mathbf{n}}$ (see fig. 3.9(a)).

The result of the rotation is to leave unchanged the component of $\mathbf{r}$

along $\hat{\mathbf{n}}$ (i.e. OX in fig. 3.9(a)). However, the component perpendicular to $\hat{\mathbf{n}}$, and which is in the plane of $\mathbf{r}$ and $\hat{\mathbf{n}}$ (XP in the figure), is reduced by a factor of $\cos\vartheta$ as compared with its original length, while a component proportional to $\sin\vartheta$ is produced out of the plane containing $\hat{\mathbf{n}}$ and $\mathbf{r}$. Thus all we have to do is to translate the last couple of sentences into vector notation, in order to have a formula for the effect of the rotation on $\mathbf{r}$. It will contain three terms, one for each of the new components.

The first is straightforward. The component of $\mathbf{r}$ along the unit vector $\hat{\mathbf{n}}$ is simply $\mathbf{r}\cdot\hat{\mathbf{n}}$, and so, as it is unaffected by the rotation, this term in the answer is $(\mathbf{r}\cdot\hat{\mathbf{n}})\hat{\mathbf{n}}$.

Now we consider the other two terms. To describe these components, we need a set of axes, of which we already have $\hat{\mathbf{n}}$. The vector $\hat{\mathbf{n}}\wedge\mathbf{r}$ is perpendicular to the plane of $\hat{\mathbf{n}}$ and $\mathbf{r}$, and so is useful for the new third component. Furthermore the sense of $\hat{\mathbf{n}}\wedge\mathbf{r}$ is into the page in fig. 3.9, and is such that the third component produced by the rotation will be positive for $\vartheta<\pi$. Finally we want an axis in the plane of $\hat{\mathbf{n}}$ and $\mathbf{r}$, that is perpendicular to the other two; this is provided by $(\hat{\mathbf{n}}\wedge\mathbf{r})\wedge\hat{\mathbf{n}}$ (compare the end of Section 3.4.2). Although $\hat{\mathbf{n}}$ is a unit vector, the other two need not be, and are of length $|\hat{\mathbf{n}}\wedge\mathbf{r}|$. The axes are shown in fig. 3.9(b).

We now return to the component of $\mathbf{r}$ which is affected by the rotation. It is $\mathbf{r}-(\mathbf{r}\cdot\hat{\mathbf{n}})\hat{\mathbf{n}}$, and is in the direction of our axis $(\hat{\mathbf{n}}\wedge\mathbf{r})\wedge\hat{\mathbf{n}}$. In fact, if we use eqn (3.34) to expand the triple vector product, we find

$$(\hat{\mathbf{n}}\wedge\mathbf{r})\wedge\hat{\mathbf{n}}=\mathbf{r}-(\mathbf{r}\cdot\hat{\mathbf{n}})\hat{\mathbf{n}} \tag{3.45}$$

Thus not only are the vectors parallel, but they are actually equal. (In addition to using (3.34), we can also check that the lengths of the two vectors in eqn (3.45) are indeed both equal to $r\sin\alpha$, where α is the angle between $\mathbf{r}$ and $\hat{\mathbf{n}}$.)

As a result of the rotation, the component $(\hat{\mathbf{n}}\wedge\mathbf{r})\wedge\hat{\mathbf{n}}$ is reduced to $(\hat{\mathbf{n}}\wedge\mathbf{r})\wedge\hat{\mathbf{n}}\cos\vartheta$, so this is our second term. The third one is in the direction of $\hat{\mathbf{n}}\wedge\mathbf{r}$, and is of length $|(\hat{\mathbf{n}}\wedge\mathbf{r})\wedge\hat{\mathbf{n}}|\sin\vartheta$. A little thought should be sufficient to convince us that a term $\hat{\mathbf{n}}\wedge\mathbf{r}\sin\vartheta$ has exactly these properties. Thus the full answer is that the effect of the rotation is to replace $\mathbf{r}$ by

$$\mathbf{r}'=(\mathbf{r}\cdot\hat{\mathbf{n}})\hat{\mathbf{n}}+\cos\vartheta(\hat{\mathbf{n}}\wedge\mathbf{r})\wedge\hat{\mathbf{n}}+\sin\vartheta\hat{\mathbf{n}}\wedge\mathbf{r} \tag{3.46}$$

It is worth checking that this gives a sensible answer for a simple case. If we choose $\hat{\mathbf{n}}$ as being along the z axis (i.e. $\hat{\mathbf{n}}=\mathbf{k}$), and write $\mathbf{r}$ as

$x\mathbf{i} + y\mathbf{j} + z\mathbf{k}$, eqn (3.46) becomes

$$\mathbf{r}' = z\mathbf{k} + (x\mathbf{i} + y\mathbf{j}) \cos \vartheta + (x\mathbf{j} - y\mathbf{i}) \sin \vartheta, \tag{3.47}$$

which is equivalent to the following relations in terms of components:-

$$\left.\begin{aligned} x' &= x \cos \vartheta - y \sin \vartheta \\ y' &= y \cos \vartheta + x \sin \vartheta \\ z' &= z \end{aligned}\right\} \tag{3.47$'$}$$

This is just as expected (compare eqns (2.11)). Clearly this does not prove eqn (3.46) is correct for all cases, but it is reassuring.

A final word of caution is needed. Rotations are not vectors. This is because they do not satisfy the rule for vectors that

$$\mathbf{a} + \mathbf{b} = \mathbf{b} + \mathbf{a}$$

i.e. the result of adding vectors is independent of their order. We can check that rotations fail this by the following simple test.

Place two books in front of you in identical orientations. Then rotate one by 90° clockwise about the vertical axis, followed by a 90° anti-clockwise rotation about a horizontal axis pointing away from you. Next perform the same operations on the second book, but in the reverse order. The books end up in different orientations from each other. This demonstrates that

$$\text{Rotation 1} + \text{Rotation 2} \neq \text{Rotation 2} + \text{Rotation 1} \tag{3.48}$$

and hence rotations are not vectors.

It turns out, however, that for small rotations, the sides of (3.48) are almost equal, and the agreement improves as the angles of rotation become smaller and smaller. It is because of this that the angular velocity $\boldsymbol{\omega}$ can be regarded as a vector, since it involves an infinitesimal rotation (divided by an infinitesimal time).

3.6 Non-orthogonal axes

We have already dealt with numerous cases of writing a vector in terms of contributions along three orthogonal axes. With

$$\begin{aligned} \mathbf{f} &= f_x\mathbf{i} + f_y\mathbf{j} + f_z\mathbf{k} \\ &= (\mathbf{f} \cdot \mathbf{i})\mathbf{i} + (\mathbf{f} \cdot \mathbf{j})\mathbf{j} + (\mathbf{f} \cdot \mathbf{k})\mathbf{k} \end{aligned}$$

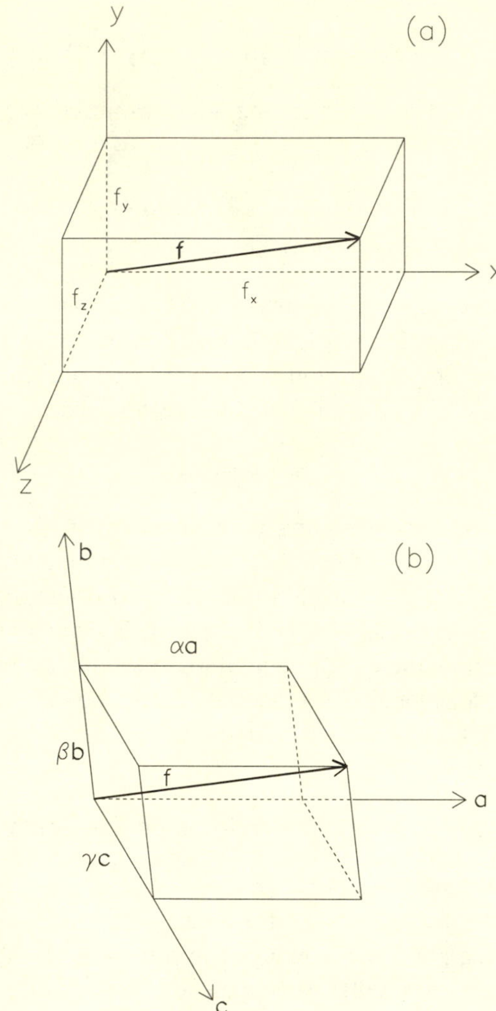

Fig. 3.10 (a) The vector **f** has components f_x, f_y and f_z along the orthogonal coordinate axes, such that **f** is the diagonal of a cuboid whose sides have lengths f_x f_y and f_z (see eqn (3.8)). (b) The same vector **f** is referred to non-orthogonal axes **a**, **b** and **c**. It can be regarded as the diagonal of the parallelepiped, whose sides have length $\alpha|\mathbf{a}|$, $\beta|\mathbf{b}|$ and $\gamma|\mathbf{c}|$ (see eqns (3.49) and (3.51)).

we are essentially regarding **f** as the diagonal of a cuboid (the three-dimensional equivalent of a rectangle) whose sides are of length f_x, f_y and f_z (see fig. 3.10(a)).

It is, however, also possible to express **f** in terms of any three non-coplanar vectors **a**, **b** and **c**. In this case we can imagine constructing a parallelepiped whose sides are in the directions of **a**, **b** and **c**, and whose diagonal is **f**. Then

$$\mathbf{f} = \alpha\mathbf{a} + \beta\mathbf{b} + \gamma\mathbf{c} \tag{3.49}$$

where the magnitude of $\alpha\mathbf{a}$ is just the length of the relevant side of the parallelepiped (see fig. 3.10(b)). For a general vector **f**, **a b** and **c** must be non-coplanar.

We now want to determine α, β and γ when **f** (and, of course, **a**, **b** and **c**) is specified. We do this by looking hard at eqn (3.49), and using a general approach which works in problems not only in the area of vectors (see, for example, Chapter 12 of Volume 2). What we do is to find something with which we can 'multiply' eqn (3.49), so that all the terms on the right-hand side vanish, except for the one containing α; we then have determined α. Clearly we must interpret 'multiply' more widely than just taking the product with a number. What is required in this case is to take the scalar product with $\mathbf{b} \wedge \mathbf{c}$. At first sight this may seem somewhat arbitrary, but with a little practice it becomes easier to spot what is likely to work.

When we take this scalar product, we obtain

$$\mathbf{f} \cdot (\mathbf{b} \wedge \mathbf{c}) = \alpha\mathbf{a} \cdot (\mathbf{b} \wedge \mathbf{c}) + \beta\mathbf{b} \cdot (\mathbf{b} \wedge \mathbf{c}) + \gamma\mathbf{c} \cdot (\mathbf{b} \wedge \mathbf{c}) \tag{3.50}$$

Once again, we are dealing with triple scalar products. The last two terms vanish because $\{\mathbf{bbc}\}$ and $\{\mathbf{cbc}\}$ are both zero. This is, of course, no coincidence; it happens because of our correct choice of the 'multiplication' factor. Then

$$\alpha = \{\mathbf{fbc}\}/\{\mathbf{abc}\} \tag{3.51}$$

and β and γ are determined similarly. We incidently note that $\{\mathbf{abc}\}$ must be non-zero in order for α to be sensibly determined by eqn (3.51); this confirms that **a**, b and **c** must not be coplanar.

There is an alternative way of using **a**, b and **c** for obtaining an expression for **f** in terms of components along specified vectors. This consists in first constructing new vectors from the original ones, namely $\mathbf{a} \wedge \mathbf{b}$, $\mathbf{b} \wedge \mathbf{c}$ and $\mathbf{c} \wedge \mathbf{a}$. In analogy with eqn (3.49), we then write

$$\mathbf{f} = \lambda\mathbf{a} \wedge \mathbf{b} + \mu\mathbf{b} \wedge \mathbf{c} + \nu\mathbf{c} \wedge \mathbf{a}$$

To determine λ, the trick this time is to take the scalar product of this equation with $\mathbf{c}$. This yields

$$\lambda = (\mathbf{f} \cdot \mathbf{c})/\{\mathbf{a}\,\mathbf{b}\,\mathbf{c}\} \tag{3.52}$$

with μ and ν being determined similarly. Thus

$$\mathbf{f} = [(\mathbf{f} \cdot \mathbf{c})\mathbf{a} \wedge \mathbf{b} + (\mathbf{f} \cdot \mathbf{a})\mathbf{b} \wedge \mathbf{c} + (\mathbf{f} \cdot \mathbf{b})\mathbf{c} \wedge \mathbf{a}] / \{\mathbf{a}\,\mathbf{b}\,\mathbf{c}\} \tag{3.53}$$

We now define

$$\left.\begin{aligned} \mathbf{a}' &= \mathbf{b} \wedge \mathbf{c}/\{\mathbf{a}\,\mathbf{b}\,\mathbf{c}\} \\ \mathbf{b}' &= \mathbf{c} \wedge \mathbf{a}/\{\mathbf{a}\,\mathbf{b}\,\mathbf{c}\} \\ \mathbf{c}' &= \mathbf{a} \wedge \mathbf{b}/\{\mathbf{a}\,\mathbf{b}\,\mathbf{c}\} \end{aligned}\right\} \tag{3.54}$$

i.e. these are just the new set of vectors we have chosen, except that they are all divided by $\{\mathbf{a}\,\mathbf{b}\,\mathbf{c}\}$. (Of course, once again these definitions make sense only if $\mathbf{a}$, $\mathbf{b}$ and $\mathbf{c}$ are not coplanar.)

The vectors $\mathbf{a}'$, $\mathbf{b}'$ and $\mathbf{c}'$ are called the reciprocal vector set of $\mathbf{a}$, $\mathbf{b}$ and $\mathbf{c}$, and the reciprocal set of $\mathbf{a}'$, $\mathbf{b}'$ and $\mathbf{c}'$ is just $\mathbf{a}$, $\mathbf{b}$ and $\mathbf{c}$ (see Problem 3.8). The reason for the convention that the first vector in eqn (3.54) is called $\mathbf{a}'$ rather than $\mathbf{b}'$ or $\mathbf{c}'$ is that two of our original three vectors occur in the numerator, and hence the unique one is the one left out (i.e. $\mathbf{a}$). Also, and perhaps more convincingly, for the case where $\mathbf{a}$, $\mathbf{b}$ and $\mathbf{c}$ are orthogonal, $\mathbf{a}'$ is parallel to $\mathbf{a}$.

With the definition of the reciprocal vector set, eqn (3.53) now becomes tidier. We obtain

$$\mathbf{f} = (\mathbf{f} \cdot \mathbf{a})\mathbf{a}' + (\mathbf{f} \cdot \mathbf{b})\mathbf{b}' + (\mathbf{f} \cdot \mathbf{c})\mathbf{c}' \tag{3.53$'$}$$

Because of the mutual relationship between reciprocal vector sets, we can also write

$$\mathbf{f} = (\mathbf{f} \cdot \mathbf{a}')\mathbf{a} + (\mathbf{f} \cdot \mathbf{b}')\mathbf{b} + (\mathbf{f} \cdot \mathbf{c}')\mathbf{c} \tag{3.53$''$}$$

With the definitions (3.54) of $\mathbf{a}'$, $\mathbf{b}'$ and $\mathbf{c}'$, eqn (3.53$''$) is identically equal to our expression (3.49), where α, β and γ are given by (3.51).

Another use of reciprocal vector sets is in determining the point of intersection of three planes (see Problem 3.9). In physics, they find applications in the field of X-ray crystallography; the original set of vectors defines the positions of ions in a crystal, and the reciprocal vector set gives the directions in which there is a large intensity of the diffracted radiation.

3.7 Solution of vector equations

We are sometimes given an equation which relates an unknown vector $\mathbf{v}$ to various given vectors $\mathbf{a}$, $\mathbf{b}$ etc. We are then required to solve this equation for $\mathbf{v}$. If $\mathbf{v}$ occurs only on its own in the equation, finding the solution is trivial. More interesting cases are where $\mathbf{v}$ is involved in vector and/or scalar products with the known vectors. An example of this type is

$$\mathbf{v} \wedge \mathbf{a} + \lambda\mathbf{v} - \mathbf{b} = 0 \tag{3.55}$$

where λ is a known scalar, and $\mathbf{a}$ and $\mathbf{b}$ are not parallel.

There are basically two methods of approach to such problems. The first requires the setting up of three axes in terms of which $\mathbf{v}$ is expanded. This involves three arbitrary constants, whose values we find from the given equation. The alternative is to be cunning, and to take scalar or vector products of the equation with suitably chosen vectors, so that $\mathbf{v}$ eventually occurs alone, to give us the solution.

First we use the axes method. Apart from $\mathbf{v}$, the problem involves the vectors $\mathbf{a}$ and $\mathbf{b}$. Since they are not parallel, we can use $\mathbf{a}$, $\mathbf{b}$ and $\mathbf{a} \wedge \mathbf{b}$ as a set of axes, and then express $\mathbf{v}$ as

$$\mathbf{v} = \alpha\mathbf{a} + \beta\mathbf{b} + \gamma\mathbf{a} \wedge \mathbf{b} \tag{3.56}$$

It does not matter that in general our axes are not orthogonal.

We now substitute our trial solution (3.56) into eqn (3.55) and obtain

$$[\alpha\mathbf{a} + \beta\mathbf{b} + \gamma(\mathbf{a} \wedge \mathbf{b})] \wedge \mathbf{a} + \lambda[\alpha\mathbf{a} + \beta\mathbf{b} + \gamma(\mathbf{a} \wedge \mathbf{b})] - \mathbf{b} = 0 \tag{3.57}$$

After some simplification involving the use of eqns (3.19) and (3.34), and collecting the terms multiplying each of the separate vectors, we obtain

$$[-\gamma(\mathbf{a} \cdot \mathbf{b}) + \lambda\alpha]\,\mathbf{a} + (\lambda\beta + \gamma\mathbf{a} \cdot \mathbf{a} - 1)\,\mathbf{b} + (-\beta + \lambda\gamma)\,\mathbf{a} \wedge \mathbf{b} = 0 \tag{3.57$'$}$$

For this to be true, and since $\mathbf{a}$ and $\mathbf{b}$ are not parallel, each of the coefficients multiplying the three separate vectors must be zero. This provides us with three simultaneous equations for our three unknowns α, β and γ. They are

$$\left.\begin{aligned} -\gamma\mathbf{a} \cdot \mathbf{b} + \lambda\alpha &= 0 \\ \lambda\beta + \gamma\mathbf{a} \cdot \mathbf{a} &= 1 \\ \text{and} \qquad -\beta + \lambda\gamma &= 0 \end{aligned}\right\} \tag{3.58}$$

Since they are linear in α, β and γ, and since they contain vectors only in scalar products, they are readily solved, to give

$$\left.\begin{aligned} \alpha &= \frac{\mathbf{a}\cdot\mathbf{b}}{(\mathbf{a}\cdot\mathbf{a}+\lambda^2)\,\lambda} \\ \beta &= \frac{\lambda}{\mathbf{a}\cdot\mathbf{a}+\lambda^2} \\ \text{and}\qquad \gamma &= \frac{1}{\mathbf{a}\cdot\mathbf{a}+\lambda^2} \end{aligned}\right\} \tag{3.59}$$

Our solution is thus

$$(\mathbf{a}\cdot\mathbf{a}+\lambda^2)\,\mathbf{v} = \frac{\mathbf{a}\cdot\mathbf{b}}{\lambda}\mathbf{a} + \lambda\mathbf{b} + \mathbf{a}\wedge\mathbf{b} \tag{3.60}$$

The form of this solution is satisfactory in that the answer $\mathbf{v}$ is not multiplied by any other vector, and all the other scalar and vector quantities in the answer are known ($\mathbf{a}$, $\mathbf{b}$, $\mathbf{a}\cdot\mathbf{a}$, $\mathbf{a}\cdot\mathbf{b}$, $\mathbf{a}\wedge\mathbf{b}$ and λ).

That method was straight-forward, but somewhat tedious. Instead we now try to be intelligent. We can try taking the scalar and/or vector products of eqn (3.55) with the vectors at our disposal, which are $\mathbf{a}$, $\mathbf{b}$ and $\mathbf{a}\wedge\mathbf{b}$. It turns out that both types of product with $\mathbf{a}$ are helpful.

Thus the scalar product gives

$$\lambda\mathbf{v}\cdot\mathbf{a} = \mathbf{b}\cdot\mathbf{a} \tag{3.61}$$

This will be useful, as it will allow us to replace any $\mathbf{v}\cdot\mathbf{a}$ term (which would be unsatisfactory in the answer) by an allowed $\mathbf{b}\cdot\mathbf{a}/\lambda$.

The vector product with $\mathbf{a}$ yields

$$(\mathbf{v}\wedge\mathbf{a})\wedge\mathbf{a} + \lambda\mathbf{v}\wedge\mathbf{a} = \mathbf{b}\wedge\mathbf{a} \tag{3.62}$$

Here we must, of course, be careful to vector multiply the equation by $\mathbf{a}$ after each term; or alternatively with $\mathbf{a}$ before each term; but not with a mixture.

We now take the terms of eqn (3.62) one by one, and see if we can make them acceptable. So far, only the $\mathbf{b}\wedge\mathbf{a}$ term is independent of $\mathbf{v}$, and hence satisfactory.

The first term can be simplified using eqn (3.34):

$$\begin{aligned} (\mathbf{v}\wedge\mathbf{a})\wedge\mathbf{a} &= (\mathbf{v}\cdot\mathbf{a})\,\mathbf{a} - (\mathbf{a}\cdot\mathbf{a})\,\mathbf{v} \\ &= \frac{\mathbf{b}\cdot\mathbf{a}}{\lambda}\mathbf{a} - (\mathbf{a}\cdot\mathbf{a})\,\mathbf{v} \end{aligned} \tag{3.63}$$

where in the last line we have made use of eqn (3.61).

Finally we consider the $\mathbf{v}\wedge\mathbf{a}$ term in eqn (3.62). Again this cannot be part of the answer, but we can make use of the original equation (3.55) to replace it by acceptable terms. We thus obtain

$$\left[\frac{(\mathbf{b}\cdot\mathbf{a})}{\lambda}\mathbf{a}-(\mathbf{a}\cdot\mathbf{a})\,\mathbf{v}\right]+\lambda\,[\mathbf{b}-\lambda\mathbf{v}]=[\mathbf{b}\wedge\mathbf{a}] \qquad (3.62')$$

where the square brackets indicate the separate vector terms in (3.62). On rearranging, (3.62′) yields our previous solution (3.60) for **v**.

In general, the second approach is intellectually more satisfying than the first, and may involve less algebra.

It is worth performing a quick check with a specific choice for the given vectors to see whether our solution is plausible. Thus for $\mathbf{a}=\mathbf{i}$ and $\mathbf{b}=\mathbf{j}$, eqn (3.55) becomes

$$\mathbf{v}\wedge\mathbf{i}+\lambda\mathbf{v}-\mathbf{j}=0 \qquad (3.55')$$

while the solution (3.60) turns into

$$\left(1+\lambda^2\right)\mathbf{v}=\lambda\mathbf{j}+\mathbf{k} \qquad (3.60')$$

Direct substitution then verifies that (3.60′) is indeed a solution of (3.55′), as required.

Finally what if **b** and **a** are parallel, so that the first method breaks down? It turns out that, in the second method, we do not need to assume at any stage that **a** and **b** are not parallel, so the solution (3.62′) is still valid. However, it simplifies. If we write $\mathbf{b}=\nu\mathbf{a}$, where ν is a constant, we quickly find

$$\mathbf{v}=\nu\mathbf{a}/\lambda=\mathbf{b}/\lambda \qquad (3.64)$$

This, in fact, is readily understood. Our original equation is now

$$\mathbf{v}\wedge\mathbf{a}+\lambda\mathbf{v}=\nu\mathbf{a} \qquad (3.55'')$$

Then the solution for **v** cannot have a component perpendicular to **a**, in a direction, say, **c**. This is because, of the various terms in eqn (3.55″), $\mathbf{v}\wedge\mathbf{a}$ would be perpendicular to both **a** and **c**, the $\lambda\mathbf{v}$ term would have a transverse component in the **c** direction, while $\nu\mathbf{a}$ has no transverse component at all. Thus there is no way that these components perpendicular to **a** can match up in eqn (3.55″), and hence there can be no such transverse component in **v**. Thus **v** must be parallel to **a**, and is given by eqn (3.64).

It is often useful to try to think about the solution of the given equation in such geometrical terms.

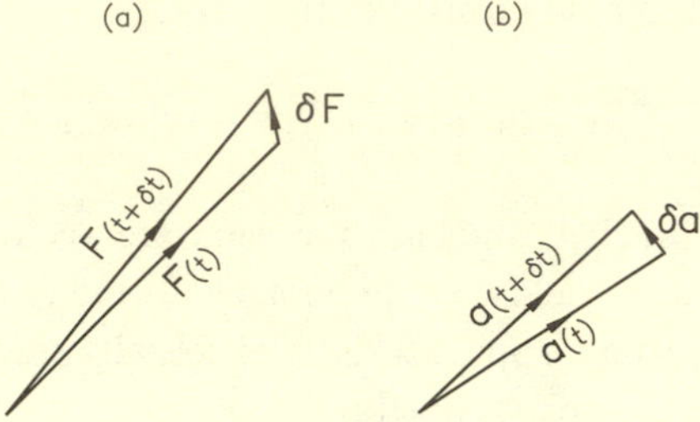

Fig. 3.11 (a) The vector **F** varies with time. At some specific time t, it is shown as $\mathbf{F}(t)$, and at a small time δt later as $\mathbf{F}(t+\delta t)$. The difference between these is given by the vector $\delta\mathbf{F}$. The time derivitive of **F** is, for small enough δt, given by the ratio $\delta\mathbf{F}/\delta t$. In general it will not be in the same direction as **F**. (b) For a vector **a** of constant length, the infinitesimal change $\delta\mathbf{a}$ (between the vector at times t and $t+\delta t$) is always perpendicular to $\mathbf{a}(t)$.

3.8 Differentiating vectors

In this section, we discuss the differentiation of vectors (and various vector expressions) with respect to time. It is also possible to consider spatial derivatives of various types; this leads to vector operators that are considered in Chapter 10 of Volume 2.

An equation containing time derivatives of vectors is, of course, a differential equation. Methods of solving such equations are beyond the scope of this book.

3.8.1 Simple examples

If we have a vector **F** which varies with time, we can obtain its derivative $d\mathbf{F}/dt$ in a similar manner to that used for a scalar. Thus

$$\frac{d\mathbf{F}}{dt} \approx \frac{\delta\mathbf{F}}{\delta t} = \frac{\mathbf{F}(t+\delta t) - \mathbf{F}(t)}{\delta t} \tag{3.65}$$

That is, we take the vector difference between **F** at two close times t and $t+\delta t$, and then divide by the time interval δt. (See fig. 3.11(a).) Thus the derivative is itself a vector.

If we express the vector **F** with respect to a set of fixed unit vectors as

$$\mathbf{F} = F_x\mathbf{i} + F_y\mathbf{j} + F_z\mathbf{k}, \tag{3.8}$$

then the derivative of **F** is given simply in terms of the derivatives of its components as

$$\dot{\mathbf{F}} = \frac{d\mathbf{F}}{dt} = \frac{dF_x}{dt}\mathbf{i} + \frac{dF_y}{dt}\mathbf{j} + \frac{dF_z}{dt}\mathbf{k} \tag{3.66}$$

This follows either from the usual rules for differentiating expressions, applied to the right-hand side of eqn (3.8) (even though it involves vectors); or by going back to first principles and using eqn (3.65) for the vector given by eqn (3.8).

It is readily verified that for scalar and vector products, the derivatives are given by

$$\frac{d}{dt}(\mathbf{a}\cdot\mathbf{b}) = \mathbf{a}\cdot\frac{d\mathbf{b}}{dt} + \mathbf{b}\cdot\frac{d\mathbf{a}}{dt} \tag{3.67}$$

and

$$\frac{d}{dt}(\mathbf{a}\wedge\mathbf{b}) = \mathbf{a}\wedge\frac{d\mathbf{b}}{dt} + \frac{d\mathbf{a}}{dt}\wedge\mathbf{b} \tag{3.68}$$

As usual we have to maintain the order of the vectors in the vector products of (3.68).

A special case of eqn (3.67) is

$$\frac{d}{dt}(\mathbf{a}\cdot\mathbf{a}) = 2\mathbf{a}\cdot\frac{d\mathbf{a}}{dt} \tag{3.69}$$

Thus for a vector which changes direction but whose length is constant, $\mathbf{a}\cdot\mathbf{a}$ does not alter and, hence,

$$\frac{d}{dt}(\mathbf{a}\cdot\mathbf{a}) = 0 = 2\mathbf{a}\cdot\frac{d\mathbf{a}}{dt} \tag{3.69$'$}$$

That is, the infinitesimal change in such a vector **a** is always perpendicular to **a** (see fig. 3.11(b)).

3.8.2 Moving axes

We now consider the case in which the unit vectors **i**, **j** and **k** are not fixed in space but are varying in direction with time. A very practical example of this is where the coordinates of an object are measured with respect to axes that rotate with the earth. That is, we measure a moving object assuming that the earth is stationary. In order to deduce the true motion of the object, it is necessary to take into account the way in which the axes themselves change.

We thus write the vector position **r** of the object as

$$\mathbf{r} = x\mathbf{i} + y\mathbf{j} + z\mathbf{k} \tag{3.70}$$

where x, y and z are the positions with respect to the moving axes defined by the vectors **i**, **j** and **k**. When we differentiate with respect to time, we obtain the true velocity of the object

$$\mathbf{v} = \frac{d\mathbf{r}}{dt} = \frac{dx}{dt}\mathbf{i} + x\frac{d\mathbf{i}}{dt} + \text{ corresponding } y \text{ and } z \text{ terms} \tag{3.71}$$

$$= \mathbf{v}_\mathrm{r} + x\frac{d\mathbf{i}}{dt} + y\frac{d\mathbf{j}}{dt} + z\frac{d\mathbf{k}}{dt} \tag{3.71$'$}$$

where

$$\mathbf{v}_\mathrm{r} = \frac{dx}{dt}\mathbf{i} + \frac{dy}{dt}\mathbf{j} + \frac{dz}{dt}\mathbf{k} \tag{3.72}$$

Thus the expression (3.71′) for the true velocity $\mathbf{v}$ contains two types of terms. The first is $\mathbf{v}_\mathrm{r}$, which is the velocity as seen with respect to the moving system of axes; the second is that due to the fact that the axes themselves are not stationary.

Now for any rotation with an angular velocity ω, the rate of change of a vector **F** is

$$\frac{d\mathbf{F}}{dt} = \omega \wedge \mathbf{F}$$

(see fig. 3.5(c)). Thus if our system of axes rotates with the earth, we then have

$$\frac{d\mathbf{i}}{dt} = \omega \wedge \mathbf{i}, \text{ etc.}$$

We can insert this in eqn (3.71) to obtain

$$\mathbf{v} = \frac{dx}{dt}\mathbf{i} + x\omega \wedge \mathbf{i} + \ldots \tag{3.73}$$

On differentiating again to obtain the acceleration, we obtain

$$\mathbf{a} = \frac{d\mathbf{v}}{dt} = \frac{d^2x}{dt^2}\mathbf{i} + \frac{dx}{dt}\frac{d\mathbf{i}}{dt} + \frac{dx}{dt}\omega \wedge \mathbf{i} + x\frac{d\omega}{dt} \wedge \mathbf{i} + x\omega \wedge \frac{d\mathbf{i}}{dt} + \ldots \tag{3.74}$$

$$= \frac{d^2x}{dt^2}\mathbf{i} + 2\frac{dx}{dt}\omega \wedge \mathbf{i} + x\omega \wedge (\omega \wedge \mathbf{i}) + x\frac{d\omega}{dt} \wedge \mathbf{i} + \ldots \tag{3.74$'$}$$

$$= \mathbf{a}_\mathrm{r} + 2\omega \wedge \mathbf{v}_\mathrm{r} + \omega \wedge (\omega \wedge \mathbf{r}) + \frac{d\omega}{dt} \wedge \mathbf{r} \tag{3.74$''$}$$

where

$$\mathbf{a}_\mathrm{r} = \frac{d^2x}{dt^2}\mathbf{i} + \frac{d^2y}{dt^2}\mathbf{j} + \frac{d^2z}{dt^2}\mathbf{k}$$

The dots in eqns (3.73), (3.74) and (3.74′) denote the corresponding y and z terms, which are implicitly included in (3.74″).

Thus the true acceleration **a** is expressed in eqn (3.74″) as the sum of several terms:

(i) $\mathbf{a_r}$, the apparent acceleration with respect to the moving axes;
(ii) $2\boldsymbol{\omega} \wedge \mathbf{v_r}$, the so-called Coriolis acceleration (see below);
(iii) $\boldsymbol{\omega} \wedge (\boldsymbol{\omega} \wedge \mathbf{r})$, the centripetal acceleration; and
(iv) $d\boldsymbol{\omega}/dt \wedge \mathbf{r}$, which is zero in the approximation where the earth rotates at a constant angular velocity.

According to Newton's Second Law of Motion, the acceleration of a body is proportional to the force acting on it. This gives the true acceleration $\mathbf{a}$. However, if we observe the body with respect to axes fixed on the earth, we measure the apparent acceleration $\mathbf{a_r}$. We then need to correct it by the extra terms (ii) and (iii) in order for Newton's Law to apply. That is, because our axes are rotating, we invent extra terms called the Coriolis and centripetal accelerations.

The latter is well known. For example, a satellite 20 000 miles above the equator appears to remain stationary above the earth. We would 'explain' this by saying that although the satellite is not moving as seen by us, we have to ascribe to it a centripetal acceleration, due to the rotating axes. This is then the acceleration caused by the gravitational attraction of the earth.

The Coriolis acceleration manifests itself in several situations. For example:

(a) In a hurricane, rapid heating of the air causes it to rise, resulting in a violent inflow of very strong winds. In the northern hemisphere, the effect of the Coriolis force is to deflect winds to the right, and thus they tend to circulate around the centre of the hurricane in an anticlockwise sense.
(b) A merry-go-round consists of a large disc rotating about its centre, with several toy horses fixed around its circumference, and on which children can ride. If a man standing at the centre of the merry-go-round throws a ball directly towards a child on one of the horses, he would see the ball apparently being deflected along its flight path, and miss the child.
(c) In a 'Tower of Pisa' experiment, a ball dropped from a great height will not fall down exactly along a stationary plumb-line (although the magnitude of the deviation is small).
(d) A pendulum which is allowed to swing for several hours will gradually have its plane of oscillation rotated.

Each of these effects is described by the invented Coriolis acceleration $2\boldsymbol{\omega} \wedge \mathbf{v_r}$ which is required in addition to the centripetal acceleration if we want to describe motion in a rotating frame. Because of the properties of vector products, the centripetal acceleration vanishes at the poles, while the Coriolis acceleration is zero if the velocity in the rotating frame is parallel to the angular velocity $\boldsymbol{\omega}$.

We conclude that these interesting and perhaps surprising effects arise simply from the differentiation of the vector position as given in eqn (3.70).

Problems

3.1 Find the angle between the directions defined by the vectors $(1,0,1)$ and $(1,-2,0)$.

3.2 By considering components of the relevant vectors, show that

$$(\mathbf{a} \wedge \mathbf{b}) \wedge \mathbf{c} = (\mathbf{a} \cdot \mathbf{c})\,\mathbf{b} - (\mathbf{b} \cdot \mathbf{c})\,\mathbf{a}$$

3.3 (i) Simplify the expression

$$\mathbf{a} \wedge (\mathbf{b} \wedge \mathbf{c}) + \mathbf{b} \wedge (\mathbf{c} \wedge \mathbf{a}) + \mathbf{c} \wedge (\mathbf{a} \wedge \mathbf{b})$$

(ii) Verify the following relationship for triple scalar products

$$\{\mathbf{a} \wedge \mathbf{b},\ \mathbf{b} \wedge \mathbf{c},\ \mathbf{c} \wedge \mathbf{a}\} = \{\mathbf{a},\ \mathbf{b},\ \mathbf{c}\}^2$$

3.4 Show that

$$(\mathbf{a} \wedge \mathbf{b}) \cdot (\mathbf{a} \wedge \mathbf{c}) = (\mathbf{b} \cdot \mathbf{c})\,(\mathbf{a} \cdot \mathbf{a}) - (\mathbf{a} \cdot \mathbf{c})\,(\mathbf{a} \cdot \mathbf{b})$$

Hence find the angle between adjacent faces of a regular tetrahedron (i.e. one which has all its edges of equal length.)

3.5 Use a vector method to show that the point on a line $\mathbf{b} + \lambda\mathbf{v}$ which is closest to a point $\mathbf{a}$ is such that the vector joining it to $\mathbf{a}$ is perpendicular to $\mathbf{v}$.

Express the square of this distance in terms of $\mathbf{a}$, $\mathbf{b}$ and $\mathbf{v}$.

3.6 Solve Problem 2.9 by vector methods.

3.7 Given a regular tetrahedron whose edges are of length l, write down vector equations for the lines corresponding to a pair of opposite edges (i.e. edges which do not meet at a corner). Then use eqn (3.44) to find the shortest distance between these edges. (Compare Problem 2.11.)

3.8 The reciprocal vectors $\mathbf{a}'$, $\mathbf{b}'$ and $\mathbf{c}'$ of an original set $\mathbf{a}$, $\mathbf{b}$ and $\mathbf{c}$ are defined by eqns (3.54). Show that

$$\{\mathbf{a}'\,\mathbf{b}'\,\mathbf{c}'\} = 1/\{\mathbf{a}\,\mathbf{b}\,\mathbf{c}\}$$

Use eqns (3.54) again to show that the reciprocal vectors of $\mathbf{a}'$, $\mathbf{b}'$ and $\mathbf{c}'$ are just equal to $\mathbf{a}$, $\mathbf{b}$ and $\mathbf{c}$ respectively.

For the special case where **a**, **b** and **c** are orthogonal, show that $\mathbf{a}'$ $\mathbf{b}'$ and $\mathbf{c}'$ are parallel to **a**, **b** and **c** respectively, and that their lengths are the reciprocals of the lengths of the original vectors.

3.9 Three planes are given by the vector equations

$$\left.\begin{array}{l}\mathbf{r}\cdot\mathbf{a}=d\\ \mathbf{r}\cdot\mathbf{b}=e\\ \mathbf{r}\cdot\mathbf{c}=f\end{array}\right\}$$

where **a**, **b** and **c** are not coplanar. Show that the point common to the three planes is given by

$$\mathbf{r} = d\mathbf{a}' + e\mathbf{b}' + f\mathbf{c}'$$

where $\mathbf{a}'$, $\mathbf{b}'$ and $\mathbf{c}'$ are the reciprocal vector set of **a**, **b** and **c**. (Because the definitions of $\mathbf{a}'$, $\mathbf{b}'$ and $\mathbf{c}'$ involve $\{\mathbf{a}\,\mathbf{b}\,\mathbf{c}\}$ in the denominators, this method breaks down if **a**, **b** and **c** are coplanar. Then the planes will not have a single common point – either they have no common point, or else they meet in a line or a plane.)

3.10 Solve the vector equation

$$\mathbf{r} = (\mathbf{b}\cdot\mathbf{r})\,\mathbf{a} + \mathbf{b}\wedge\mathbf{r} + \mathbf{c}$$

where $\mathbf{a}\cdot\mathbf{b} = 0$.

3.11 The position of a body moving in a plane about the origin is given by polar coordinates r and θ. At any moment, the unit vectors $\hat{\mathbf{r}}$ and $\hat{\boldsymbol{\theta}}$ are along the radial and tangential directions respectively. Show that $\dot{\hat{\mathbf{r}}} = \dot{\theta}\hat{\boldsymbol{\theta}}$, and find the corresponding equation for $\dot{\hat{\boldsymbol{\theta}}}$. Show that the velocity $\dot{\mathbf{r}}$ and the acceleration of $\ddot{\mathbf{r}}$ of the body are given by

$$\dot{\mathbf{r}} = \dot{r}\hat{\mathbf{r}} + r\dot{\theta}\hat{\boldsymbol{\theta}}$$

and

$$\ddot{\mathbf{r}} = \left(\ddot{r} - r\dot{\theta}^2\right)\hat{\mathbf{r}} + \frac{1}{r}\frac{d}{dt}\left(r^2\dot{\theta}\right)\hat{\boldsymbol{\theta}}$$

4
Complex numbers

If you dial a wrong number, a certain telephone exchange has a recorded message which announces 'You have reached an imaginary number. If you require a real number, please rotate the phone by 90°, and try again.'

(Unsubstantiated rumour)

This chapter first deals with what complex numbers are and how we manipulate them and understand their properties; and then goes on to describe how useful they are in helping to solve a variety of problems. The case of a resonant system will be dealt with in detail. Because the use of complex numbers makes the solution of this type of problem trivial, we will be able to concentrate on the form of the results, rather than on the details of how to get there.

Complex numbers are also helpful in describing waves of all sorts (see Chapter 14 of Volume 2). They are particularly useful in situations involving the diffraction and interference of light waves. In quantum mechanics, it is not so much that they are a calculational aid, but rather that the wave function used to describe objects like electrons involves complex numbers in an essential manner. Neither of these last two applications is described here.

4.1 What are complex numbers?

Mathematicians are adept at inventing new types of concepts. Thus while the use of positive numbers to represent the magnitude of actual quantities in the real world gives them an immediate significance, this is not so obvious for negative numbers. However, just as we can think of

+3 as the solution of the equation

$$x - 3 = 0 \tag{4.1}$$

so we can invent −3 to be the solution of the not-too-different-looking

$$x + 3 = 0 \tag{4.2}$$

Thus we (or rather they – the mathematicians) can develop a set of rules for the arithmetic of negative numbers.

If we now consider

$$x^2 - 9 = 0 \tag{4.3}$$

we find that its solutions are $x = \pm 3$. In a similar vein, we can ask what values of x satisfy

$$x^2 + 9 = 0 \tag{4.4}$$

We find that we are in a somewhat similar situation to what we encountered when we first faced eqn (4.2) – nothing we know up till now is a solution of eqn (4.4). So the same device is adopted: a new type of number is invented, such that it provides a solution of this equation. Just as eqn (4.2) needed the introduction of an extra concept or symbol, so here we meet the idea of *imaginary* numbers.† The basic imaginary number is denoted by the symbol i and is defined to have the property that

$$i^2 = -1 \tag{4.5}$$

and so the solutions of eqn (4.4) are $x = \pm 3i$.

In general, a quadratic equation

$$ax^2 + 2bx + c = 0 \tag{4.6}$$

will have roots

$$x = (-b \pm \sqrt{b^2 - ac})/a \tag{4.7}$$

Then if $b^2 - ac$ is negative, the square root is imaginary. The solution x then has a conventional part $-b/a$ (which is called *real*) and an imaginary part

$$\sqrt{b^2 - ac}/a = i\sqrt{ac - b^2}/a \tag{4.8}$$

† The term 'imaginary' is a little lurid, and gives the feeling that such numbers are unreal, especially when compared with ordinary numbers, which in this context are called 'real'. This is a little unfortunate. We will see later that the 'complex' numbers we deal with in this chapter can be usefully applied to very pratical situations.

where $\sqrt{ac-b^2}/a$ is real. A number which consists of a real and an imaginary part is called complex. The simplest form is

$$x = p + iq$$

where p and q are real numbers (like $3, -7.1, \sqrt{2}, \pi^2/15, \sin(0.18\pi)$, etc.).

Now all this might seem like a fairly meaningless exercise. In particular, we might guess that if we were to continue the sequence started in eqns (4.1)–(4.4), we would find that we need to invent even more complicated objects than the complex numbers we have just introduced. Thus, for example, what are the solutions of the equation

$$x^2 - 9i = 0 \tag{4.9}$$

This requires us to calculate $\sqrt{i}$. It is a big surprise for most people that in fact we do not need any new types of numbers for this; the solution involves only the complex numbers we have already met, and will very shortly come to know and love. Similarly, complex numbers provide all the solutions of any polynomial equation for x, where the coefficients of the various terms in x^m in the equation are either real or complex. In this case, complex numbers provide a very natural extension of the real ones, so that all polynomial equations of the same degree have the same number of solutions.

Three points are to be noted in passing. First, in simple cases, some of the complex solutions may reduce to real or to imaginary numbers. Secondly, it may turn out that the exact solution of the equation is impossible to obtain analytically. This, however, does not negate the principle of the existence of complex solutions, whose values can if necessary be determined numerically to any desired accuracy. Finally, some of the solutions may be equal to each other.

4.2 Arithmetic of complex numbers

4.2.1 Addition

The first operation we might want to perform on complex numbers is to add them up. The rule is very simple. If we have n complex numbers

$$z_n = a_n + ib_n \tag{4.10}$$

their sum z is such that its real part is the sum of the real parts, and similarly for the imaginary part i.e.

$$z = \Sigma z_n = \Sigma a_n + i\Sigma b_n \tag{4.11}$$

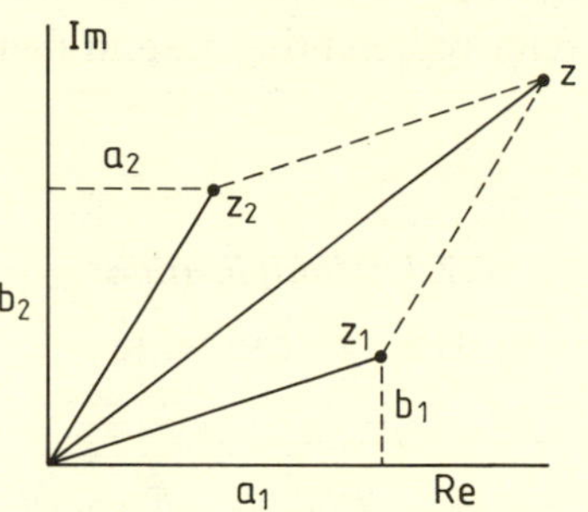

Fig. 4.1 Argand diagram illustrating the addition of two complex numbers. The sum $z = z_1 + z_2$ is the point at the end of the diagonal of the parallelogram, two of whose sides start at the origin and end at z_1 and z_2 respectively.

In this respect, complex numbers remind us of vectors in two dimensions. Thus if we regard each complex number z_n as being replaced by a vector $\mathbf{z}_n$ with components (a_n, b_n), then the vector sum

$$\mathbf{z} = \Sigma \mathbf{z}_n = (\Sigma a_n, \Sigma b_n) \tag{4.12}$$

This suggests that we could draw a diagram in which a complex number is represented by a point whose x and y coordinates are respectively the real and imaginary parts of thc complex number. Such a picture is known as an Argand diagram. The sum of two complex numbers can then be read off from the corner of the parallelogram diagonally opposite the origin, as shown in fig. 4.1. Alternatively, it can be thought of as the point at the far end of a vector $\mathbf{z}'_2$, which is parallel to $\mathbf{z}_2$ and of length equal to it, but which starts at the point represented by $\mathbf{z}_1$ in our Argand diagram, rather than at the origin (see fig. 4.1). Because of the properties of parallelograms, another possibility is to add the vector $\mathbf{z}'_1$ onto $\mathbf{z}_2$. This confirms that

$$z_1 + z_2 = z_2 + z_1 \tag{4.13}$$

4.2.2 Subtraction

Subtraction really involves nothing new. Thus

$$z_3 - z_4 = z_3 + (-z_4) \tag{4.14}$$

where $-z_4$ is such that when it is added to z_4, the answer is zero, i.e.

$$-z_4 = -a_4 - ib_4 \tag{4.15}$$

Hence the operation (4.14), as performed in the Argand diagram, consists in first reversing the vector representing z_4, and then adding it to that of z_3.

4.2.3 Multiplication

Next comes multiplication. Here we can write

$$\begin{aligned} z_5 z_6 &= (a_5 + ib_5)(a_6 + ib_6) \\ &= a_5 a_6 + i(b_5 a_6 + a_5 b_6) + i^2 b_5 b_6 \\ &= (a_5 a_6 - b_5 b_6) + i(b_5 a_6 + a_5 b_6) \end{aligned} \tag{4.16}$$

where we have made use of our basic definition (4.5). We see that the product is a new complex number, but its value does not seem particularly easy to interpret.

In order to achieve a better understanding of such products, we digress to remind ourselves that e^x can be written as an infinite series

$$e^x = 1 + x + x^2/2! + x^3/3! + \ldots x^n/n! + \ldots \tag{4.17}$$

Up till now, we would, of course, have chosen some real value for x, but since we have recently expanded our horizons, we might be brave enough to try something imaginary. So let us put

$$x = i\theta \tag{4.18}$$

where θ is real.† (The reason for choosing the variable as θ will soon become apparent). Then if we remember that $i^2 = -1, i^3 = -i, i^4 = +1$, etc, the infinite series reduces to two sets of terms, one of which contains a single factor of i, and the other which does not; these are respectively the imaginary and the real parts of our answer. These give us

$$e^{i\theta} = (1 - \theta^2/2! + \theta^4/4! - \ldots) + i(\theta - \theta^3/3! + \theta^5/5! - \ldots) \tag{4.19}$$

At this stage, we are supposed to have a flash of recognition as we see that these are the series for $\cos\theta$ and for $\sin\theta$. Thus we can write

$$e^{i\theta} = \cos\theta + i\sin\theta \tag{4.20}$$

This is a fantastically useful result, which we will use frequently.

† We might be a little worried about using the series (4.17) for an imaginary exponent. It turns out that everything is all right, mainly because there is no alternative definition of $e^{i\theta}$ which could conflict with what we are doing here. The even more adventurous might like to try replacing x by a complex number, instead of simply an imaginary one. The choice $x = \ln(r) + i\theta$ then leads directly to the result (4.25).

From eqn (4.20) we can obtain some surprising deductions. A particularly amazing one follows by setting $\theta = \pi$. Then

$$e^{i\pi} = -1 \tag{4.21}$$

This relates some of the fundamental constants of mathematics in a particularly simple way. It is also somewhat unexpected that when we raise e to this particular imaginary power, the answer is the real number -1.

Similarly we can deduce

$$e^{i\pi/2} = i \tag{4.22}$$

and

$$e^{2\pi i} = 1 \tag{4.23}$$

We leave it as an exercise for the student to derive the even more remarkable result

$$i^i = e^{-\pi/2} \tag{4.24}$$

Thus we have raised the square root of -1 to an imaginary power, and obtained a real answer. I am afraid I can throw no enlightenment on what deep significance this has. Numerically, $e^{-\pi/2} \sim 0.208$. Amusingly enough, for a long period in the 1930s, this was the rate of exchange between English pounds and American dollars.

If we return to eqn (4.20) and multiply both sides by r, we obtain

$$\begin{aligned} re^{i\theta} &= r\cos\theta + ir\sin\theta \\ &= a + ib \end{aligned} \tag{4.25}$$

where

$$\left.\begin{aligned} a &= r\cos\theta \\ \text{and } b &= r\sin\theta \end{aligned}\right\} \tag{4.26}$$

This demonstrates that $a + ib$ and $re^{i\theta}$ are two equivalent alternative representations of a complex number z. In particular, eqns (4.26) show that r and θ are the radius and the polar angle of the point z in the Argand diagram (see fig. 4.2).

Two 'other' concepts related to our complex number z are frequently used. The first is the length of the corresponding vector $\mathbf{z}$; it is denoted by $|z|$, and is called the modulus of z. The other refers to the direction

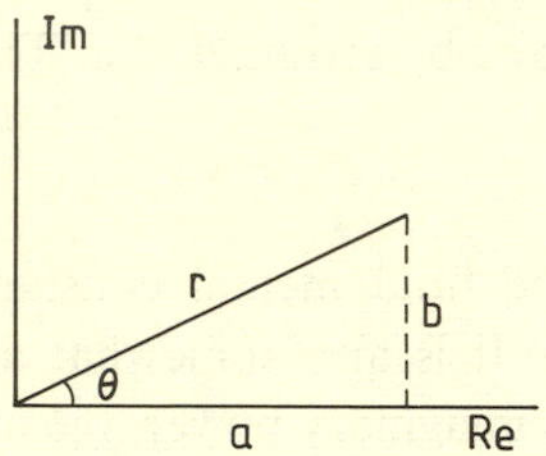

Fig. 4.2 Illustration of the alternative representations of a complex number $z = a + ib$ or $z = re^{i\theta}$. The modulus of z is r, and its argument is θ.

of z in the Argand diagram; it is called the argument of z, or arg(z) for short. In terms of components,

$$\left.\begin{aligned} |z| &= \sqrt{a^2 + b^2} \\ \text{and } \arg(z) &= \tan^{-1}(b/a) \end{aligned}\right\} \tag{4.27}$$

In fact the polar coordinate description is more appropriate, since $|z|$ and arg(z) are simply r and θ respectively.

We are now ready to return to our multiplication sum of eqn (4.16), but instead of representing the complex numbers by their real and imaginary parts, instead we will use the description in terms of r and θ.

Then

$$\begin{aligned} z_5 z_6 &= (r_5 e^{i\theta_5})(r_6 e^{i\theta_6}) \\ &= r_5 r_6 e^{i(\theta_5 + \theta_6)} \end{aligned} \tag{4.28}$$

This is easier to remember and to understand than eqn (4.16). The product is a new complex number whose modulus $r_5 r_6$ is the product of r_5 and r_6, the lengths of z_5 and z_6 respectively. On the other hand, the argument of the product is the sum of the arguments of the individual complex numbers.

For those who are suspicious about the validity of multiplying the polar form of complex numbers as if they were ordinary real numbers, it is possible to go back to the product in terms of the real and imaginary parts as shown in eqn (4.16), to calculate its modulus and its argument, to compare these with those of the individual complex numbers z_5 and z_6, and finally to deduce that the last two sentences of the previous paragraph are correct. The main point to emerge from this exercise is that, as far as products of complex numbers are concerned, the polar form makes calculations considerably easier than does the component representation.

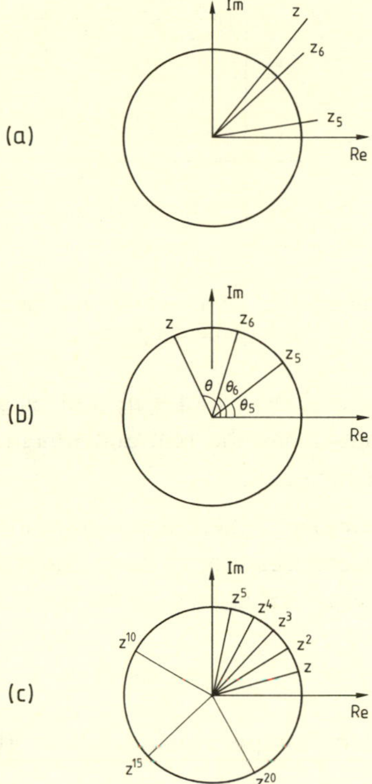

Fig. 4.3 Multiplication of numbers, as seen in the Argand diagram. The curve in the three diagrams is the unit circle. A product $z = z_5z_6$ has its modulus equal to the product of the moduli of z_5 and z_6, while its argument is the sum of their arguments. (a) Two numbers outside the unit circle have a product even further away from the origin. (b) The product of two numbers on the unit circle is also on the unit circle, and at an angle $\theta = \theta_5 + \theta_6$ (c) Raising a complex number z on the unit circle to the nth power simply results in a new point on the unit circle with argument n times that of z.

The operation of multiplication is demonstrated in the Argand diagram of fig. 4.3. The circle drawn there has unit radius ($r = 1$). Thus the product of two zs outside the unit circle will be a complex number even further away from the origin (fig. 4.3(a)). On the other hand, if both complex numbers are on the unit circle, so will be their product (4.3(b)). Finally in fig. 4.3(c), we show the result of raising a complex number on the unit circle to various powers.

A particularly simple special case of the product z_5z_6 occurs when one of the numbers is real, e.g. $z_5 = a_5$. Then the result is just z_6

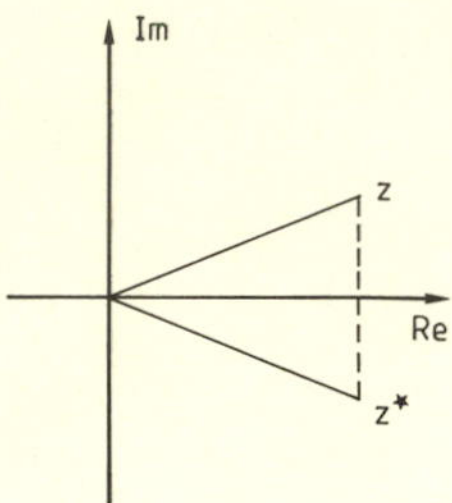

Fig. 4.4 The complex conjugate z^* of a complex number z is simply given by its reflection in the real axis of the Argand diagram.

with its modulus increased by a factor a_5 and its argument unchanged. Alternatively, the product has the real and imaginary parts of z_6 both increased by the same factor a_5.

For any complex number z, there is always another complex number z^* called 'the complex conjugate of z', such that the lengths of z and z^* are equal, and the product zz^* is real. If we write z in the now familiar polar form $re^{i\theta}$, then

$$z^* = re^{-i\theta} \tag{4.29}$$

Alternatively, in terms of components $z = a + ib$; then

$$z^* = a - ib \tag{4.29'}$$

The relation of z and z^* in the Argand diagram is shown in fig. 4.4. The complex conjugate is simply the reflection of z in the real axis.

4.2.4 Division

Finally we come to division, which (just like the relation of subtraction to addition) is really no more complicated than multiplication. Thus if

$$z = z_7/z_8 \tag{4.30}$$

it is simplest to write the numbers in their polar form. Then

$$\begin{aligned} z = re^{i\theta} &= (r_7e^{i\theta_7})/(r_8e^{i\theta_8}) \\ &= (r_7/r_8)e^{i(\theta_7-\theta_8)} \end{aligned} \tag{4.31}$$

That is, the modulus of z is the *ratio* of the moduli of z_7 and z_8, while its argument is the *difference* of their arguments.

An alternative approach can be used for z_7 and z_8 in component form.

The idea is that we try to avoid dividing by a complex number, since we are not yet quite sure how to do it. Thus we multiply the numerator and denominator of eqn (4.30) by something in order to make the numerator real. (Presumably we still remember how to divide by a real number.) An appropriate choice for 'something' is z_8^* since, from the definition of a complex conjugate, $z_8 z_8^*$ is real.† Thus

$$\begin{aligned} z &= \frac{a_7 + ib_7}{a_8 + ib_8} \\ &= \frac{(a_7 + ib_7)(a_8 - ib_8)}{(a_8 + ib_8)(a_8 - ib_8)} \\ &= [(a_7 a_8 + b_7 b_8) + i(a_8 b_7 - a_7 b_8)]/(a_8^2 + b_8^2) \end{aligned} \tag{4.31$'$}$$

With some effort, we can show that this is equivalent to (4.31).

As a final method, we can write the answer

$$z = a + ib$$

and then rewrite (4.30) as

$$zz_8 = z_7 \tag{4.30$'$}$$

We next perform the multiplication on the left-hand side of this equation, and equate the product to z_7 in order to determine a and b, and hence the ratio z. The reader might at first be curious how the *one* equation (4.30′) can be solved to find *two* quantities a and b. The important point is that a complex equation is satisfied only if both the real and the imaginary parts of the two sides of the equation match up separately, i.e.

$$p + iq = 0 \tag{4.32}$$

implies

$$\left.\begin{aligned} p &= 0 \\ \text{and } q &= 0 \end{aligned}\right\} \tag{4.32$'$}$$

Again this is in line with our representing complex numbers in the Argand diagram as two-dimensional vectors; an equation relating vectors in n-dimensional space is the same as n relations between their separate components.

Just as in the case of multiplication, it is well worth remembering that the algebra involved in dividing complex numbers is simpler if they are in the polar form $re^{i\theta}$, rather than when we write them as $a + ib$. Thus if we are interested in the modulus and/or the argument of a ratio, it

† This is reminiscent of the method used to convert, say, $1/(5+\sqrt{3})$ into $(5-\sqrt{3})/(25-3)$

will almost invariably be much easier† to deduce it from eqn (4.31) than by extracting the answers via eqn (4.31′). We will come across some examples of this later in this chapter.

4.2.5 Square root of *i*

Now that we have learnt the basic operations that can be performed with complex numbers, we return to the question of the square root of i, that was raised in Section 4.1. That is, we are trying to solve the equation

$$z^2 = i \tag{4.33}$$

We shall do this by three different methods.

The first, and most boring, is the same as the last approach we used for dividing. We thus assume that a solution in which z is a complex number is waiting to be found, and we write this as

$$z = a + ib$$

We then square it, substitute into eqn (4.33), and equate real and imaginary parts in order to find a and b. We discover

$$a = b = \pm 1/\sqrt{2}$$

so that the solutions are

$$z = \pm(1 + i)/\sqrt{2} \tag{4.34}$$

Thus, as promised, the square root of i is a complex number, and there is no need to continue the sequence of inventing all kinds of new numbers.

The second method begins by writing the i of eqn (4.33) in (r, θ) form as

$$i = e^{i\theta/2}$$

This we have done (i) by remembering eqn (4.22); or (ii) by reading off $r = 1$ and $\theta = \pi/2$ for the point in the Argand diagram corresponding to the 'complex' number $z = i$; or (iii) by inverting eqns (4.26) (and remembering that for $z = i$, a is zero and b is unity).

Then since from eqn (4.33)

$$z^2 = e^{i\pi/2} \tag{4.33′}$$

† Also the chance of making a trivial algebraic error will be correspondingly reduced as well.

we have

$$\begin{aligned} z &= (e^{i\pi/2})^{\frac{1}{2}} \\ &= e^{i\pi/4} \\ &= 1/\sqrt{2} + i/\sqrt{2} \end{aligned} \tag{4.35}$$

This agrees with one of the solutions given in (4.34). We will shortly return to the question of where the other one has disappeared.

What we have done here is in fact to use a specific case of the more general de Moivre's Theorem. This states that

$$(\cos\theta + i\sin\theta)^n = \cos n\theta + i\sin n\theta \tag{4.36}$$

which follows immediately from the fact that

$$(e^{i\theta})^n = e^{in\theta} \tag{4.36$'$}$$

Our specific example requires us to set $\theta = \pi/2$ and $n = 1/2$, whence

$$i^{\frac{1}{2}} = \cos\pi/4 + i\sin\pi/4 \tag{4.37}$$

in agreement with (4.35). We return to de Moivre's Theorem in the next section.

Now for our final method, which uses the Argand diagram. We have already discovered when we were considering how to multiply complex numbers that when we raise a number on the unit circle to the nth power, we simply multiply its argument (i.e. its angular position) by n – see fig. 4.3(c). A corollary of this is that the nth root of a number on the unit circle is obtained by dividing its angle by n. Thus square roots are obtained by halving the angle. For the particular case of i, which is on the unit circle at an angle of $\pi/2$, the square root will also be on the unit circle at an angle of $\pi/4$ (see fig. 4.5(a)). Thus once again the answer is $1/\sqrt{2} + i/\sqrt{2}$. In fact this method is merely a visual way of applying de Moivre's Theorem.

We now return to the question of the second solution for $\sqrt{i}$. When we give a bit more thought to where i is on the Argand diagram, we may realise that whereas the value of the radius $r = 1$ is uniquely defined, the argument is ambiguous and can have an integral multiple of 2π added to it. Thus values of

$$\ldots -7\pi/2, \quad -3\pi/2, \quad +\pi/2, \quad +5\pi/2, \quad +9\pi/2, \quad \ldots$$

are all equally valid. The answer is determined simply by halving this angle. Now $-7\pi/2$, $+\pi/2$, $+9\pi/2$, etc. yield respectively $-7\pi/4$, $+\pi/4$,

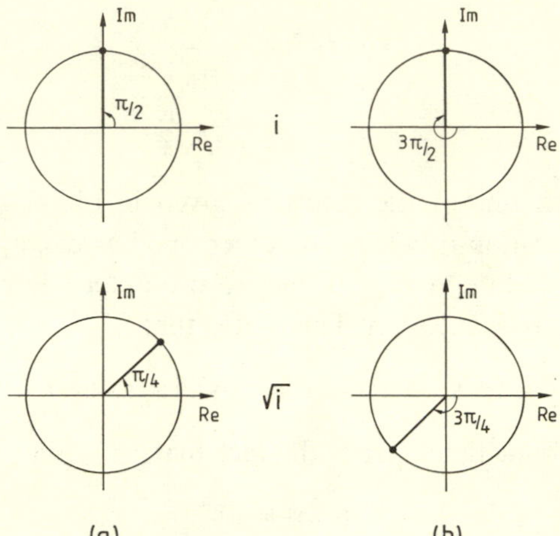

Fig. 4.5 The square root of i is on the unit circle with argument half that of i. (a) Since i is at an angle of $\pi/2$, $\sqrt{i}$ is at $\pi/4$ and hence is $1/\sqrt{2} + i/\sqrt{2}$ (b) Alternatively i can be regarded as being at $-3\pi/2$, whence $\sqrt{i}$ is $-(1/\sqrt{2} + i/\sqrt{2})$.

$+9\pi/4, \ldots$; these are all identical in position to the value for $\sqrt{i}$ displayed in fig. 4.5(a), as they differ from each other by $2n\pi$. However, the other set of possibilities yield $\ldots -3\pi/4,\ +5\pi/4, \ldots$, which are all equivalent to $+5\pi/4$. The important point is that this is meaningfully different from $+\pi/4$, and so we have a second solution (see fig. 4.5(b)). In component form, this second solution is $-1/\sqrt{2} - i/\sqrt{2}$, which fortunately agrees with eqn (4.34).

The same basic reason provides the second solution for our second method. Thus, since $e^{2\pi mi} = 1$, we could have written

$$i = e^{i\pi/2} \cdot e^{2\pi mi}$$

rather than simply as $e^{i\pi/2}$. Then when we halve the argument to obtain the solution in eqn (4.35), we obtain

$$z = e^{i\pi/4} \cdot e^{\pi mi} \tag{4.35$'$}$$

for any integer m. Of course only two values of m are required to produce the different solutions for z, since all even values of m result in our obtaining identical zs (and similarly for odd m).

The fact that the argument of a complex number is ambiguous and results in n different solutions for the nth root of a complex number is a general feature of this type of problem.

We thus conclude that our various approaches all provide the correct number of solutions for $\sqrt{i}$, and agree as to what their values are. We reiterate that $\sqrt{i}$ is merely a complex number.

4.3 Using complex numbers

In this section, we discuss various applications of complex numbers. We start by describing the use of de Moivre's Theorem for solving some polynomial equations. Section 4.3.2 deals with summation of series involving trigonometric functions. Simultaneous differential equations are considered briefly in Section 4.3.3. The large subject of second order differential equations, with its many applications and possibilities, is dealt with on its own in Section 4.4. We regard as beyond the scope of this book the subject of mapping, in which a function $z(w)$ is used to transform points in the complex w plane into corresponding points in the complex z plane.

4.3.1 Polynomial equations

We have already met de Moivre's Theorem briefly in Section 4.2 (eqn (4.36)). Here we apply it to polynomial equations. We start with a specific example.

The equation

$$x^7 = 1 \tag{4.38}$$

should have seven roots, of which one is clearly unity. We find all of them by rewriting x as $r(\cos\theta + i\sin\theta)$, and then using de Moivre's Theorem to obtain

$$r^7(\cos 7\theta + i\sin 7\theta) = \cos 2m\pi \tag{4.38$'$}$$

where we have remembered that the '1' that appears on the right-hand side of eqn (4.38) has an argument ambiguity of $2m\pi$. Comparison of the modulus and argument then yields

$$\left.\begin{aligned} r = 1, \quad \theta = 2\pi m/7 \\ \text{or } x = \cos 2m\pi/7 + i\sin 2m\pi/7 \end{aligned}\right\} \tag{4.39}$$

Values of m from 1 to 7 yield different solutions, with $m = 7$ producing the $x = 1$ answer that we already knew.

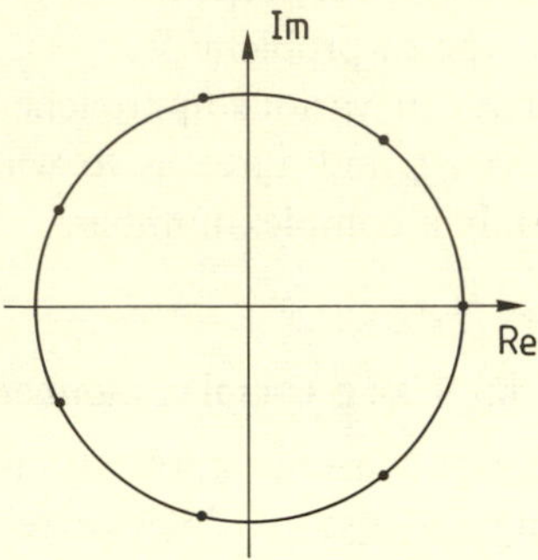

Fig. 4.6 The seventh roots of 1, as given by eqn (4.39). One is at unity and the other six are equally spaced around the unit circle.

These solutions are best displayed on an Argand diagram (from which we could in this case have deduced the answers more quickly in the first place). Thus we see from fig. 4.6 that the seven solutions are equally spaced round the unit circle, with one solution at $x = 1$. It is a general feature that the nth roots of unity divide the unit circle equally in this way.

The general polynomial equation, for which de Moivre's Theorem is useful for finding the solution, is one which can be reduced to the form

$$[f(x)]^n = c \tag{4.40}$$

where c is a constant which can, of course, be complex, and $f(x)$ is a simple function of x, by which we mean that we can solve the equation

$$f(x) = c' \tag{4.41}$$

where c' is another constant.

We first express

$$f(x) = re^{i\theta}$$
$$\text{and } c = r_c e^{i\theta_c} e^{2im\pi}$$

Substituting these in eqn (4.40) yields

$$f(x) = r_c^{1/n} e^{i(\theta_c + 2\pi m)/n} \tag{4.42}$$

The right-hand side of eqn (4.42) is a set of m different constants, and so this equation is of the same form as (4.41), and hence can be solved for x. This then provides our solutions. (See Problem 4.5.)

Sometimes equations which are not at first sight of the form (4.40) can be reduced to it with a little cunning. Thus, for example

$$x^6 - 15x^4 + 15x^2 - 1 = 0 \tag{4.43}$$

can be rewritten as

$$(x+i)^6 + (x-i)^6 = 0 \tag{4.44}$$

or

$$\left(\frac{x+i}{x-i}\right)^6 = -1 \tag{4.45}$$

which possesses the required structure. We could have made the sequence of operations required for solving this problem less obvious by writing our equation as

$$y^3 - 15y^2 + 15y = 1 \tag{4.43$'$}$$

This is identical to (4.43), and hence the solutions can be found in the same way, provided $y = x^2$.

4.3.2 Summing trigonometric series

Below are two different series. Which is it easier to sum?

$$C_1 = 1 + \cos\theta + \cos 2\theta + \cos 3\theta + \ldots + \cos l\theta \tag{4.46}$$

$$C_2 = 1 + \cos\theta + \cos^2\theta + \cos^3\theta + \ldots + \cos^l\theta \tag{4.47}$$

The answer is that C_2 is simply a geometric series, whose sum can be written down immediately as

$$C_2 = \frac{1 - \cos^{l+1}\theta}{1 - \cos\theta} \tag{4.48}$$

The first looks more complicated. However, we remember that de Moivre's Theorem gives us a recipe for converting the sines and cosines of multiple angles (like 3θ) to the corresponding powers. Thus the problem of summing C_1 would have been easier if it had been coupled with that of summing

$$S_1 = \sin\theta + \sin 2\theta + \sin 3\theta + \ldots + \sin l\theta \tag{4.46$'$}$$

since in that case we can multiply S_1 by i and add it to C_1 to obtain

$$\begin{aligned} C_1 + iS_1 &= 1 + (\cos\theta + i\sin\theta) + (\cos 2\theta + i\sin 2\theta) + \ldots (\cos l\theta + i\sin l\theta) \\ &= 1 + z + z^2 + \ldots + z^l \end{aligned}$$

where $z = \cos\theta + i\sin\theta$, and we have made use of de Moivre's Theorem.

Thus we have obtained $C_l + iS_l$ as a geometric series which we can sum to give

$$\begin{aligned} C_l + iS_l &= (1 - z^{l+1})/(1 - z) \\ &= (1 - e^{i(l+1)\theta})/(1 - e^{i\theta}) \\ &= \frac{e^{i(l+1)\theta/2}\sin[(l+1)\theta/2]}{e^{i\theta/2}\sin(\theta/2)} \end{aligned} \tag{4.49}$$

where we have again used de Moivre's Theorem. Finally we extract C_l and S_l by comparing the real and the imaginary parts of eqn (4.49) and obtain

$$C_l = \cos(l\theta/2)\sin[(l+1)\theta/2]/\sin(\theta/2) \tag{4.50}$$

and

$$S_l = \sin(l\theta/2)\sin[(l+1)\theta/2]/\sin(\theta/2) \tag{4.50$'$}$$

Thus what is needed is the imagination to see that, if we are presented with the series C_l, which involves only real terms, we can make it easier by inventing the series S_l and turning the problem into one containing complex numbers.

Enough algebra was involved in deducing our solutions (4.50) and (4.50′) that it is worthwhile to check them for simple cases. For example, it is encouraging to see that for $l = 1$, our answers are obviously correct. With a hand calculator, we could also try out some small value of l (e.g. 3) and a specific choice for θ. Such simple tests could help reveal any algebraic slips in our working.

In our series for C_l above, the coefficients of the various terms $\cos k\theta$ were all unity. If the coefficients instead are some simple function of k, we may still be able to add up the series for z, and hence solve the problem. An example of this type appears in Problem 4.4.

A very slightly different approach to summing the series (4.46), and which has something in common with what we shall be doing in Section 4.4, is to write each term in the series as

$$\cos k\theta = \mathrm{Re}[e^{ik\theta}] \tag{4.51}$$

where Re[x] implies the real part of x (and similarly Im[x] would be its imaginary part). Then

$$C_l = \mathrm{Re}[1 + e^{i\theta} + e^{2i\theta} + \ldots e^{li\theta}] \tag{4.52}$$

which then yields the same result as before. In this case we do not explicitly introduce the sine series, and instead recognise cosines or sines as being respectively the real or imaginary parts of $e^{i\theta}$.

Thus, for example, the series

$$M = 1 - \sin\theta - \frac{1}{2}\cos 2\theta + \frac{1}{3}\sin 3\theta + \frac{1}{4}\cos 4\theta - \ldots \qquad (4.53)$$

can be written as

$$\mathrm{Re}[1 + iz + (iz)^2/2 + (iz)^3/3 + (iz)^4/4 + \ldots] \qquad (4.53')$$

which again is a summable series. The answer follows by much the same steps as used above.

4.3.3 Simultaneous differential equations

There is a special class of simultaneous differential equations which, because they have the correct relationships between certain of the coefficients, can be solved most easily by using complex numbers. An example is provided by the equations:

$$\left.\begin{aligned} \frac{d^2x}{dt^2} + 3\frac{dy}{dt} - 2x = f(t) \\ \frac{d^2y}{dt^2} - 3\frac{dx}{dt} - 2y = g(t) \end{aligned}\right\} \qquad (4.54)$$

These are simultaneous equations for x and y, which we have to obtain as functions of t.

The trick, which by now should be easy to guess, is to multiply the second equation by i, to add them, and to replace $x + iy$ by z:

$$\frac{d^2z}{dt^2} - 3i\frac{dz}{dt} - 2z = f(t) + ig(t) \qquad (4.54')$$

This is a second order differential equation for the one variable z which we can solve by the standard methods. Then we extract x and y by taking the real and imaginary parts of our solution.

This works only when the corresponding terms in the two equations have the same magnitude and the correct relative signs. Otherwise, it is necessary to use one of the other methods for solving simultaneous differential equations (see Section 5.15).

Of course this method can be applied not only to simultaneous *differential* equations, but also to ordinary simultaneous ones. It is, however, a rather heavy-handed technique to use for a few special cases of such simple problems.

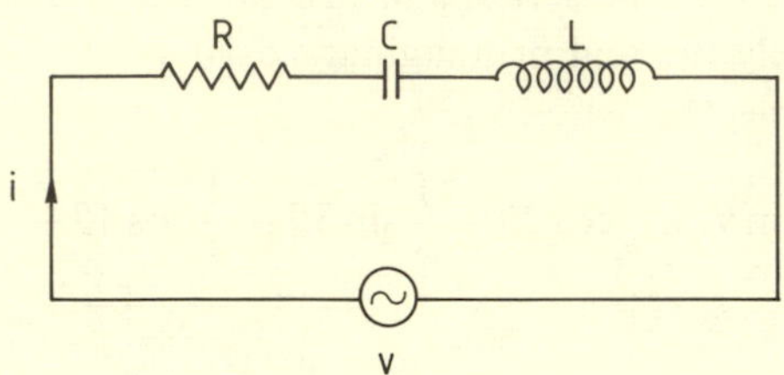

Fig. 4.7 A simple electrical circuit consisting of a resistor R, a capacitor C and an inductor L, connected in series to an external voltage source V of frequency ω. The current in the circuit is i, and the charge on the capacitor q.

4.4 Second order differential equations, physics and complex numbers

A multitude of physical situations are to a good approximation described by the differential equation

$$a\frac{d^2x}{dt^2} + b\frac{dx}{dt} + cx = f(t) \tag{4.55}$$

where a, b and c are constants, and $f(t)$ is a function of time. In this section, we shall generally be concerned with the case where $f(t)$ is harmonically varying.

For example, in the electrical circuit of fig. 4.7, the charge q on the capacitor satisfies

$$L\frac{d^2q}{dt^2} + R\frac{dq}{dt} + \frac{q}{C} = V(t) \tag{4.56}$$

We can regard $V(t)$ as an externally applied influence, and q as describing how our system responds.

Alternatively, x can represent the position of a mass m on a light spring obeying Hooke's Law, such that the restoring force is proportional to the displacement. In the presence of an external force $F(t)$ acting on the mass, and a resistive force $-k\,dx/dt$ proportional to the mass's velocity, the differential equation for x is

$$m\frac{d^2x}{dt^2} + k\frac{dx}{dt} + \lambda x = F(t) \tag{4.57}$$

which has exactly the same structure as (4.55) and (4.56).

In all these problems we are interested in solving our differential equation for x, when the system is subject to the given external influence $f(t)$. Now the solutions of differential equations of this type contain two parts, a complementary function and a particular integral (see Section 5.10). The former is defined to be the solution of the equation when modified to have $f(t) = 0$, and is thus obviously independent of the external influence $f(t)$. In contrast, the particular integral depends on the specific form of $f(t)$, and it is this that we will be especially interested in finding.

Another way of seeing this is to realise that the complementary function describes the 'transient' behaviour of the system for the specific initial conditions, while the particular integral gives the 'steady state' solution.† For example, in our electric circuit, there may well be an initial surge of current when the circuit is first connected because someone had recently charged up the capacitor. For our spring, the mass could be initially moving because you had previously given it a kick, and it would continue to move and perhaps oscillate for a while afterwards. Such oscillation is at the so-called natural frequency of the system. These effects almost always die away with time, owing to dissipative effects in the system; in the circuit, this is caused by the resistor R and for the spring by the resistive force $-k\,dx/dt$. The system will after a short while settle down to the steady state response. Thus the charge on the capacitor will oscillate with a constant amplitude in response to an applied harmonic voltage, and similarly for the mass on the spring when there is an external harmonic force.

It is the particular integral which describes this steady state behaviour. Because after a short while the transients associated with the initial conditions have died out while the steady state solution persists, we are generally more interested in that part of the solution corresponding to the particular integral. Thus, in practical terms, if we use an oscilloscope to monitor the charge on the capacitor, we will usually either be uninterested in or perhaps even fail to notice the way the trace may initially give a slight jump when the circuit is first connected.

Later we shall justify mathematically our statement that the complementary function corresponds to transient effects. First, however, we shall consider the more-important steady state particular integral for the case where $f(t)$ is harmonic.

† The complementary function exhibits this transient behaviour when b/a is positive. When $b = 0$, it oscillates with constant amplitude. In this case, if the driving force is of the same frequency as those of these natural oscillations, the 'steady state' solution grows with time; thus again, after some initial interval, the particular integral dominates over the complementary function. In realistic physical systems, the resistive effects can be small, but usually b is not zero.

The case of b/a being negative corresponds to a system which oscillates spontaneously, and is not dealt with here.

We are going to make use of complex numbers for doing this. The basic reason why they are appropriate is that the x oscillations need not have the same phase as the harmonic driving force $f(t)$. Thus the oscillations require an amplitude and a phase to describe them, and this corresponds to the number and type of variables that we need for a complex number. At a practical level, we shall see that the use of complex numbers makes the solution of such problems very much simpler, once the somewhat unusual technique has been mastered.

Now if †

$$f(t) = f_0 \cos \omega t \tag{4.58}$$

we will be interested in finding harmonic solutions at the same frequency,‡ i.e.

$$x(t) = x_0 \cos(\omega t + \phi) \tag{4.59}$$

where ϕ is the phase difference between x and f. (If you do not believe there is a phase difference, set $\phi = 0$, and substitute (4.58) and (4.59) into (4.55). You will discover that in general this is not a valid solution).

It looks a little curious, but it is clearly valid, to write

$$\begin{aligned} f(t) &= f_0 \cos \omega t \\ &= \mathrm{Re}[f_0 e^{i\omega t}] \end{aligned} \tag{4.58$'$}$$

and similarly the solution

$$\begin{aligned} x(t) &= x_0 \cos(\omega t + \phi) \\ &= \mathrm{Re}[x_0 e^{i\omega t} e^{i\phi}] \\ &= \mathrm{Re}[\chi e^{i\omega t}] \end{aligned} \tag{4.59$'$}$$

where the complex number

$$\chi = x_0 e^{i\phi} \tag{4.60}$$

We now need to substitute (4.58′) and (4.59′) into our differential equation (4.55) in order to determine χ, and hence $x(t)$. This requires us to differentiate expressions like (4.59′). We might guess that, since χ is a constant (still to be determined and admittedly complex), when we

† In general we might have expected $f(t)$ to have a phase associated with it (i.e. $\cos(\omega t + \phi')$), but we can eliminate ϕ' by a suitable choice of our origin of time, without significantly affecting the problem.

‡ Throughout this book, ω is often referred to simply as the frequency, rather than the more accurate but cumbersome 'angular frequency'.

differentiate $x(t)$ of eqn (4.59′) with respect to time, because the only place that t occurs is in the exponent, the answer will be

$$\frac{dx}{dt} = \text{Re}[i\omega\chi e^{i\omega t}] \tag{4.61}$$

Our only doubt about this might be that perhaps the factor $i\omega$ should be placed before the 'Re'. That, however, would give us an imaginary answer for dx/dt which is the current in the circuit or the mass's velocity, in our electrical or mechanical examples respectively. This sounds highly undesirable. Furthermore, if we differentiate $x_0 \cos(\omega t + \phi)$ and compare it with

$$\text{Re}[i\omega\chi e^{i\omega t}] = \text{Re}[i\omega x_0 \{\cos(\omega t + \phi) + i\sin(\omega t + \phi)\}] = -\omega x_0 \sin(\omega t + \phi)$$

we see that our guess (4.61) is indeed correct. Thus differentiating an expression like (4.59′) simply has the effect of bringing down a factor of $i\omega$ from the exponent.

What we are capable of doing once we can also succeed in performing a second time, and hence

$$\frac{d^2x}{dt^2} = \text{Re}[(i\omega)^2\chi e^{i\omega t}] \tag{4.62}$$

On substituting these into the differential equation (4.55), we obtain

$$\begin{aligned}\text{Re}[f_0 e^{i\omega t}] &= a\text{Re}[-\omega^2\chi e^{i\omega t}] + b\text{Re}[i\omega\chi e^{i\omega t}] + c\text{Re}[\chi e^{i\omega t}] \\ &= \text{Re}[(-\omega^2 a + i\omega b + c)\chi e^{i\omega t}] \end{aligned} \tag{4.63}$$

When we examine this equation, we see that on both sides we have a 'Re', and similarly within the square brackets there is an $e^{i\omega t}$ on each side. It would be convenient if we could regard these as ordinary factors and cancel them out. Then we would be left with

$$f_0 = (-\omega^2 a + i\omega b + c)\chi \tag{4.63′}$$

which immediately yields our solution

$$\chi = \frac{f_0}{(c - \omega^2 a) + i\omega b} \tag{4.64}$$

This is, of course, a complex number, from which we can recover the usual real solution as

$$x(t) = \text{Re}[\chi e^{i\omega t}] \tag{4.59′}$$

If we prefer to write x explicitly as

$$x(t) = x_0 \cos(\omega t + \phi) \tag{4.59}$$

then, as we have already seen from eqn (4.60),

$$\left.\begin{aligned} x_0 &= |\chi| \\ \text{and}\ \tan\phi &= \text{Im}[\chi]/\text{Re}[\chi] \end{aligned}\right\} \tag{4.65}$$

With a little bit of practice and familiarity, we will be prepared to regard the complex χ as our solution, rather than having to write it explicitly in the form (4.59).

We now return to the step where we merely dropped the 'Re' and $e^{i\omega t}$ from the two sides of the equation. It is *not* valid to say that if

$$\text{Re}[u] = \text{Re}[v]$$

then

$$u = v$$

The example

$$\left.\begin{aligned} u &= 1 + i \\ v &= 1 + 2i \end{aligned}\right\}$$

clearly invalidates this. However, what saves us in this case is that we have something of the form

$$\text{Re}[ue^{i\omega t}] = \text{Re}[ve^{i\omega t}] \tag{4.66}$$

which we require to be true *at all times t*. In that case it is true that

$$u = v \tag{4.67}$$

This can be verified by writing u and v each as consisting of a real part and an imaginary part, substituting into eqn (4.66), and then ensuring that the equation is true at all t. This requires that the real parts of u and of v are equal, and similarly for their imaginary parts. This demonstrates that eqn (4.67) is valid, and hence that we were in fact justified in dropping the 'Re' and the $e^{i\omega t}$ from both sides of eqn (4.63).

Now it really seems as if we have taken longer in deducing the solution than we would have done by a more conventional method. This is, of course, because we have gone over each step in great detail. So what we are going to do now is: (i) to recall the mathematical tricks that we are going to employ; and then (ii) to solve the equation at a faster pace, in order to demonstrate how easy and quick it actually can be.

The main mathematical tools are as follows:

(a) When we have Re and $e^{i\omega t}$ on both sides of the equation, we can in effect cancel them, and hence, if we want to save time, we can leave them out altogether. We should remember, however, that at all stages down to the equivalent of eqn (4.63), they really are present, even though in future we will not bother to write them down.

(b) Differentiation brings down a factor of $i\omega$. This we saw in eqn (4.61), which can be written as

$$\frac{d}{dt}\mathrm{Re}[\chi e^{i\omega t}] = \mathrm{Re}[i\omega\chi e^{i\omega t}] \tag{4.61$'$}$$

and hence with our new-found convention becomes

$$\frac{d\chi}{dt} = i\omega\chi$$

where χ is any complex number representing a variable with a harmonic time variation. Since this is true for any χ, we can rewrite it as

$$\frac{d}{dt} \equiv i\omega \tag{4.61$''$}$$

which is an operator equation, in that it makes sense only when some harmonically varying complex quantity appears on each side of the equation.

Before we apply our technique to the specific case of the electrical circuit, it is vital to remember how to tell the difference between a mathematician and a scientist. When dealing with electrical problems, mathematicians use i as the symbol for $\sqrt{-1}$, and j for electric current; scientists use the opposite convention. Thus for the rest of this section, we shall change from i to j for $\sqrt{-1}$, and keep i for electric current.

We are now going to find a harmonic solution for the charge on the capacitor of fig. 4.7, which satisfies the equation

$$L\frac{d^2q}{dt^2} + R\frac{dq}{dt} + \frac{q}{C} = V_0\cos\omega t \tag{4.56$'$}$$

We replace the real voltage $V_0\cos\omega t$ by the 'complex' voltage $\mathscr{V}$, which happens to be real because the phase angle of the voltage was zero. We represent the charge that we want to find by the complex quantity $\mathscr{Q}$. Substitution of $\mathscr{Q}$ into eqn (4.56′) yields

$$(-\omega^2L + j\omega R + 1/C)\mathscr{Q} = \mathscr{V}$$

or

$$\mathcal{Q} = \frac{\mathcal{V}}{(1/C - \omega^2 L) + j\omega R} \tag{4.68}$$

This is our answer; the problem is solved and it has taken two lines of algebra.

Of course, $\mathcal{Q}$ is complex. If we want to know the amplitude of the charge oscillations, we must calculate $|\mathcal{Q}|$. Similarly for the phase of these oscillations (with respect to the voltage as defining zero phase) we need $\arg(\mathcal{Q})$. In order to obtain these, we might think it is necessary to make the denominator D of expression (4.68) real by multiplying both numerator N and D by $(1/C - \omega^2 L) - j\omega R$, which is the complex conjugate of D. There is nothing actually wrong with this, but it is not necessary, wastes time and gives us an extra opportunity of making an arithmetic slip. Instead we need to remember the rules about the modulus and argument of the ratios of complex numbers as given in Section 4.2.4. Then

$$\begin{aligned} |\mathcal{Q}| &= \frac{|\mathcal{V}|}{|\,1/C - \omega^2 L + j\omega R\,|} \\ &= V_0/\sqrt{(1/C - \omega^2 L)^2 + (\omega R)^2} \end{aligned} \tag{4.69}$$

and

$$\tan\phi = -\omega R/(1/C - \omega^2 L) \tag{4.70}$$

If we really insist, we can finally write out our solution in conventional form as

$$q = |\mathcal{Q}| \cos(\omega t + \phi) \tag{4.71}$$

with $|\mathcal{Q}|$ and ϕ as given above. Very soon, however, we shall find that we are quite happy stopping at eqn (4.68).

Often in electrical problems of this sort, we are interested in the current i, rather than the charge on the capacitor. These are related by

$$i = dq/dt \tag{4.72}$$

and so we could deduce the current by differentiating eqn (4.71). However, it is more in the spirit of our new approach, and it also obviates the need for steps (4.69)–(4.71), to use eqn (4.61″) to write the complex current as $\mathcal{I}$

$$\begin{aligned} \mathcal{I} &= j\omega\mathcal{Q} \\ &= \frac{j\omega\mathcal{V}}{(1/C - \omega^2 L) + j\omega R} \end{aligned} \tag{4.73}$$

This is, of course, a complex number like, for example, $3 + 4j$ amperes. In that case, what would an RMS ammeter in the circuit actually read? Is it 3 amperes, $3/\sqrt{2}$, $3/2$ or something else? The reader should think about this before continuing; the answer is given shortly.

An insight into the significance of this complex number technique can be gained by anyone familiar with the 'phasor diagram' approach, which is sometimes used for electrical problems, but in fact can in principle be used for any system represented by our standard second order differential equation (4.55) with a harmonic driving effect. The input (4.58) is represented by the projection onto the horizontal axis of a vector of length f_0 rotating at constant angular speed ω in the $x - y$ plane. (See fig. 4.8(a)). If the vector is along the horizontal axis at $t = 0$, then the projection is $f_0 \cos \omega t$, as required.

The solution $x_0 \cos(\omega t + \phi)$ is similarly denoted by the projection of another vector of length x_0, advanced in angle by ϕ with respect to the present position of **f**, and of course rotating also with angular speed ω. In fig. 4.8(b) they have been plotted in the same diagram, although it must be remembered that **f** and **x** have different dimensions. The pair of vectors then rotate together, and at any instant the respective projections give the input and the response. Now we know that both of these, of course, vary with frequency ω, but we usually are interested only in their amplitude and perhaps in their phase relationship. This corresponds to ignoring the fact that the vectors are rotating and forgetting about the need to take their projections onto the horizontal axis† but regarding the essence of the solution as being encapsulated in the vectors **f** and **x** at a fixed time, which by convention can be chosen such that **f** is horizontal (see fig 4.8(c)).

In our complex number method, we are essentially using the philosophy of fig. 4.8(c), with the complex χ and (the conventionally real) f_0 telling us all we want to determine about the solution. Of course, if we really want to determine the value of x at any given instant, we reinsert the $e^{i\omega t}$ and the 'Re', which correspond to letting the vectors rotate together by the relevant amount, and taking their horizontal projections.

Now we return to the question of what our ammeter reads when the complex current $\mathscr{I}$ is $3 + 4j$ amperes. Although our actual current is derived from $\mathscr{I}$ by at some stage taking the real part, the answer is *not* related to the 3 in the $3 + 4j$ amperes. This is because, although the current at any moment is $\text{Re}[\mathscr{I} e^{i\omega t}]$, the amplitude is given by $|\mathscr{I}|$,

† These correspond respectively to our not bothering to write down $e^{i\omega t}$ and 'Re' in the complex number approach.

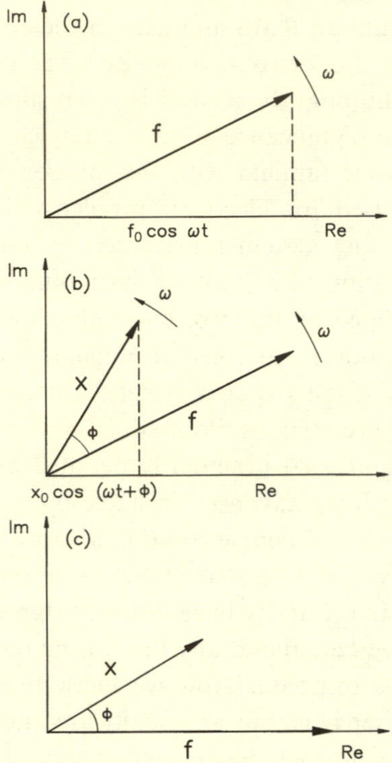

Fig. 4.8 The phasor diagram approach for displaying solutions of second order differential equations. (a) The driving force $f_0 \cos \omega t$ is given by the projection onto the horizontal axis of a vector **f** rotating with constant angular velocity ω, and which coincided with the horizontal axis at $t = 0$. (b) The solution $x_0 \cos(\omega t + \phi)$ is given by the projection onto the horizontal axis of a vector **x** rotating with angular velocity ω, and at a phase angle ϕ ahead of **f**. The vectors **f** and **x** thus rotate together. (c) The essence of the solution is contained in the relationship of the vectors **f** and **x**. By convention, the external force **f** is usually displayed as horizontal.

and not by Re[$\mathscr{I}$] [compare eqns (4.65)]. Thus an RMS ammeter, which shows $1/\sqrt{2}$ of the peak value, will give a reading of $5/\sqrt{2}$ amperes.

Finally, we return to the statement made much earlier in this section that the complementary function usually is a transient. In the spirit of this chapter, we will look for a solution of the equation

$$L\frac{d^2q}{dt^2} + R\frac{dq}{dt} + q/C = 0 \qquad (4.56'')$$

which is of the form

$$q = \mathrm{Re}[\mathscr{P}e^{jpt}] \tag{4.74}$$

where $|\mathscr{P}|$ is the amplitude and p the angular frequency of the free oscillations of the circuit. When we substitute this in eqn (4.56″), we find

$$-Lp^2 + jpR + 1/C = 0 \tag{4.75}$$

which yields as solution

$$p = (-jR \pm \sqrt{4L/C - R^2})/2L \tag{4.76}$$

Our first observation is that $\mathscr{P}$ is undefined. That is, the system is capable of sustaining free oscillations of any magnitude and phase, these being determined only by the initial conditions of any given problem (e.g. the charge on the capacitor and the current in the circuit at $t = 0$). The second point is that the frequency p is complex. This at first sight looks disastrous. However, we keep calm, and remember that the real charge is given by eqn (4.74). Thus the frequency p occurs in the term e^{jpt} which, with our solution for p, becomes

$$e^{jpt} = e^{-Rt/2L}e^{\pm j\sqrt{4L/C-R^2}t/2L} = e^{-Rt/2L}e^{\pm jgt} \tag{4.77}$$

where

$$g^2 = 1/LC - R^2/4L^2 \tag{4.78}$$

Thus the imaginary part of the frequency becomes an exponential damping term in the expression for the time dependence. It is this factor which makes the complementary function in general a transient. It is only in the case where the resistance of the circuit is zero (e.g. when we are dealing with a superconductor) that the damping term is absent and the free oscillations can continue for ever.

For the case where $R^2 < 4L/C$ (i.e. g^2 of eqn (4.78) is positive), the second exponential factor in eqn (4.77) with its imaginary argument as usual corresponds to oscillations. If in doubt about this, we can most easily convince ourselves by realising that the two solutions implicit in eqns (4.77) have arbitrary complex coefficients, which as before we determine from the initial conditions of the circuit. When we take the real part of their sum, as required from eqn (4.74), we can write the complementary function as

$$e^{-Rt/2L}(A\cos gt + B\sin gt) \tag{4.79}$$

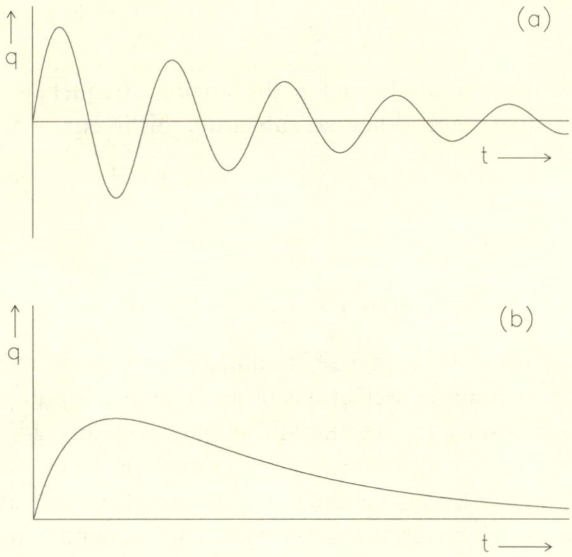

Fig. 4.9 Complementary functions for the second order differential equation (4.56″) for the charge q on the capacitor for the circuit of fig. 4.7. At $t = 0$, there is a current flowing in the circuit, but the charge on the capacitor is zero. (a) For low resistance R, the solution corresponds to a damped oscillation. The envelope of the oscillations is given by $e^{-Rt/2L}$. (b) For large resistance, the system is overdamped, and no oscillations occur.

where A and B are our arbitrary constants. For small R, this will correspond to slowly damped waves, as shown in fig. 4.9(a). Thus the circuit is such that the charge on the capacitor, or more or less equivalently the current in the circuit, can oscillate with an overall slowly decreasing envelope corresponding to the damping. Similarly, if we initially displace the mass in our mechanical problem and then release it but do not apply any time-varying force, it also can oscillate with gradually decreasing amplitude.

For the case of large damping so that $R^2 > 4L/C$, g^2 is negative and the exponentials are indeed exponentials rather than harmonic. We can now write our complementary function as

$$A\,e^{-(R/2L+\sqrt{-g^2})t} + B\,e^{-(R/2L-\sqrt{-g^2})t} \tag{4.80}$$

Since $-g^2$ is smaller than $(R/2L)^2$, the coefficients $-(R/2L \pm \sqrt{-g^2})$ are both negative, and the complementary function again corresponds to a decaying solution (see, for example, fig. 4.9(b)).

This is the overdamped situation, in which there are no oscillations. The easiest example to imagine is our displaced mass on a weak spring, with the resistance increased by submerging the system in thick honey. When released, the mass will slowly return to its initial position.

The critically damped solution is when $R^2 = 4L/C$. Then g is zero, and the complementary function is

$$(A + Bt)e^{-Rt/2L} \tag{4.81}$$

(compare Section 5.11.4.) Again, there is no oscillation, and the solution dies away exponentially.

Thus we have justified our statement that in general the complementary function is a transient, and the long-term behaviour of the system is given by the more interesting particular integral.

4.5 Other electrical circuits

Not all electric circuits look like the series circuit of fig. 4.7. In principle we can write down the required (probably coupled) differential equations in each case we encounter, and solve them anew for every problem by the techniques we developed in Section 4.4. An easier approach, however, is to associate with each basic circuit element a complex impedence; and then to combine the impedences for elements in parallel or in series by the same rules as if they were resistors. The basic impedences are

$$\left.\begin{aligned} \text{Resistor} : \mathscr{Z}_R &= R \text{ (real)} \\ \text{Capacitor} : \mathscr{Z}_C &= 1/j\omega C \text{ (imaginary)} \\ \text{Inductor} : \mathscr{Z}_L &= j\omega L \text{ (imaginary)} \end{aligned}\right\} \tag{4.82}$$

Thus for the series circuit of fig. 4.7, the total impedence

$$\begin{aligned} \mathscr{Z} &= \mathscr{Z}_R + \mathscr{Z}_L + \mathscr{Z}_C \\ &= R + j(\omega L - 1/\omega C) \end{aligned}$$

and the complex current $\mathscr{I}$ produced by the alternating voltage $V \cos \omega t$ is

$$\mathscr{I} = \frac{V}{R + j(\omega L - 1/\omega C)} \tag{4.83}$$

which agrees with eqn (4.73)

Some other examples are shown in fig. 4.10. We see that the simple expedient of employing a complex impedence for each element enables us to write down the impedence of the circuit as easily as for the

D.C.		A.C.	
Circuit	$I = V/R$	Circuit	$I = V/Z$
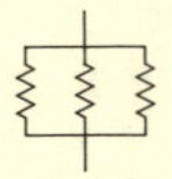	$\frac{1}{R} = \frac{1}{R_1} + \frac{1}{R_2} + \frac{1}{R_3}$	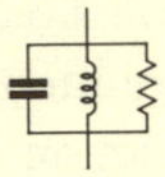	$\frac{1}{Z} = \frac{1}{R} + \frac{1}{j\omega L} + j\omega C$
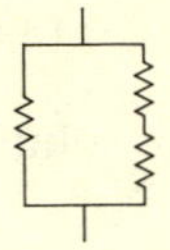	$\frac{1}{R} = \frac{1}{R_1} + \frac{1}{R_2 + R_3}$	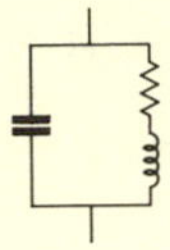	$\frac{1}{Z} = \frac{1}{j\omega L + R} + j\omega C$
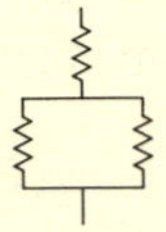	$R = R_1 + \left(\frac{1}{R_2} + \frac{1}{R_3}\right)^{-1}$	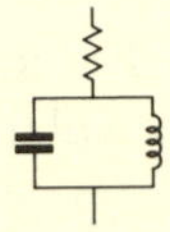	$Z = R + j\omega L/(1 - \omega^2 LC)$
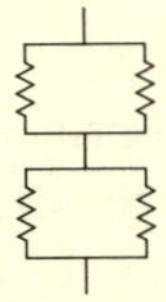	$R = \left(\frac{1}{R_1} + \frac{1}{R_2}\right)^{-1} + \left(\frac{1}{R_3} + \frac{1}{R_4}\right)^{-1}$	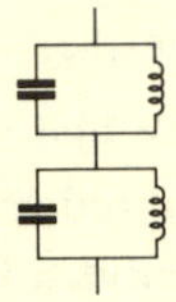	$Z = j\omega\left(\frac{L_1}{1 - \omega^2 L_1 C_1} + \frac{L_2}{1 - \omega^2 L_2 C_2}\right)$

Fig. 4.10 Relationship between the resistance of various circuits consisting of resistors, and the impedence to AC voltages of the corresponding circuits with resistors, inductors and capacitors.

corresponding DC circuit with resistors. Of course, our impedence is frequency dependent, and so will be meaningful only when the input waveform is harmonic. For other types of repetitive input, it is necessary to employ Fourier series in order to decompose the voltage waveform into its harmonic components (see Chapter 12 of Volume 2).

4.6 Power dissipation in AC circuits

Just to demonstrate that complex numbers do not make all problems simpler, we here discuss the power P dissipated in our electric circuit of fig. 4.7. For a given time interval from t_1 to t_2, the power is defined by

$$P = \frac{1}{t_2 - t_1} \int_{t_1}^{t_2} vidt \tag{4.84}$$

where v and i are the real, time-dependent voltage and current respectively. With

$$\left.\begin{aligned} v &= v_0 \cos \omega t \\ \text{and } i &= i_0 \cos(\omega t + \phi) \end{aligned}\right\} \tag{4.85}$$

and for a time interval that covers a whole number of cycles, explicit evaluation of the integral gives

$$\begin{aligned} P &= \frac{1}{2} v_0 i_0 \cos \phi \\ &= v_{\mathrm{RMS}} i_{\mathrm{RMS}} \cos \phi \end{aligned} \tag{4.86}$$

The formula for the power contains a factor of $\cos \phi$, the cosine of the phase angle between the voltage and the current. If ϕ is zero, as is the case for a resistor, then our formula for the power gives the well known result $v_{\mathrm{RMS}}\ i_{\mathrm{RMS}}$. Alternatively for a perfect inductor or capacitor, ϕ is $\pm\pi/2$, and the power consumption is zero; again our formula gives the correct answer.

For most practical situations, we are interested in the power dissipation in a time interval which is very much longer than the period (i.e. $\omega(t_2 - t_1) >> 1$). Thus, for example, the time interval could be seconds, and the period 1/50 seconds. Then the requirement for eqn (4.86) that the time interval be an exact number of periods is not too important. That is, the fractional error caused by using a slightly different time interval t ($= 2\pi n/\omega + \Delta t$) will be of order

$$\Delta t / t << 1$$

and hence can generally be neglected.

We obtained the formula (4.86) by using good old-fashioned real variables. Now if we want to we can, of course, express the power in terms of our complex voltage and current $\mathscr{V}$ and $\mathscr{I}$. It becomes

$$\begin{aligned} P &= \frac{1}{2} \mathrm{Re}[\mathscr{V} \mathscr{I}^*] \\ &= (\mathscr{V} \mathscr{I}^* + \mathscr{I} \mathscr{V}^*)/4 \end{aligned} \tag{4.86$'$}$$

where as usual the star denotes the complex conjugate. However, there is no direct way of deriving the complex expression (4.86′) from eqn (4.84). This is because the power involves integration of the product of the instantaneous values of the current and voltage, while the complex quantities give only their amplitudes and their relative phase.

The general rule is that complex notation works well in *linear* situations. Thus, for example, our original differential equation (4.56′) is linear in q (or equivalently in i) and in v, but the power involves their *product*. In these situations, it is best to return to the simple description of i and v as real harmonic quantities.

If we want to, we can of course calculate the power from eqn (4.86′) just about as easily as from (4.86). As an example for practice, we can calculate the power dissipated in the circuit of fig. 4.7. We can do this from formula (4.86), where i_0 and $\cos\phi$ are derived from the expression (4.73) for the complex current. Alternatively, we can use (4.73) directly in the first line of eqn (4.86′). We can compare these with

$$P_R = \frac{1}{2}i_0^2 R \tag{4.87}$$

which is the power dissipated in the resistor alone. If we have not made any mistakes, we should find that all three answers are equal. This implies that P, which is strictly speaking the power delivered by the source of the applied voltage, is equal to the power P_R lost by the resistor. Since no net power is dissipated by either the ideal inductor or the capacitor in our circuit, this simply reduces to an expression of energy conservation. It is to be noted that these are time-averaged expressions. It is not true that at every instant $P(t) = P_R(t)$, since energy is stored both on the capacitor and in the inductor. As these energies vary during the cycle, $P(t) \neq P_R(t)$. Over a whole number of cycles, however, the energies of the capacitor and inductor return to their initial values, and so they do not contribute to the time-averaged power balance.

In the corresponding mechanical problem described by eqn (4.57), the work done by the external force (over a whole number of cycles) equals the energy lost by pushing the mass backwards and forwards against the resistive force $-k\,dx/dt$.

4.7 Resonance in the series electrical circuit

4.7.1 Resonance frequency

Having used complex numbers so effectively to find the particular integral of the second order differential equation (4.55), we end this chapter with a discussion of the phenomenon of resonance. This utilises the solution we derived, but in fact makes no more use of complex numbers. For ease of visualisation, we begin with the example of the electrical circuit, but later generalise the treatment to include several other branches of physics. From the point of view of the examination-oriented student, this clearly is a subject worth mastering, since it provides a fruitful area for examiners searching for problems in second order differential equations, AC theory, mechanics, atomic physics, nuclear and elementary particle physics, sound, absorption of electromagnetic waves, etc., etc.

For the series circuit of fig. 4.7 and eqn (4.56′), we found that the amplitude of the current as deduced from eqn (4.73) is

$$i_0 = \frac{V_0}{\sqrt{(1/\omega C - \omega L)^2 + R^2}} \tag{4.88}$$

Usually we tend to think of the voltage source as variable in amplitude, but fixed in frequency. Now, however, we are going to consider the opposite situation. That is, we are interested in how the current in the circuit varies as the frequency changes, with V_0 kept constant.

First let us determine when i_0 is a maximum.† The sledgehammer approach is to calculate $di_0/d\omega$, and find when this is zero. This, however, is not too intelligent. The angular frequency ω occurs inside an unpleasant looking square root sign, so it is certainly better to consider

$$i_0^2 = \frac{V_0^2}{(1/\omega C - \omega L)^2 + R^2} \tag{4.88′}$$

and then evaluate $d(i_0^2)/d\omega$, since (apart from possible minor complications at $i_0 = 0$), i_0 and i_0^2 will have stationary values at the same ω. The next stage of sophistication leads us to make use of the even simpler derivative $d(1/i_0^2)/d\omega$, since a maximum of i_0^2 will correspond to a minimum of $1/i_0^2$. Thus

$$\frac{d(1/i_0^2)}{d\omega} = 2\left(\frac{1}{\omega C} - \omega L\right)\left(\frac{1}{-\omega^2 C} - L\right)/V_0^2 \tag{4.89}$$

and we set this to zero to obtain the value of ω which gives us the stationary value of i_0 (and i_0^2 and $1/i_0^2$). This yields the well known result

$$\omega_R = 1/\sqrt{LC} \tag{4.90}$$

for considerably less algebra as compared with starting with $di_0/d\omega$.

Now in fact you should have realised that the whole of the discussion of the last paragraph was a waste of time. The maximum of i_0 in eqn (4.88) will occur when the denominator is a minimum. Within the square root sign, there are two terms both of which are squares (and hence

† It is important to realise that we want to find the frequency at which the *amplitude* i_0 of the current is a maximum, rather than when the current $i(t)$ is a maximum as a function of time. The latter occurs once every period $2\pi/\omega$ of the alternating current. That is not what interests us. We want the single angular frequency ω for which we have the maximum response of the system.

greater than or equal to zero) but only one of which depends on our variable ω. The minimum of the denominator will thus be obtained when the variable term is zero. This immediately gives us our condition (4.90). The moral of this is that it often pays to think for a moment, before plunging in and using some standard mathematical procedure.

4.7.2 Width of resonance

Having found that i_0 is a maximum for $\omega = \omega_R$, we are now interested in how the current amplitude i_0 varies as we change the frequency ω.

We first rewrite eqn (4.88) as

$$i_0 = \frac{V_0/L}{\sqrt{(\omega_R^2/\omega - \omega)^2 + (R/L)^2}} \tag{4.91}$$

Thus we see that the functional form depends on three parameters, which can be chosen as V_0/R, ω_R and R/L. The first determines the overall height of the distribution and ω_R determines where the maximum occurs; we shall see shortly that R/L determines the width of the peak.

In fig. 4.11(a) we plot i_0 as a function of ω for a given set of values of the parameters. The peak in the distribution for $\omega = \omega_R$ is clearly visible. Another feature of the graph is that i_0 tends to zero for ω very small or very large. This is because the impedence of the capacitor or of the inductor respectively becomes very large in these limits, and prevents the flow of current.

Fig. 4.11(b) contains another plot of i_0 as a function of ω for a slightly different set of parameters. As compared with fig. 4.11(a), the resistance has been increased by a factor of 2. Clearly R affects the width (and the height) of the peak. We return to a more precise statement of this relationship shortly.

When the peaks are relatively narrow, as in figs. 4.11(a) or (b), the curve is called a resonance curve, and ω_R is the resonant frequency. For our circuit, the response (i.e. the current) is very much larger when the input is at or near the resonance frequency. The behaviour of the tuning circuit in a radio very much follows this pattern. We want the signal that we hear to be strong when we have our radio tuned to the desired station, but negligible when we are off it, since otherwise we would hear all possible programmes simultaneously. This phenomenon of resonance

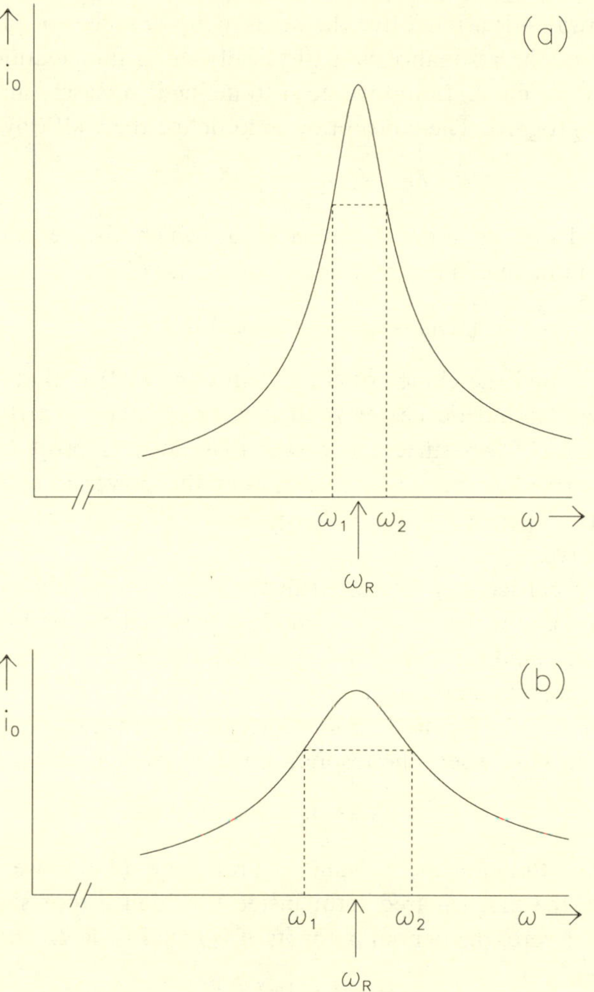

Fig. 4.11 The resonance curve for the response of the series circuit of fig. 4.7 as a function of the variable frequency of the external source of alternating voltage (but which is of constant amplitude). The resulting current oscillations have maximum amplitude at an angular frequency ω_R, and fall off to $1/\sqrt{2}$ of this value at frequencies ω_1 and ω_2. The curve (b) corresponds to a resistance twice as big as in (a), and results in a resonance curve twice as wide, and only half as high at the maximum. The quality factor Q of a resonance can be defined as $\omega_R/(\omega_2 - \omega_1)$, and is twice as big for the narrow resonance (a) as for the wider one (b).

– a strong response close to a particular frequency,† and a much weaker one further away – is a repetitive theme in many branches of physics.

The width of the resonance is a physically important feature of the response curve of fig. 4.11, and we need to define it precisely in order to make further progress. The convention is to define the width by

$$\delta\omega = (\omega_2 - \omega_1)/2 \tag{4.92}$$

where ω_1 and ω_2 are those frequencies at which the response i_0 is decreased from its maximum by a factor of $\sqrt{2}$ (see fig. 4.11), i.e.

$$i_0(\omega_1) = i_0(\omega_2) = i_0(\omega_R)/\sqrt{2} \tag{4.93}$$

The logic behind the choice of the $\sqrt{2}$ in eqn (4.93) is that it makes the subsequent arithmetic neater than if we had taken something else (like $2, e, \pi$, etc.). Also since the power dissipated is proportional to i^2, the frequencies ω_1 and ω_2 are such that the power is half that at the maximum. Then $\delta\omega$ is the half-width at half-height of the power dissipation curve.

We shall be particularly interested in the case of a narrow resonance (i.e. $\delta\omega << \omega_R$), in which case to a good approximation the frequencies ω_1 and ω_2 are equidistant from ω_R, and the shape of the resonance is reasonably symmetric about ω_R.

Next we shall find a formula for $\delta\omega$ in terms of the values of the circuit elements, in the case where the resonance is narrow. We then write

$$\omega_{1,2} = \omega_R \pm \delta\omega \tag{4.94}$$

and remember that $\delta\omega/\omega_R$ is small. From eqn (4.91), we see that at resonance, the first squared term inside the square root sign in the denominator is zero, the denominator itself is equal to R/L, and

$$i_0(\omega_R) = V_0/R \tag{4.95}$$

This is not too surprising, since the resonance frequency is such that the impedence of the capacitor and inductor cancel, and the impedence of the circuit at this frequency is simply R.

As we move away from ω_R, the first term in the denominator stops being zero, and hence grows. In order to make i_0 smaller than $i_0(\omega_R)$ by a factor of $\sqrt{2}$, the frequencies $\omega_R \pm \delta\omega$ must be such that this term

† This is the behaviour of the series resonant circuit we have just discussed. For the parallel resonant circuits of fig. 4.10, however, the impedence has a maximum at the resonant frequency, and hence the current will exhibit a minimum as the frequency is varied at fixed input voltage. The basic ideas are, however, similar.

has become equal to the second (constant) term in the denominator, i.e. $(R/L)^2$. Thus †

$$\frac{\omega_R^2}{\omega_R + \delta\omega} - (\omega_R + \delta\omega) = \pm R/L \tag{4.96}$$

This equation is in fact exact, and does not require the assumption that $\delta\omega << \omega R$. It is quadratic for $\delta\omega$, and so we could solve it to find the two values, one positive (which corresponds to $-R/L$ in the above equation) and the other negative (with $+R/L$). (The bright student is invited to consider why there are not four solutions, two for the quadratic equation with $+R/L$, and two with $-R/L$)

For the case of a narrow resonance, however, it is more instructive to multiply both sides of eqn (4.96) by $\omega_R + \delta\omega$:

$$2\omega_R\delta\omega + \delta\omega^2 = \mp R(\omega_R + \delta\omega)/L \tag{4.96$'$}$$

and to make use of the fact that $\delta\omega << \omega_R$ in order to ignore the second term on each side of the equation as compared with their respective first terms:

$$2\omega_R\delta\omega = \mp R\omega_R/L \tag{4.96$''$}$$

or

$$\delta\omega = \mp R/2L \tag{4.97}$$

This thus justifies our earlier claim that for a narrow resonance the two values of $\delta\omega$ are (approximately) equal in magnitude.

This then is our fundamental result for the half-width of a narrow resonance, in terms of the parameters of the circuit. We see that it is proportional to the resistance of the circuit. If we have an ideal case with $R = 0$, then $\delta\omega = 0$ and the resonance is simply a sharp line at $\omega = \omega_R$. As R increases, however, the resonance curve widens, in the manner shown in fig. 4.11. At the same time, the height of the curve at the maximum decreases (see eqn (4.95)), in such a way that the area under the curve becomes smaller.

4.7.3 Quality factor

For resonant curves it is common to define what is known as the quality factor Q (not to be confused with the charge, which also is represented

† We are just about to change our convention slightly as compared with eqn (4.94). There $\delta\omega$ was assumed to be positive, whereas here we allow it to take either sign, and hence write $\omega_R + \delta\omega$, rather than $\omega_R \pm \delta\omega$.

by the letter Q). There are several different but equivalent definitions, of which any can be chosen and then the others can be derived from it. An easily visualisable definition is

$$Q = \frac{\omega_R}{2\delta\omega} \tag{4.98}$$

i.e. it is the ratio of the resonant frequency to the full width at half maximum of the power curve. It is dimensionless, and for a narrow resonance is a large number. In our electrical circuit, because of eqn (4.97),

$$Q = \frac{\omega_R L}{R} = \frac{1}{R}\sqrt{\frac{L}{C}} = \frac{1}{\omega_R CR} \tag{4.99}$$

Thus Q is also the ratio of the impedence at the resonant frequency of either the inductor or the capacitor (since they are equal there) to that of the resistor. From this it follows that Q is also 2π times the ratio of the stored energy in the capacitor and inductor, to that dissipated in the resistor during one cycle of the applied alternating voltage. We shall discover some other properties of Q later.

We should note that the exact value of the resonant frequency for a given circuit depends slightly on how we choose to define 'resonance'. Up till now, we have used the maximum amplitude of the current oscillations. However, we could alternatively have used the maximum of the amplitude of the oscillations of the charge on the capacitor (see eqn (4.69)). Because of the ω factor in the $(\omega R)^2$ term, $|\mathcal{Q}|$ maximises at a frequency ω' which is not exactly equal to our previous ω_R. It will, however, be quite close to it since the fast variation of the denominator of $|\mathcal{Q}|$ occurs in a narrowish range of frequency around ω_R, for which the ωR term remains approximately constant. If ωR were exactly constant, then the maximum would require

$$\frac{1}{C} - \omega^2 L = 0$$

or

$$\omega' = \sqrt{1/LC} = \omega_R$$

The exact position of the maximum is given by

$$\begin{aligned}\omega'^2 &= \frac{1}{LC}\left(1 - \frac{R^2C}{2L}\right)\\ &= \omega_R^2(1 - 1/2Q^2)\end{aligned} \tag{4.100}$$

For a narrow resonance, Q is large and

$$\begin{aligned}\omega' &\sim \omega_R\left(1 - \frac{1}{4}Q^2\right)\\ &\sim \omega_R - \delta\omega/2Q \qquad (4.100')\end{aligned}$$

Thus ω' is lower than ω_R by only a small fraction of the width of the resonance curve.

4.7.4 Phase of output

So far we have concentrated on the amplitude of the current in our circuit, as a function of frequency. Now we turn to its phase, relative to that of the input voltage. From eqn (4.83), we have that

$$\begin{aligned}\tan\phi &= -\frac{\omega L - 1/\omega C}{R}\\ &= \frac{L}{R}\left(\frac{\omega_R^2}{\omega} - \omega\right)\\ &= Q\left(\frac{\omega_R}{\omega} - \frac{\omega}{\omega_R}\right) \qquad (4.101)\end{aligned}$$

Thus we can immediately see that $\phi = 0$ when $\omega = \omega_R$, and tends to $+\pi/2$ or $-\pi/2$ as ω becomes very small or very large respectively. This is all very sensible, since the circuit behaves as a simple resistor, capacitor or inductor respectively in these three cases.

What also emerges from eqn (4.101) is that the phase varies rapidly in the neighbourhood of the resonance. Thus the phase is $\pm\pi/4$ at the same frequencies at which the amplitude has decreased by the standard factor of $\sqrt{2}$. The variation of phase with frequency for a typical resonance is shown in fig. 4.12(b).

It is instructive to display the alternating current and alternating applied voltage on a double trace oscilloscope, which is triggered by the voltage (i.e. the voltage trace begins at the same position on the screen throughout the display). As we vary the frequency from below to above the resonance, we see that not only does the size of the current trace at first increase and then decrease rapidly in magnitude, but also its peak moves from the left to the right of the voltage peak, coinciding at the resonant frequency (see fig. 4.13).

An alternative reinterpretation of our definition of Q in eqn (4.98) is thus that ω_R is the frequency at which the current is in phase with

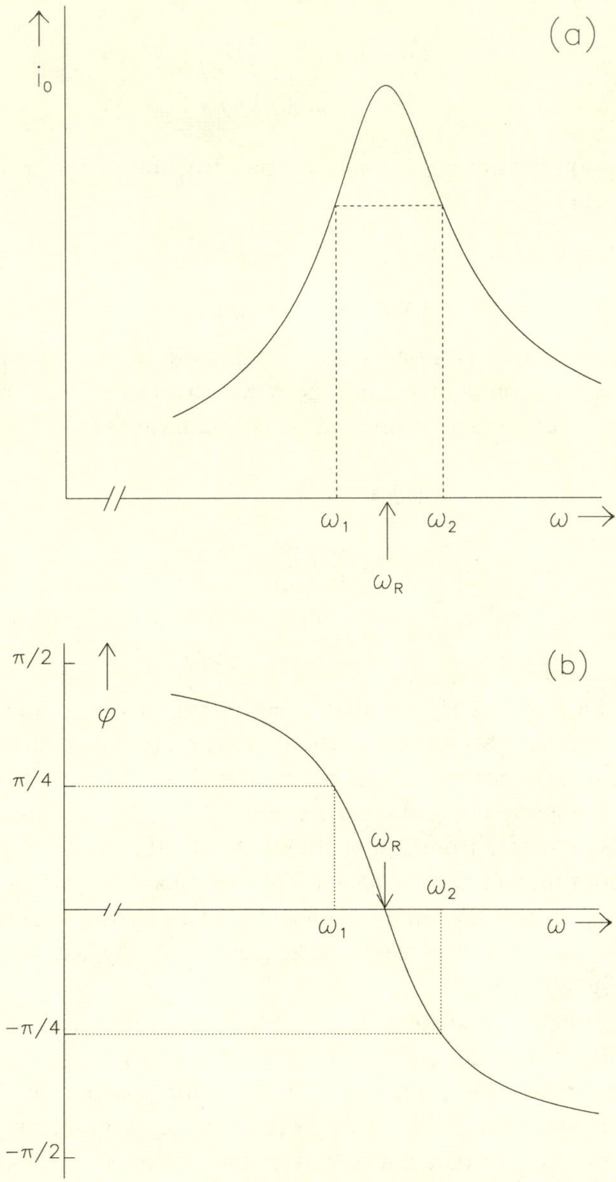

Fig. 4.12 (a) A resonance curve, showing the current response for the series circuit of fig. 4.7. (b) The phase ϕ of the current as a function of frequency, for the same circuit used in (a). At zero frequency, ϕ is $\pi/2$, and remains close to it for a large frequency range. As the resonance is approached, the phase suddenly changes rapidly, and passes through zero at the resonance frequency. It has values of $\pm\pi/4$ at the same frequencies ω_1 and ω_2 where i_0 is down by a factor $1/\sqrt{2}$ from its maximum value. For very large frequencies, the impedence is dominated by that of the inductor, and the phase is close to $-\pi/2$.

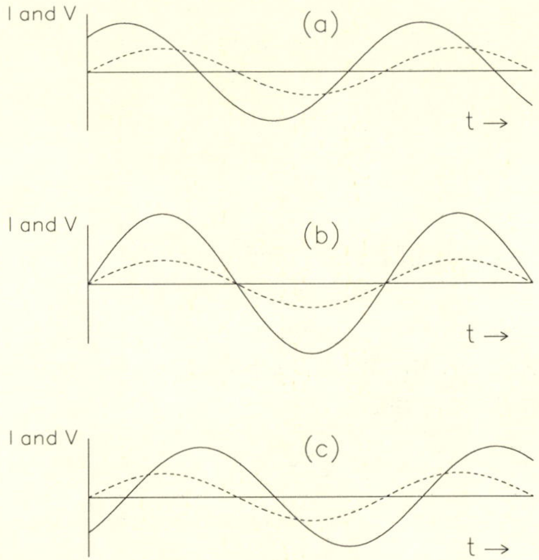

Fig. 4.13 The relationship between the current (solid curve) and voltage (dashed) as a function of time. The graph (b) is for the resonance frequency ω_R, where the current is in phase with the voltage, and is largest; (a) is for the frequency ω_1 below the resonance at which the phase of the current with respect to the voltage is $+\pi/4$, and the current amplitude is 0.707 times that at the resonance. Similarly in (c), the frequency is ω_2, the phase difference is $-\pi/4$, and the amplitude is as in (a). The slight differences in frequencies ω_1, ω_R and ω_2 are not visible on these diagrams.

the applied voltage, while $\delta\omega$ is the change required to produce a phase difference of $\pi/4$.

4.7.5 Relation between free and forced oscillations

We now return to the free oscillations of the circuit, as given by the complementary function (see eqn (4.79) and fig. 4.9(a)). The amplitude of these oscillations, which in fact are of the charge on the capacitor, falls off by a factor of e after a time interval

$$T = 2L/R \tag{4.102}$$

since when this value of T is substituted into eqn (4.79), the exponential factor, which for smallish R determines the amplitude of the oscillations, has fallen to a value of $1/e$ (as compared with unity at $t = 0$).

Eqn (4.102) provides us with yet another interpretation of Q. By comparison with eqn (4.99), we see that

$$T = 2Q/\omega_R \tag{4.103}$$

or alternatively

$$\begin{aligned} Q &= \omega_R T/2 \\ &\sim \pi N \end{aligned} \tag{4.104}$$

where N is the number of free oscillations of the system† in the time T. Thus Q is equal to π times the number of free oscillations that occur before the amplitudes of the current or charge oscillations have decreased by a factor of e. Alternatively, as the power in the circuit is proportional to the square of the current,

$$Q \sim 2\pi N' \tag{4.104$'$}$$

where N' is now the number of oscillations for the power to decrease by a factor of e.

If we take the at-first-sight surprising step of combining eqns (4.97) and (4.102), we obtain the suggestive-looking relationship

$$T\delta\omega = 1 \tag{4.105}$$

That is, we obtain unity when we multiply

(i) the time it takes the free oscillations of the circuit to fall off significantly, and
(ii) the width of the resonance curve for forced oscillations.

The fact that the answer is exactly 1 is perhaps a bit of a coincidence. It certainly depends on our somewhat arbitrary choices of how to define the width of the resonance, and what constitutes a reasonable fall-off in the free oscillations. The really important point is that these two physical quantities are inversely proportional to each other.

A physical reason behind eqn (4.105) is that as, for example, the value of the resistance in our circuit is increased, the resonance becomes wider and also the free oscillations die out faster. The decrease in T cancels the increase in $\delta\omega$, and their product remains constant.

From a mathematical viewpoint, the explanation is as follows. The free oscillations describe the *time* variation of the response of the system to

† From eqn (4.78), the angular frequency of the free oscillations is given by $g^2 = \omega_R^2(1 - 1/4Q^2)$ and hence, for a narrow resonance with large Q, is very close to ω_R. (Compare the discussion above eqn (4.100) relating to the frequencies for the maximum amplitude of the current and of the charge for forced oscillations.)

an impulse, while the resonance curve gives its *frequency* response. These are the Fourier transforms of each other. Eqn (4.105) is thus simply an expression of the fact that the Fourier transform of a pulse lasting for a time t has a width γ in angular frequency such that $t\gamma$ is of order unity (see Section 12.12 of Volume 2).

We can convert eqn (4.105) into something that looks remarkably like Heisenberg's Uncertainty Principle, if we multiply both sides of the equation by $h/2\pi$, where h is Planck's constant. According to quantum theory, the energy E associated with a frequency ν comes in units of $h\nu$, or $h\omega/2\pi$. Thus eqn (4.105) becomes

$$T\delta E = h/2\pi \tag{4.106}$$

where δE is the uncertainty in energy which results from the fact that our wave lasts only for a time or order T.

Of course this 'derivation' should not be taken too seriously, since eqn (4.105) applies to a very classical electric circuit, for which the effects of quantum mechanics are not too relevant. The reason we can apparently obtain (4.106) from (4.105) is that they are both expressions of a basic result relating the product of the widths of Fourier transforms (see Section 4.8.2).

4.8 Resonances in other systems

4.8.1 Mechanical

The mechanical system described by eqn (4.57) also exhibits resonance phenomena, as indeed it must since (4.57) is in fact identical in structure to eqn (4.56) for the electrical circuit, with only the symbols differing. We took eqn (4.57) as applying to a mass on the end of a spring, but it is perhaps a little more intuitive to think instead of a child on a swing. For small horizontal displacements x, eqn (4.57) applies with λ replaced by mg/l, where l is the length of the swing and m the mass of the child.

If the swing is either given a single impulse, or else it is displaced from the vertical position and then released, it will undergo free oscillations at an angular frequency very close to $\sqrt{g/l}$ (compare the footnote below eqn (4.104)). These will gradually die out. The quality factor Q of the system can be deduced immediately via eqn (4.104) from the number of swings of the free oscillations before the amplitude has decreased by a factor of e. Clearly the smaller the frictional effects, the longer the swing will continue to oscillate, and the larger Q will be.

We have, of course, assumed above that the resistive force $-k\,dx/dt$ is small enough for the complementary function of the differential equation to correspond to damped oscillations (compare eqn (4.79)). Unless the swing is exceptionally rusty, or it is submerged in water, this is very likely to be the case in any practical situation.

Next we turn to the forced oscillations. The term $F(t)$ in eqn (4.57) now represents someone pushing the swing. We are interested in how the amplitude of these forced oscillations varies with the frequency of the applied force $F(t)$. The answer is obvious to anyone who has ever tried pushing a swing: in order to produce a significant amplitude, one pushes at the natural frequency of the swing.

We can also deduce the 'width of the resonance' $2\delta\omega$ (which is the difference between the frequencies of $F(t)$ for which the amplitude is smaller than the resonant one by a factor of $\sqrt{2}$) from our previous observation of the swing's free oscillations, since the Q that we can obtain from their decay is also equal to $\omega_R/2\delta\omega$.

4.8.2 Atoms

The electrons in an atom can exist in various quantised states. Usually they arrange themselves in the state of lowest energy. However, if the atom absorbs electromagnetic radiation, they can be raised to states of higher energy. The simplistic view is that, in order to do this, the radiation has to be of exactly the correct frequency so that its quantised energy matches exactly the increase in energy of the atom, i.e.

$$h\omega/2\pi = E_f - E_i \tag{4.107}$$

where ω is the angular frequency of the incident radiation, and E_i and E_f are the energies of the lower state of the atom and its relevant excited one.† At this degree of approximation, when the electron returns from the upper level to the ground state the atom emits monochromatic radiation whose frequency is again given by eqn (4.107).

The reason the emitted radiation is not perfectly monochromatic is that this would imply it was of the form

$$A\sin(\omega t + \phi) \tag{4.108}$$

† Throughout this section, we ignore the fact that the recoil of the atom (or nucleus in Section 4.8.3) absorbs a small amount of energy, so that the incident frequency must be slightly larger than would be given by eqn (4.107), and the emitted frequency on de-excitation is slightly lower.

for *all* values of t (compare Chapter 12 of Volume 2). Instead, when the atom is given an 'impulse' of energy and then left to itself, it will undergo 'free oscillations' and emit radiation, but these do not continue for ever. This is because the excited state of the atom has only a finite lifetime, and after several lifetimes there will be hardly any excited atoms left to emit radiation. Rather than eqn (4.108), a better description of the emitted radiation is thus provided by the damped oscillations which are the complementary function of our second order differential equation (compare fig. 4.14(a) and eqn (4.79)). This then is the time dependence of the radiation.

Now a *modulated* sine wave does not have a unique frequency, but can be composed of a range of monochromatic waves. For a non-repetitive form like fig. 4.14(a), the procedure for finding the amplitudes of the various components is that of the Fourier transform, and we need a continuous range of frequencies in the neighbourhood of ω_R. The Fourier transform thus provides us with the frequency content of the radiation.

Since energy and frequency are simply related via Planck's constant, the Fourier transform gives also the energy spectrum of the radiation (see fig. 4.14(b) and (c)).

A related way of looking at the problem is by the Uncertainty Principle. This states that, because the radiation is emitted only for a short time of order T, the energy of the system cannot be determined to an accuracy better than ΔE, where

$$T\Delta E \geq O(h/2\pi) \tag{4.109}$$

We can thus think of the upper energy level as being not so much a line on an energy plot, but rather as a smeared out region, whose width is given by eqn (4.109) (see fig. 4.15). Thus when we expose the atom to radiation, its energy does not have to correspond to some unique value of the energy difference between the states. If its frequency is slightly lower, it will suffice to raise the system into the lower tail of the upper state; and correspondingly to the higher side of the state if its frequency is larger. The exact shape of the profile of the level is given by the square† of the Fourier transform of the decaying exponential, and is known as the Lorentz line shape:

$$L(\omega) \sim \frac{1}{(\omega - \omega_R)^2 + (\gamma/2)^2} \tag{4.110}$$

† The square is involved because the problem is to do with the *intensity* of the radiation, which is proportional to the square of the *amplitude*.

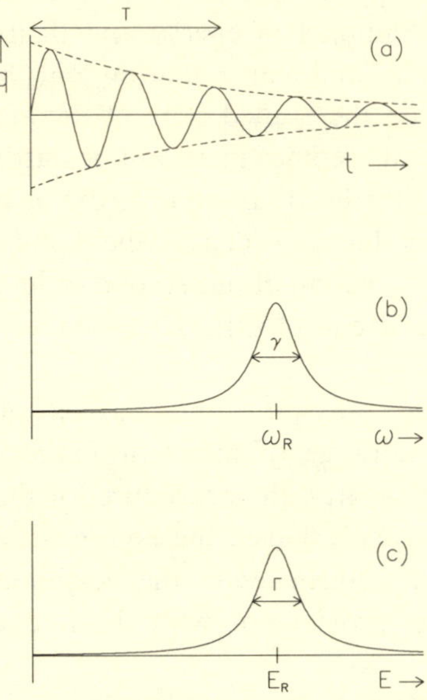

Fig. 4.14 (a) The damped free oscillations of a swing, or the charge on the capacitor in a series electric circuit, or an atom emitting radiation, as a function of time. The amplitude decreases by a factor of e in a time T. (b) The frequency response ($L(\omega)$ of eqn (4.110)) of the system, whose free oscillations are shown in (a). The shape of the curve is obtained from (a) by Fourier analysis, and can be interpreted as giving the intensities of the different frequencies of harmonic waves necessary to produce the pattern (a). From the properties of Fourier transforms, it follows that $T\gamma \sim 1$. (c) The same graph as in (b), except that the horizontal axis has been multiplied by $h/2\pi$ (where h is Planck's constant) to turn it into an energy scale. This would then give us the energy spectrum of the radiation shown in (a). It is such that $T\Gamma \sim h/2\pi$.

Here ω_R is the resonant frequency corresponding to the exact energy difference, and γ is the full width at half-intensity of the line (see fig. 4.14(b)).

After our experience in analysing electric circuits, we of course recognise (4.110) as having the same form as the frequency response of the forced oscillations of our series circuit, whose free oscillations look like the graph of fig. 4.14(a).

4.8.3 Nuclear reactions

When nuclear beams are projected onto other nuclei, the probability of scattering depends on the incident energy. For many reactions, this probability exhibits characteristic resonance peaks. A typical example

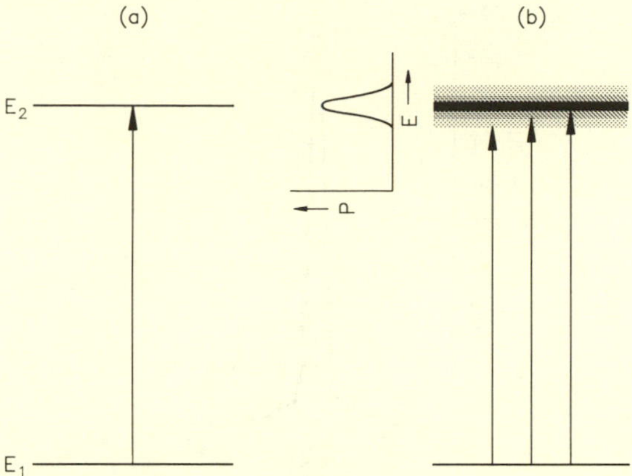

Fig. 4.15 Schematic diagram of the energy levels for the ground state E_1 and an excited state of a system. In (a), the excited state is assumed to exist at a unique energy E_2. (b) is more realistic in that the upper level is depicted as having a non-zero intrinsic width, because of its finite lifetime and the Uncertainty Principle. The arrows show radiation of different energies, which are all capable of exciting the system because of the finite width of the upper level. The inset shows the probability P of exciting the system as a function of the incident energy E.

is shown in fig. 4.16. In analogy with the atomic case, this is ascribed to our producing an excited state of the combined system, which can then decay back to the constituents that produced it (or perhaps to other decay products as well). This decay process is characterised by a specific lifetime, and hence once again the energy of the upper state is not uniquely defined. That is why a range of energies of the incident particle are capable of exciting the system into its upper state.

This time the shape of the resonance is known as the Breit–Wigner curve:

$$BW(E) \sim \frac{1}{(E - E_0)^2 + (\Gamma/2)^2} \tag{4.111}$$

where E is the beam energy, E_0 the resonant energy and Γ the full width at half-height of the graph of scattering probability plotted against beam energy. We see that the Breit–Wigner nuclear curve is of exactly the same form as the Lorentz line shape of atomic physics or the frequency response of the series electric circuit.

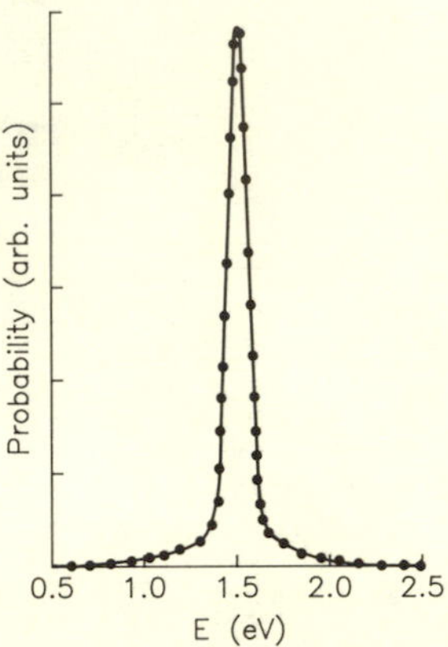

Fig. 4.16 Probability of absorption of neutrons by indium, as a function of the neutrons' energy. Neutrons are captured by the isotope ^{115}In, and a specific excited state of ^{116}In is produced. The data are shown as dots, and the curve is of the form (4.111). (From *Phys. Rev.* **98** 1267.)

4.8.4 Elementary particle reactions

Fig. 4.17(a) shows the probability of scattering of π^+ mesons of various energies when they are beamed at protons. (More precisely, the experiment consists of allowing π^+ mesons to pass through a hydrogen target. The scattering by the protons is readily distinguishable from the effect of the electrons which are also in the target.)

Once again we see the typical resonance phenomenon associated with the enhanced absorption of pions around a certain energy. This produces an excited state of the combined system; this one is known as the Δ^{++}. The process is equivalent to the forced response of the system to an external influence of variable energy.

The excited state will decay back to the particles that formed it, i.e.

$$\Delta^{++} \to p + \pi^+ \tag{4.112}$$

in a time of about 10^{-23} seconds; this corresponds to the free oscillations of the disturbed system. By the Uncertainty Principle, such a short lifetime implies a large width for the energy of the Δ^{++} state. This can be seen from fig. 4.17(a), where the upper scale gives the mass of the combined $p\pi^+$ system when the kinetic energy of the π^+ is as shown on the lower scale. In this case, the calculation must be performed relativistically, and the correction for the recoil of the target proton is significant and has to be incorporated. Thus we read off the mass of the

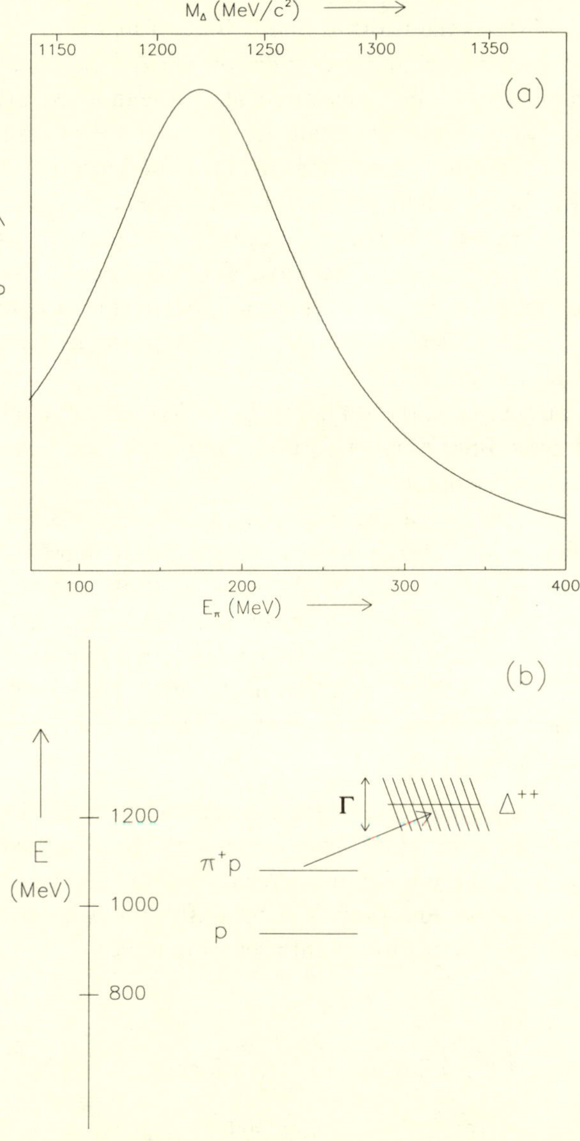

Fig. 4.17 (a) Probability P (in arbitrary units) of a π^+ scattering from a proton in a hydrogen target, as a function of the kinetic energy E_π of the pion beam. The peak is typical of resonance phenomena. The top scale gives the mass M_Δ of the combined π^+p system, as a function of E_π. (b) Energy level diagram. The lower line corresponds to the rest mass of a proton. The line marked π^+p gives the energy of a π^+ plus a proton when both are at rest; this is simply equivalent to the sum of their masses. A heavier state Δ^{++} of mass $\sim$ 1220 MeV/c^2 can be excited by pions of the correct energy. Because of its fast decay to $\pi^+ + p$, the Δ^{++} has a significant uncertainty in mass; its width Γ is $\sim$120 MeV/c^2. Thus the incident pion can in fact produce the Δ^{++} over a range of energies; this explains the width of the resonance curve in (a). The calculation of the kinetic energy E_π needed to produce the Δ^{++} state must be performed relativistically, and includes the effect of the proton recoil.

Δ^{++} as about 1220 MeV/c^2, and Γ, its full width at half-maximum, as $\sim$120 MeV/c^2.† Thus the width is about 10% of its mass.

In fact the lifetime of 10^{-23} seconds for the Δ^{++} is much too short to be determined by direct measurement. Instead it is inferred from the observed width of the Δ^{++} state from experimental results such as that displayed in fig. 4.17, and then invoking the Uncertainty Principle. (Compare the electric circuit in Section 4.7.5, where we made use of the equivalent relation, but in the opposite sense, in order to deduce the width of the frequency response of its resonance for forced oscillations, from the the observed lifetime of its free-oscillation mode.) This technique enables the range of lifetime determinations to be extended into an otherwise unmeasurable domain.

As in the nuclear case, the shape of the resonance curve of fig. 4.17(a) is described by a Breit–Wigner formula, although here relativistic and other corrections are needed.

The common feature of the examples of Sections 4.8.2–4.8.4 is that the atom, nucleus and 'elementary' particles all have substructure. These are respectively: the nucleus plus electrons; protons and neutrons; and quarks. These exist in various quantum states, and upward transitions can be produced by suitable incoming beams, and downward ones occur spontaneously. The emitted radiation in the latter case decays exponentially as a function of time; the excitation curve for the former has the typical resonance shape.

Resonance phenomena occur over a much wider range of phenomena than we have discussed. Our examples should be regarded only as a small sampling of the possibilities. What we have aimed to convey is that resonance phenomena provide a beautiful link across many of the conventional divisions of physics into separate topics.

Problems

4.1 (i) Convert the following into (r, θ) form:

$$4 + 3i$$

$$4 - 3i$$

$$-4 + 3i$$

$$-4 - 3i$$

† These mass units are commonly used in nuclear and elementary particle physics. On the same scale, the mass of the proton is 938 MeV/c^2.

(ii) Calculate the amplitude and phase of $(2+i)/(2-i)$. Did you do this in the most efficient manner?

4.2 Find the values of $(-1)^{2/5}$. Plot them on an Argand diagram.

4.3 Use complex numbers to evaluate $\int e^{ax}\cos bx\,dx$.

4.4 Use complex numbers to find the sum of the $n-1$ terms of the series

$$S = 2\sin\theta + 3\sin 2\theta + 4\sin 3\theta + \ldots n\sin(n-1)\theta$$

Show that, if $\theta = 2\pi/n$, then $S = -\frac{1}{2}n\cot\theta/2$.

4.5 Show that, for integral n, the solutions of the equation

$$(x+i)^n - (x-i)^n = 0$$

are given by $x = \cot(r\pi/n)$. What values of r give different solutions? How many solutions do you expect? By a suitable choice of n, show that

$$\cot^2(\pi/5) + \cot^2(2\pi/5) = 2$$

4.6 The charge q on the capacitor in a series R–L–C circuit in the presence of an external harmonic voltage obeys eqn (4.56′). The angular frequency ω_R is that for which the amplitude of the steady state current oscillations is a maximum. Given that R is very much smaller than $\sqrt{L/C}$, find (approximately) the change in frequency $\delta\omega$ such that at $\omega_R \pm \delta\omega$, the amplitude of the current oscillations is reduced from its maximum by a factor of $\sqrt{5}$.

4.7 An electric circuit contains a resistor R, a capacitor C and an inductor L each in parallel with each other. Write down the value of $1/\mathscr{Z}$, where $\mathscr{Z}$ is the complex impedence of this combination. For what frequency of the externally applied voltage is (a) the impedence real, and (b) its modulus a stationary value? As the frequency is changed from the resonance value in (b) above, does the impedence rise or fall? If $\omega_R \pm \delta\omega$ is the frequency at which the impedence is a factor of $\sqrt{2}$ different from its value at ω_R, show that $\omega_R/2\delta\omega \sim R\sqrt{C/L}$. How does this compare with the formula for the Q value in terms of the components of a series R–L–C circuit? In particular, why is the functional dependence on R different?

4.8 The position x of a mass on a spring satisfies eqn (4.57), with the force $f(t)$ given by eqn (4.58). The values of the constants, in suitable units, are

$$m = 1,\ k = 2,\ \lambda = 10^4,\ f_0 = 1$$

Use the complex number technique to solve the differential equation for χ, the 'complex' position. Find the resonant frequency ω_R, the half-width of the resonance $\delta\omega$, and the Q value. At frequency values $\omega_R \pm 5\delta\omega$, $\omega_R \pm 3\delta\omega$, $\omega_R \pm 2\delta\omega$, $\omega_R \pm \delta\omega$, $\omega_R \pm 0.5\delta\omega$ and ω_R, plot on a graph the following:

(i) The amplitude of the oscillations against ω.

(ii) The phase of χ against ω. What is the value of the phase at ω_R, and at what frequencies does it differ from this by $\pm\pi/4$?

(iii) An Argand diagram of $\mathrm{Re}(\chi)$ against $\mathrm{Im}(\chi)$, with each point being labelled by its value of ω. Note the approximate form of this curve, and also the relative speeds at which χ traverses the various parts of its trajectory (assuming that ω is increased at a constant rate). Remember these results when you study the subject of scattering amplitudes in the neighbourhood of resonances in quantum mechanics.

5
Ordinary differential equations

'Decrease in rate of inflation flattens off.'
(Newspaper headline)

5.1 What are differential equations?

In many situations in physics and in mathematics, relations among the different variables relevant to a problem involve derivatives. We give some simple examples.

5.1.1 Geometry

We may be interested in finding curves that have the property that their gradient dy/dx at any point (x, y) is perpendicular to the direction from a fixed point (a, b) to (x, y). Then

$$\frac{dy}{dx} = -\left(\frac{x-a}{y-b}\right) \tag{5.1}$$

We have thus converted the requirement about the curves into a differential equation involving dy/dx, x and y.

The curves that have the desired property are circles of arbitrary radius, centred on (a, b). i.e.

$$(x-a)^2 + (y-b)^2 = R^2 \tag{5.2}$$

It can be verified by differentiating (5.2) and substituting the derivative into (5.1) that these circles do indeed satisfy the differential equation.

5.1.2 Radioactive decay

The probability of a given radioactive nucleus decaying in any small time interval δt is constant, independent of how long the nucleus has already lived, how many other nuclei are around, etc. If the source is strong enough so that there are a large number D of decays during δt, we then expect

$$D/N = \lambda \delta t \tag{5.3}$$

where N is the number of original nuclei present at that time, and λ is a constant which specifies the rate at which these particular nuclei decay. Since the decays deplete the number of these nuclei, $D = -\delta N$. Thus eqn (5.3) becomes a differential equation

$$\frac{dN}{dt} = -\lambda N \tag{5.4}$$

describing how the number of non-decayed nuclei changes with time.

The solution of this differential equation is

$$N = N_0 e^{-\lambda t} \tag{5.5}$$

where N_0 is the number of nuclei present initially. Thus we expect an exponential decrease with time. (Of course in any actual experiment, there will be statistical fluctuations around this prediction.)

5.1.3 Statics

If a heavy string has its ends fixed, the effect of gravity is to make the string assume a certain shape. This is determined by a differential equation, derived by considering a short portion of the string between nearby points, marked A and B in fig. 5.1.

If the string makes a small angle θ with the horizontal, there is a downward force $g\rho\delta x$ (where ρ is the linear density of the string), and this must be balanced by a net upward force due to the tension in the string. Thus

$$g\rho\delta x = (T\sin\theta)_{\mathrm{B}} - (T\sin\theta)_{\mathrm{A}} \tag{5.6}$$

For small values of θ, the tension in the string is the same everywhere, in order to ensure that there is no horizontal motion. Also $\sin\theta \sim dy/dx$, whence

$$g\rho\delta x = T\left[\left(\frac{dy}{dx}\right)_{\mathrm{B}} - \left(\frac{dy}{dx}\right)_{\mathrm{A}}\right] \tag{5.7}$$

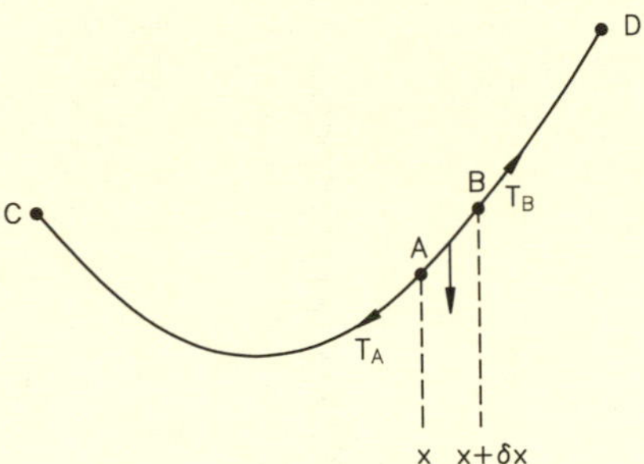

Fig. 5.1 A string of linear density ρ is held between two fixed points C and D. The equilibrium of the short length of string between A and B is determined by the three forces acting on it: T_A and T_B, the tensions in the string at the two points, and the gravitational force $g\rho\delta x$ on the string segment. This results in a differential equation for the shape of the string. If the deviations of the string from the horizontal are small everywhere, the eqn (5.9) results.

But from Taylor's Theorem applied to dy/dx (see eqn (7.2))

$$\begin{aligned}\left(\frac{dy}{dx}\right)_B &= \left(\frac{dy}{dx}\right)_A + \frac{d}{dx}\left(\frac{dy}{dx}\right)\delta x \\ &= \left(\frac{dy}{dx}\right)_A + \left(\frac{d^2y}{dx^2}\right)_A \delta x \qquad (5.8)\end{aligned}$$

Thus eqn (5.7) becomes (after cancelling out the δx)

$$g\rho = T\frac{d^2y}{dx^2} \qquad (5.9)$$

and this is our differential equation for the shape of the string.

The solution of this equation is the parabola

$$y = \frac{1}{2}\frac{g\rho}{T}x^2 + Cx + D \qquad (5.10)$$

where C and D are constants whose values are determined by the conditions that we want the parabola to pass through the two end-points where the string is fixed.

If we had not made the assumption that θ is small everywhere, the solution of the exact differential equation would have been a catenary.

5.1.4 Dynamics : A spring

Consider the vertical motion of a mass suspended on a spring. (see fig. 5.2) The forces to which it is subjected are gravity, the restoring force

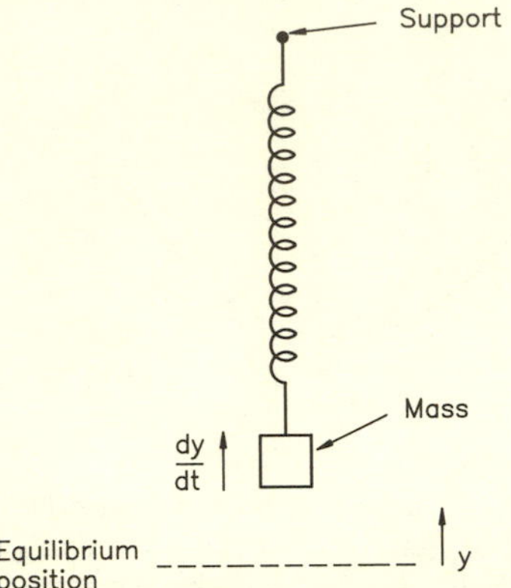

Fig. 5.2 A mass on a spring. It is currently above the equilibrium position, and moving vertically upwards. The retarding force is thus in the negative y direction, as are the gravitational force and that due to the compression of the spring. This ensures that, for the case shown, the signs of eqn (5.11′) are correct.

due to the spring, and air resistance. As these in general do not cancel, their net effect is to produce an acceleration.

If the spring obeys Hooke's Law, such that the force is proportional to its extension, the combined effect of gravity and the spring is to produce a restoring force towards the equilibrium position (i.e. where the mass would be in static equilibrium under the influence of the spring and gravity), which is proportional to the mass's upward displacement y from there. We write this as $-cy$, where the minus sign is needed because the force is a restoring one; this works for y either positive or negative.

We further assume that the force due to the air resistance is proportional to the velocity of the mass dy/dt. Because the force is retarding, we write it as $-k\,dy/dt$; this is correct for upward or downward motion (i.e. dy/dt positive or negative).

Thus Newton's Second Law applied to the mass gives us our differential equation

$$m\frac{dy^2}{dt^2} = -cy - k\frac{dy}{dt} \tag{5.11}$$

This is more usually written as

$$m\frac{d^2y}{dt^2} + k\frac{dy}{dt} + cy = 0 \tag{5.11′}$$

We note in passing that the identical equation would have been obtained if we had defined positive y as being downwards. For small enough

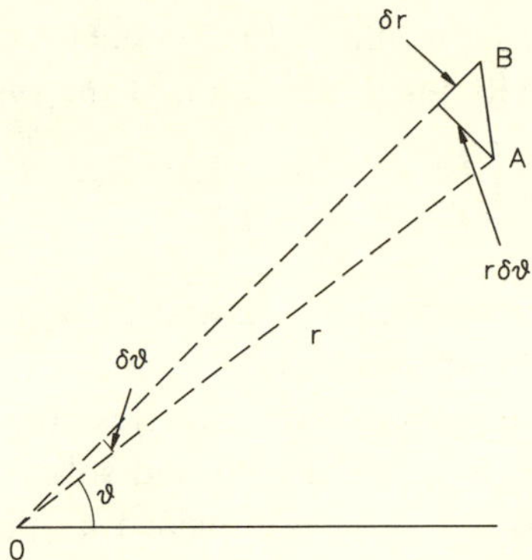

Fig. 5.3 A planet's position is described by polar coordinates r and ϑ. In moving from A to B, its radial distance changes by δr and its tangential one by $r\delta\vartheta$. Its velocity thus has components dr/dt and $rd\vartheta/dt$. The components of its acceleration are $d^2r/dt^2 - r(d\vartheta/dt)^2$ radially, and $rd^2\vartheta/dt^2 + 2(dr/dt)d\vartheta/dt$ tangentially.

damping (i.e. k not too large), the solution of eqn (5.11) corresponds to the mass undergoing damped oscillations.

If some time-dependent external force $F(t)$ is also applied to the mass (for example, if it were charged, and in a varying external electric field), then the equation becomes

$$m\frac{d^2y}{dt^2} + k\frac{dy}{dt} + cy = F(t) \tag{5.12}$$

Differential equations of the above form occur in a very wide variety of physics phenomena. We will have a lot to say about solving eqns (5.11′) and (5.12) later in this chapter.

5.1.5 Dynamics : Planets

If we ignore the gravitational attraction of everything except the sun, a planet moves in a plane containing the sun, in an elliptical orbit. We want to derive the differential equation, of which the solution is an ellipse.

The planet's polar coordinates with respect to the sun are r and ϑ, both of which can vary with time (see fig. 5.3). Then its radial acceleration with respect to the sun is given by $d^2r/dt^2 - r(d\vartheta/dt)^2$. (This is simply derived in books on mechanics or on vectors – see also Problem 3.11.) It can be understood as being composed of d^2r/dt^2, which depends on how the radial distance changes with time; and $-r(d\vartheta/dt)^2$, which gives the centripetal acceleration, and which is important even if the radial distance r is constant, as it would be for a circular orbit. This acceleration is provided by the gravitational attraction of the sun, so that

$$m\left[\frac{d^2r}{dt^2} - r\left(\frac{d\vartheta}{dt}\right)^2\right] = -\frac{GmM}{r^2} \tag{5.13}$$

where G is the gravitational constant, and M and m are the masses of the sun and the planet respectively.†

This is a differential equation, but it involves both r and ϑ as functions of time. In order to hope to be able to find solutions, we need another equation relating r, ϑ and t. This is provided by the fact that, since the gravitational force on the planet is directed towards the sun, its angular momentum L is constant. Thus

$$L = mr^2\frac{d\vartheta}{dt} = \text{constant} \tag{5.14}$$

Thus (5.13) and (5.14) are two simultaneous differential equations for r and ϑ.

In fact we can simply use (5.14) to eliminate $d\vartheta/dt$ from equation (5.13), to obtain

$$\frac{d^2r}{dt^2} - \frac{L^2}{m^2r^3} = -GM/r^2 \tag{5.15}$$

as a differential equation for r in terms of t.‡ Its solution tells us how the distance of the planet from the sun varies with time.

However, if we want the orbit of the planet, we need a differential equation relating r and ϑ. To obtain this we use the substitutions

$$\frac{dr}{dt} = \frac{dr}{d\vartheta}\frac{d\vartheta}{dt}$$

$$\text{and} \quad \frac{d\vartheta}{dt} = \frac{L}{mr^2}$$

to eliminate all reference to the variable t in eqn (5.15).

This yields

$$\frac{d}{d\vartheta}\left(\frac{1}{r^2}\frac{dr}{d\vartheta}\right) - \frac{1}{r} = -\frac{GMm^2}{L^2} \tag{5.16}$$

which is the differential equational we want. For small enough values of L, the solutions correspond to ellipses (see Problem 5.10).

† We assume that $m << M$, so that we do not have to consider the difference between the centre of mass of the sun plus the planet, and the centre of the sun.

‡ It is usually a little more difficult than this to reduce simultaneous differential equations to separate ones, each in terms of a single dependent variable.

At this point, it is worth noting a general feature about differential equations. Their solutions are definite functions relating the variables involved in the problem, but inevitably contain arbitrary constants of integration. The values of these constants are determined by the particular conditions relevant to the given problem.

Thus ellipses are solutions to eqn (5.16), but this ellipse could correspond to that of the earth's orbit, or Jupiter's, or some other possible but non-existent planetary orbit. Which turns out to be the solution for a specific problem depends on, for example, the velocity and the radial distance of the planet when $\vartheta = 0$. With this extra information, the orbit is completely specified. Without this, it is, of course, completely reasonable that all we can deduce from eqn (5.16) is that 'planets move in elliptical orbits around the sun'.

Another illustration is provided by the radioactive decay example of Section 5.1.2. The solution is an exponential decrease in the number of nuclei with time. The overall constant N_0 multiplying this exponential (see eqn 5.5)) is determined by how many radioactive nuclei were present at some specified time.

5.1.6 Coupled differential equations

In some circumstances, we obtain differential equations for two (or more) variables that are related to each other, e.g.

$$\left.\begin{aligned} \frac{dy}{dt} + 3x - y = 4 \\ \frac{d^2x}{dt^2} + 7\frac{dx}{dt} + y = e^{-t} \end{aligned}\right\} \tag{5.17}$$

The unknowns are x and y, which we want to determine as functions of time. Not surprisingly, to determine these two unknowns, we need two simultaneous equations. Each equation is a function of both x and y, and their derivatives.

We already had an example in the previous section, where for planetary orbits, the polar coordinates r and ϑ are given by eqns (5.13) and (5.14). Another example is provided by radioactive decay series, such as

$$\left.\begin{aligned} &A_1 \to A_2 + \alpha \\ &A_2 \to A_3 + e^- + \bar{\nu} \\ &A_3 \to A_4 + \alpha \end{aligned}\right\} \tag{5.18}$$

where $A_1, \ldots A_4$ denote various atomic nuclei, and the emitted decay

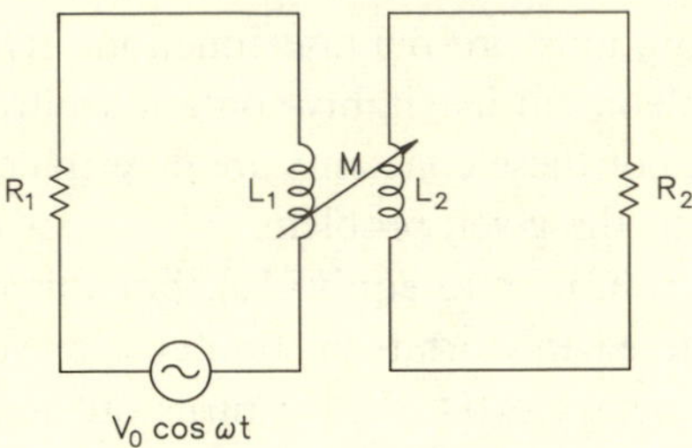

Fig. 5.4 Two coupled electrical circuits. The one on the left has an alternating voltage source, a resistor R_1 and an inductor L_1. Because the alternating current i_1 in L_1 produces a changing magnetic flux which passes through the inductor L_2 in the right-hand circuit, a voltage is induced there which is determined by the mutual inductance M.

particles are either alphas, or electrons and neutrinos. If the decay rates for the various processes are given by λ_1, λ_2 and λ_3 respectively and assuming that A_4 is stable, then the numbers $n_1, \ldots n_4$ of the different nuclei satisfy

$$\left.\begin{aligned} \frac{dn_1}{dt} &= -\lambda_1 n_1 \\ \frac{dn_2}{dt} &= \lambda_1 n_1 - \lambda_2 n_2 \\ \frac{dn_3}{dt} &= \lambda_2 n_2 - \lambda_3 n_3 \\ \text{and } \frac{dn_4}{dt} &= \lambda_3 n_3 \end{aligned}\right\} \tag{5.19}$$

Yet another example is the case of the two electrical circuits shown in fig. 5.4. The differential equations for the currents are

$$\left.\begin{aligned} R_1 i_1 + L_1 \frac{di_1}{dt} &= M \frac{di_2}{dt} + V_0 \cos \omega t \\ \text{and } R_2 i_2 + L_2 \frac{di_2}{dt} &= M \frac{di_1}{dt} \end{aligned}\right\} \tag{5.20}$$

In biology, the predator–prey situation results in coupled differential equations. Two populations of animals inhabit a common area. The rate of growth of the predator population depends, among other factors, on how much of the other there is to eat; this in turn reduces the number of prey. The result is that the differential equations governing the numbers of the two species are coupled to each other.

Methods of solving such differential equations are described in Section 5.15.

5.1.7 Partial differential equations

In the previous section, we had examples with two dependent variables as functions of one independent variable (which was time t). If we had a transversely vibrating stretched string, its displacement y would be a function of two independent variables, the distance x along the string and the time t. Newton's Second Law again would give a differential equation, but in this case it involves the partial derivatives of y with respect to x and with respect to t. For small displacements, it is

$$\rho \frac{\partial^2 y}{\partial t^2} = T \frac{\partial^2 y}{\partial x^2} \tag{5.21}$$

where T and ρ are the tension along the string, and its linear density.

Such partial differential equations are the subject of Chapter 11 of Volume 2. Here we deal with ordinary differential equations involving only one independent variable. For most of the rest of this chapter, we consider cases where there is only one dependent variable as well.

5.1.8 Definitions

Differential equations obviously involve derivatives like dy/dx. We describe y as the dependent variable, and x as the independent one. The solution of such an equation should be some functional relationship between y and x, without any derivatives. In general it is not necessary to solve this relationship explicitly, in order to write y as a function of x. Thus

$$4e^{xy} + x^2 - xy^3 = 2$$

is a perfectly satisfactory form of solution to a differential equation.

Differential equations can involve not only derivatives like dy/dx, but also higher ones such as d^2y/dx^2, d^3y/dx^3, etc. If the highest derivative that occurs in the equation is d^ny/dx^n, its *order* is said to be n.

The solution of a differential equation of order n in general involves n arbitrary constants. Thus

$$\frac{d^3y}{dx^3} = 24$$

has the solution

$$y = 4x^3 + Ax^2 + Bx + C$$

where A B and C are arbitrary. Their values may be determined by other information we have concerning the solution (e.g. that y and its first two derivatives, all evaluated at $x = a$, have specific values).

The *degree* of a differential equation is the power to which the highest order derivative is raised, after the equation has been rationalised so that none of the derivatives occurs raised to a fractional power. Thus

$$\frac{d^3y}{dx^3} = \sqrt{1 + \left(\frac{dy}{dx}\right)^3}$$

is of second degree.

A differential equation is said to be *linear* if each term in it is such that the dependent variable or its derivatives occurs only once, and only to the first power. Thus

$$x^3\frac{d^3y}{dx^3} + e^x \sin x\frac{dy}{dx} + y = \ln x$$

is linear, but

$$\frac{d^3y}{dx^3} + y\frac{dy}{dx} = 0$$

is not. If in a linear differential equation there are no terms independent of y, the equation is also said to be homogeneous†; this would have been true for the first equation above if the $\ln x$ term on the right-hand side had been replaced by zero.

A very important property of linear homogeneous equations is that, if we know two solutions y_1 and y_2, we can construct others as linear combinations of them, i.e.

$$y = \alpha y_1 + \beta y_2 \tag{5.22}$$

is also a solution, where α and β are any constants. Thus if such an equation has e^x and e^{-x} as solutions, then

$$\left.\begin{aligned} &e^x + 2e^{-x} \\ &\sinh x \\ \text{and } &-3\cosh x \end{aligned}\right\} \tag{5.23}$$

must also be solutions of the same equation.

The terms 'complementary function' and 'particular integral' are discussed in Section 5.10.

† In fact, 'homogeneous' as applied to differential equations has two separate meanings (see also Section 5.5).

5.2 How to solve differential equations

Most differential equations are very difficult to solve. This has the advantage for students that, as far as examinations are concerned, the equations have to be chosen to be particularly simple, so that it is possible to find the solution in a short time. This, in turn, means that there are only a relatively small number of types of differential equation that you have to know about.

The basic method is thus to prepare for yourself a short list of the various types of differential equation, and of the approach to be used for solving each of them. This is to be committed to memory. Then, when you have a specific problem to solve, you check which of the types in your list it matches, and apply the appropriate method.

Sometimes differential equations can look unfamiliar because of the letters used for the variables. It is remarkable how a trivial change of variables (such as replacing θ by t; or interchanging x and y) can reduce a seemingly impossible question into one whose type is readily recognisable.

The next few sections describe the various types of differential equations, and how to solve them. We start with equations of first order.

5.3 Separable variables

Consider the differential equation

$$\frac{dy}{dx} = -y^2 e^x \tag{5.24}$$

We can rewrite this with all the y terms on one side of the equation, and the x ones on the other, thus:

$$-\frac{dy}{y^2} = e^x dx \tag{5.24$'$}$$

(Rigorous mathematicians may be unhappy about treating dy/dx as if it were the ratio of dy and dx, which can be manipulated independently. We, however, are scientists. If necessary we could justify it by considering dy and dx to represent small finite changes δy and δx, before we have actually gone to the limit where each becomes infinitesimal.) Our reason for rewriting the equation as in (5.24′) is to make it clear that we have actually separated the variables.

Then each side of the equation can be integrated separately to give us the solution of the differential equation. For our particular example, this is straightforward, and we obtain

$$\frac{1}{y} = e^x + C \tag{5.25}$$

where C is a constant. As expected, our solution for a first order differential equation contains one arbitrary constant.

For other separable equations, the actual integrations after the variables have been separated may be difficult, or even impossible in terms of standard functions. However, we have, at least in principle, a method of solving this type of differential equation. Furthermore, examples you are likely to come across will almost certainly involve integrals that are not too complicated to perform.

The general form of a separable first order differential equation can be written as

$$\frac{dy}{dx} = \frac{f(x)}{g(y)} \tag{5.26}$$

where f and g are any arbitrary functions of the relevant single variable. A counter example would be

$$\frac{dy}{dx} = x + y \tag{5.27}$$

5.4 Almost separable

The equation

$$\frac{dy}{dx} = 8x + 4y + (2x + y - 1)^2 \tag{5.28}$$

contains x and y on the right-hand side only in terms of a single combination

$$w = 2x + y \tag{5.29}$$

If we substitute this into the differential equation, we obtain

$$\frac{dy}{dx} = 4w + (w - 1)^2 \tag{5.30}$$

which is more complicated than (5.28) in the sense that it now contains three variables, x, y and w. We thus need to differentiate (5.29) with respect to x, in order to express dy/dx in terms of dw/dx, thus:

$$\frac{dw}{dx} = 2 + \frac{dy}{dx} \tag{5.31}$$

Thus our differential equation becomes

$$\frac{dw}{dx} + 2 = 4w + (w - 1)^2 \tag{5.32}$$

On taking the 2 to the other side of the equation, we then see that this is separable in terms of w and x, and hence can be solved by the method of Section 5.3.

The general form of differential equation amenable to this approach is

$$\frac{dy}{dx} = f(ax + by) \tag{5.33}$$

where f is an arbitrary function, and a and b are constants. Another specific example is eqn (5.27). However,

$$\frac{dy}{dx} = x(2x + y) + \ln(2x + y) \tag{5.34}$$

is not of this type because of the extra factor of x in the first term on the right-hand side.

5.5 Homogeneous equations†

This type is illustrated by the example

$$\frac{dy}{dx} = \frac{y^2 + xy}{x^2} \tag{5.35}$$

The right-hand side can be rewritten as $(y/x)^2 + (y/x)$, and hence is a function of only the single variable

$$v = y/x \tag{5.36}$$

rather than of the two variables x and y. In this sense it is similar to the examples of the previous section, which also involved functions of only a single combination of x and y.

We thus use the substitution (5.36) both for simplifying the right-hand side of our equation, and also for rewriting dy/dx in terms of v and x, thus

$$\begin{aligned} \frac{dy}{dx} &= \frac{d}{dx}(xv) \\ &= v + x\frac{dv}{dx} \end{aligned} \tag{5.37}$$

† See footnote in Section 5.1.8.

Then eqn (5.35) becomes

$$v + x\frac{dv}{dx} = v^2 + v \tag{5.38}$$

which immediately yields the separated form

$$\frac{dv}{v^2} = \frac{dx}{x} \tag{5.38$'$}$$

The solution of this is readily obtained as

$$-\frac{1}{v} = \ln x + C \tag{5.39}$$

or

$$x = Ae^{-x/y} \tag{5.39$'$}$$

where C and $A(= e^{-C})$ are constants.

When dealing with this type of problem, after using the substitutions (5.36) and (5.37) for y/x and dy/dx respectively, we must end up with a differential equation for x and v which is separable in the variables. If not, we have made a mistake.

The general form of this equation is

$$\frac{dy}{dx} = f(y/x) \tag{5.40}$$

where f is any function of the variable y/x. Thus the differential equation

$$\frac{dy}{dx} = \frac{y^2 + xy}{x} \tag{5.41}$$

is not homogeneous.

A comment about acceptable forms of solution for differential equations is in order. (See also the first paragraph of Section 5.1.8.) We had earlier obtained our solution as (5.39), involving x and v. Since v did not occur in the original differential equation, we used the definition (5.36) for v in order to rewrite the solution in terms of the original variables x and y; this gave us eqn (5.39′). This was not difficult. However, in more complicated problems where several consecutive changes of variables have been used, a lot of straight-forward but tedious algebra could be involved in returning to the original variables of the problem. In general this is not worth doing. It is sufficient to write the solution as

$$x = Ae^{-\frac{1}{v}}, \text{ where } v = y/x$$

(or alternatively with a whole series of 'where's if necessary).

5.6 Nearly homogeneous equations

The equation

$$\frac{dy}{dx} = \frac{y + x - 5}{y - 3x - 1} \tag{5.42}$$

would be homogeneous if it were not for the constants −5 and −1 in the numerator and denominator respectively. However, we can simply eliminate them by a suitable redefinition of the variables x and y:

$$\left.\begin{aligned} x' &= x + \alpha \\ y' &= y + \beta \end{aligned}\right\} \tag{5.43}$$

where α and β are constants specially chosen in order to make the equation become

$$\frac{dy'}{dx'} = \frac{y' + x'}{y' - 3x'} \tag{5.44}$$

(Note that for the definitions (5.43) of the new variables in terms of the old ones, $dy'/dx' = dy/dx$.)

Trivial algebra then yields

$$\left.\begin{aligned} \alpha &= -1 \\ \beta &= -4 \end{aligned}\right\} \tag{5.45}$$

so that in terms of the variables

$$\left.\begin{aligned} x' &= x - 1 \\ y' &= y - 4 \end{aligned}\right\} \tag{5.43'}$$

the differential equation reduces to the form

$$\frac{dy'}{dx'} = \frac{v + 1}{v - 3}$$

where $v = y'/x'$. It is thus now homogeneous, and can be solved as described in Section 5.5.

A variant of this method is needed for the differential equation

$$\frac{dy}{dx} = \frac{y - 3x - 5}{2y - 6x - 1} \tag{5.46}$$

in which the ratio of the coefficients of x and y in the numerator is the same as that in the denominator. Then no choice of α and β in eqn (5.43) will make the constant terms −5 and −1 both disappear. However, in this case the equation is of the type discussed in Section 5.4. We thus use the substitution

$$u = y - 3x \tag{5.47}$$

to obtain

$$\frac{du}{dx} + 3 = \frac{u-5}{2u-1} \tag{5.46$'$}$$

This then becomes a simple example of a separable differential equation in u and x, and is readily solved.

Thus the general examples of these types of differential equation are

$$\frac{dy}{dx} = \frac{a_1x + b_1y + c_1}{a_2x + b_2y + c_2} \tag{5.48}$$

Then provided $a_1/b_1 \neq a_2/b_2$, we use linear transformations (5.43) of x and of y separately, in order to eliminate c_1 and c_2. The resulting equation is then homogeneous. For the special case $a_1/b_1 = a_2/b_2$, we use the substitution

$$u = a_1x + b_1y$$

which reduces the differential equation (5.48) to one which is separable in x and u.

5.7 Exact equations

We now turn to equations of a different type: an example is provided by

$$x^2 \cos y \frac{dy}{dx} + 2x \sin y = 0 \tag{5.49}$$

This may look complicated, but if we rewrite it† as

$$x^2 \cos y dy + 2x \sin y dx = 0, \tag{5.49$'$}$$

we might recognise this as the derivative of $x^2 \sin y$. This is readily verified by carrying out the differentiation. Thus the solution of our differential equation can immediately be written down as

$$x^2 \sin y + c = 0 \tag{5.50}$$

where c is the usual arbitrary constant of integration.

This type of differential equation, which is the derivative of some function of the two variables involved, is known as 'exact'. Since the solution can be written down almost by inspection, it is a good idea to be able to recognise them. We shall write the differential equation we want to test as being exact or not in the form

$$f(x,y)dy + g(x,y)dx = 0 \tag{5.51}$$

† Again we adopt the not strictly rigorous approach of manipulating dy and dx separately. This is done merely to make the derivatives more readily apparent.

Now it must be possible to rewrite an exact differential equation as

$$d[h(x, y)] = 0, \tag{5.52}$$

since by definition we require it to be the derivative of some function $h(x, y)$ of x and y. On performing the differentiation indicated in (5.52), we obtain

$$\frac{\partial h}{\partial x}dx + \frac{\partial h}{\partial y}dy = 0 \tag{5.53}$$

Here $\partial h/\partial x$ and $\partial h/\partial y$ are the partial derivatives of the function h with respect to x, and with respect to y. (This simply means that for evaluating $\partial h/\partial x$ we regard y as a constant, while for $\partial h/\partial y$, x is treated as a constant. For more details, see Chapter 6.) Both $\partial h/\partial x$ and $\partial h/\partial y$ are functions of x and y.

It is a general property of partial derivatives of any reasonable function that

$$\frac{\partial}{\partial y}\left(\frac{\partial h}{\partial x}\right) = \frac{\partial}{\partial x}\left(\frac{\partial h}{\partial y}\right) \tag{5.54}$$

(see eqn (7.45)). Now if our differential equation (5.51) is of the form (5.52), we must be able to identify

$$\left.\begin{aligned} f(x, y) &= \frac{\partial h}{\partial y} \\ \text{and } g(x, y) &= \frac{\partial h}{\partial x} \end{aligned}\right\} \tag{5.55}$$

Then it follows from (5.54) that

$$\frac{\partial g}{\partial y} = \frac{\partial f}{\partial x} \tag{5.56}$$

This is the important condition that we need to check for any differential equation of the form (5.51), in order to test whether it is exact.

Thus for our eqn (5.49) or (5.49′),

$$\left.\begin{aligned} &\frac{\partial h}{\partial y} = f(x, y) = x^2 \cos y \\ \text{and} \quad &\frac{\partial h}{\partial x} = g(x, y) = 2x \sin y \end{aligned}\right\} \tag{5.57}$$

We then evaluate the required partial derivatives to check (5.56), remembering that for $\partial f/\partial x$ we regard y as a constant (and correspondingly for $\partial g/\partial y$). Thus

$$\left.\begin{aligned} \frac{\partial f}{\partial x} &= 2x\cos y \\ \text{and} \quad \frac{\partial g}{\partial y} &= 2x\cos y \end{aligned}\right\} \tag{5.58}$$

Since these are equal, our differential equation (5.49) satisfies the test for being exact.

Having found that our equation is exact, we next need its solution. In simple cases, we can write down by inspection the solution $h(x, y) =$ constant. Alternatively we integrate the first of eqns (5.57), to obtain

$$h(x, y) = x^2 \sin y + w(x) \tag{5.59}$$

where $w(x)$ is an arbitrary function of x. This expression for $h(x, y)$ may require a little thought. The first term arises because we are integrating $x^2 \cos y$ with respect to y, treating x as a constant. The second term is our 'constant' of integration, except that since we are integrating with respect to y at constant x, any function of x will be a constant as far as this is concerned. If we are still doubtful, we can evaluate $\partial h/\partial y$ for eqn (5.59), and discover that we obtain our aimed for $f(x, y)$, as given in eqn (5.57).

Similarly we integrate $g(x, y)$ of eqn (5.57), and obtain

$$h(x, y) = x^2 \sin y + v(y) \tag{5.59$'$}$$

where $v(y)$ is an arbitrary function of y.

We finally have to make (5.59) and (5.59′) consistent. (If we find it impossible to do this, we have made an error somewhere. Either we were wrong in concluding that the differential equation was exact; or we have made a mistake in integrating $f(x, y)$ and $g(x, y)$.) We can achieve this by choosing $w(x)$ and $v(y)$ both being equal to a constant c. Our final solution to the differential equation is thus

$$x^2 \sin y + c = 0,$$

as already stated at the beginning of this section.

In this particular example, $w(x)$ and $v(y)$ both were simply constants. Thus the discussion below eqn (5.59) (about $w(x)$ being a function of x, even though it is a 'constant' of integration) may appear a little unnecessary. This is not so. In general, we may well require $w(x)$ and $v(y)$ to vary with their respective variables in order to achieve identity of the two forms of $h(x, y)$ (see Problem 5.2).

Thus in summary we first use eqn (5.56) to check if our differential equation (5.51) is exact. If so, we then integrate $\partial h/\partial y = f(x,y)$ and $\partial h/\partial x = g(x,y)$ to obtain two forms for $h(x,y)$. Finally we choose the relevant 'constants' of integration to make these forms of h identical, and our solution is $h(x,y) = c$.

In our original equation (5.49), had we changed the factor 2 in the second term to any other constant, we would have found that the differential equation was no longer exact.

It is interesting to consider differential equations of the type

$$f(x,y)\frac{dy}{dx} + g(x,y) = k(x) \tag{5.60}$$

where the left-hand side is an exact differential $(d/dx)[h(x,y)]$, and $k(x)$ on the right-hand side is a function of x only. Then the solution of the differential equation can be written as

$$h(x,y) = \int k(x)dx \tag{5.61}$$

Alternatively eqn (5.60) can be rearranged as

$$f(x,y)\frac{dy}{dx} + [g(x,y) - k(x)] = 0 \tag{5.60$'$}$$

Now since the left-hand side of (5.60) is exact

$$\frac{\partial f}{\partial x} = \frac{\partial g}{\partial y} \tag{5.56}$$

Then eqn (5.60′) is exact as well. This is because the test for exactness for (5.60′) requires

$$\frac{\partial}{\partial x}[f(x)] = \frac{\partial}{\partial y}[g(x,y) - k(x)] \tag{5.62}$$

This follows directly from (5.56), since $\partial k/\partial y = 0$. This is because k is a function of x only, and is independent of y. Thus eqn (5.60′) satisfies the necessary requirement for being exact.

We can thus write its solution as

$$H(x,y) = c \tag{5.61$'$}$$

where

$$\left.\begin{aligned} \frac{\partial H}{\partial y} &= f(x,y) \\ \text{and}\quad \frac{\partial H}{\partial x} &= g(x,y) - k(x) \end{aligned}\right\}$$

Of course, the solution (5.61′) must agree with (5.61) (see Problem 5.2).

5.8 Integrating factors

The differential equation

$$\frac{dy}{dx} + \frac{y}{x} = \ln x \tag{5.63}$$

is such that its left-hand side is not exact. However, if we adopt the simple expedient of multiplying the equation throughout by x, we obtain

$$x\frac{dy}{dx} + y = x\ln x \tag{5.63$'$}$$

Now the left-hand side is exact. It satisfies the test (5.56), and it is the derivative with respect to x of xy. Thus the solution of our differential equation is

$$\begin{aligned} xy &= \int x\ln x dx \\ &= \frac{x^2}{4}(2\ln x - 1) + c \end{aligned} \tag{5.64}$$

The factor x which turned the left-hand side of the differential equation into an exact derivative is known as an integrating factor.

Thus we have solved the differential equation (5.63). The only problem is that we need some procedure for finding the integrating factor, rather than just trial and error.

A prescription exists for finding an integrating factor for first order differential equations of the form

$$\frac{dy}{dx} + f(x)y = g(x) \tag{5.65}$$

where f and g are arbitrary functions of x. We assume that there exists an integrating factor $F(x)$ such that when eqn (5.65) is multiplied through by it, the new left-hand side is exact. Thus

$$F(x)\frac{dy}{dx} + F(x)f(x)y = F(x)g(x) \tag{5.65$'$}$$

is such that

$$\frac{\partial}{\partial x}[F(x)] = \frac{\partial}{\partial y}[F(x)f(x)y] \tag{5.66}$$

Now $\partial/\partial x$ operates on $F(x)$, which is a function of x only, and so this is completely equivalent to d/dx. Thus

$$\frac{dF}{dx} = Ff \tag{5.66$'$}$$

where on the right-hand side we have evaluated the partial derivative with respect to y, remembering that $F(x)$ and $f(x)$ are here to be treated as constants.

Eqn (5.66′) is separable in the variables F and x. Thus

$$\frac{dF}{F} = f(x)dx$$

and hence

$$\ln F = \int f(x)dx$$

or

$$F = e^{\int f(x)dx} \tag{5.67}$$

This then is the expression for the integrating factor $F(x)$ for the differential equation (5.65).

Finally the solution of (5.65′) is

$$F(x)y = \int F(x)g(x)dx \tag{5.68}$$

where $F(x)$ is given by (5.67).

We can see how this works for eqn (5.63). There

$$f(x) = \frac{1}{x} \tag{5.69}$$

and so

$$\begin{aligned} F(x) &= e^{\int dx/x} \\ &= e^{\ln x} \\ &= x \end{aligned} \tag{5.70}$$

This was the factor that we merely stated earlier was needed in order to make eqn (5.63) exact.

After having evaluated an integrating factor explicitly, it is worth checking that the modified eqn (5.65′) is indeed exact. If the left-hand side is not the derivative of Fy, we have made some mistake in calculating F, and should do it again. (For example, had we made an error in sign in performing the integral in (5.70), we would have obtained an incorrect integrating factor of $1/x$, rather than x.) It is worth doing this check before evaluating the integral on the right-hand side of eqn (5.68), required for the complete solution.

We can also look back at our exact equation

$$x^2 \cos y \frac{dy}{dx} + 2x \sin y = 0 \tag{5.49}$$

of Section 5.7. It would be very reasonable to divide through by x, and perhaps also by $\cos y$, to obtain

$$x\frac{dy}{dx} + 2\tan y = 0 \tag{5.49''}$$

This is no longer exact, in that it does not satisfy eqn (5.56). However, it clearly can be made so by multiplying back by the factor $x \cos y$, to reobtain eqn (5.49). Thus $x \cos y$ is the integrating factor for eqn (5.49″). It cannot, however, be deduced by the regular procedure described earlier in this section, since that works only for equations of the form (5.65). Eqn (5.49″) is not of that form in that it contains a $\tan y$ term, which is not allowed in (5.65). Hopefully we will not be presented with these non-standard examples of integrating factors, without being given a very good hint on how to find them. Otherwise inspired guesswork is required.

5.9 Bernoulli's Equation

Another example which is not standard for the integrating factor approach of Section 5.8 is

$$\frac{dy}{dx} + f(x)y = g(x)y^n \tag{5.71}$$

It differs from (5.65) in having the extra factor of y^n on the right-hand side. This type of equation is named after Jacob Bernoulli, who studied it around 1700.

The trick is to use the substitution

$$z = y^\alpha \tag{5.72}$$

with α suitably chosen in order to reduce the new differential equation between z and x to the form (5.65). This can be achieved if $\alpha = 1 - n$, i.e.

$$z = 1/y^{n-1} \tag{5.72'}$$

(see Problem 5.3).

Thus, for example, if $n = 4$, the required substitution is

$$z = 1/y^3$$

and hence

$$\frac{dz}{dx} = -\frac{3}{y^4}\frac{dy}{dx}$$

Thus (5.71) becomes

$$-\frac{y^4}{3}\frac{dz}{dx} + f(x)y = g(x)y^4$$

or

$$-\frac{1}{3z}\frac{dz}{dx} + f(x) = g(x)/z$$

This can be rewritten as

$$\frac{dz}{dx} - 3f(x)z = -3g(x)$$

which is identical in form to (5.65).

We can then determine the integrating factor which makes this differential equation for z and x exact, and hence solve it.

5.10 Second order equations : General properties

In this section we consider more general properties of second order linear differential equations with constant coefficients. They are thus of the form

$$a\frac{d^2y}{dt^2} + 2b\frac{dy}{dt} + cy = f(t) \tag{5.73}$$

where a, b and c are the constant coefficients, and $f(t)$ is an arbitrary function of the independent variable t. We thus see that we are very much restricting the range of possible second order linear equations, since we could imagine a, b and c being replaced by general functions of t.

Nevertheless, our eqn (5.73) has widespread applicability in physics. In Section 5.1.4, we derived the differential equation (5.12) obeyed by a mass on a spring. There the constant a represents the mass, $2b$ is the coefficient of the frictional term and c is the constant of proportionality in Hooke's Law for the spring; $f(t)$ is some time-dependent external force acting on the mass. It is convenient to keep this particular case in mind when considering the solutions.

Equally, eqn (5.73) applies to an electric circuit consisting of an inductor (inductance $= a$), resistor (resistance $= 2b$) and a capacitor (capacitance $= 1/c$) in series. Then y is the charge on the capacitor and $f(t)$ is a varying external voltage applied to the circuit. This case is considered in more detail in Section 4.4.

A general property of linear differential equations emerges when we consider the homogenous equation derived from (5.73) by setting $f(t)$ on the right-hand side equal to zero, i.e.

$$a\frac{d^2y}{dt^2} + 2b\frac{dy}{dt} + cy = 0 \tag{5.74}$$

The solution of this differential equation is known as the complementary function $y_c(t)$, whereas any solution of (5.73) is known as a particular integral $y_p(t)$. Because our eqn (5.73) is linear, we can make new solutions by adding the complementary function to a particular integral, i.e.

$$y(t) = y_c(t) + y_p(t) \tag{5.75}$$

is also a solution.

Clearly the complementary function is independent of $f(t)$, and hence has nothing to do with the behaviour of the system in response to the externally applied influence (i.e. the external force in the spring situation, or the applied voltage for the electrical circuit.) What it does represent is the free oscillations† of the system. Thus even without external forces applied, the spring can oscillate, because of any initial displacement and/or velocity. Similarly had the capacitor been charged already at $t = 0$, the circuit will subsequently display current oscillations even though there is no applied voltage.

A specific example helps to illustrate these generalities. We examine solutions of the differential equation

$$\frac{d^2y}{dt^2} + 0.2\frac{dy}{dt} + 0.37y = \cos t \tag{5.76}$$

Setting the right-hand side of this to zero, we obtain

$$\frac{d^2y}{dt^2} + 0.2\frac{dy}{dt} + 0.37y = 0 \tag{5.77}$$

The general solution of this is the complementary function, and is

$$y_c(t) = e^{-0.1t}\,(A\sin 0.6t + B\cos 0.6t) \tag{5.78}$$

For a derivation of this type of solution, see Section 5.11 below. The complementary function is thus a decaying oscillation, suitable to describe the free oscillations of the spring (see fig. 5.5(a)).

Now for the particular integral. From the methods described in Sections 5.12 or 5.13, this can be obtained as

$$y_p(t) = 1.51\cos(t - 162°) \tag{5.79}$$

† We describe the free motion of the system as 'oscillations'. However, as we shall shortly see, if the damping term $2b$ is too large, the system will exponentially decay rather than oscillate.

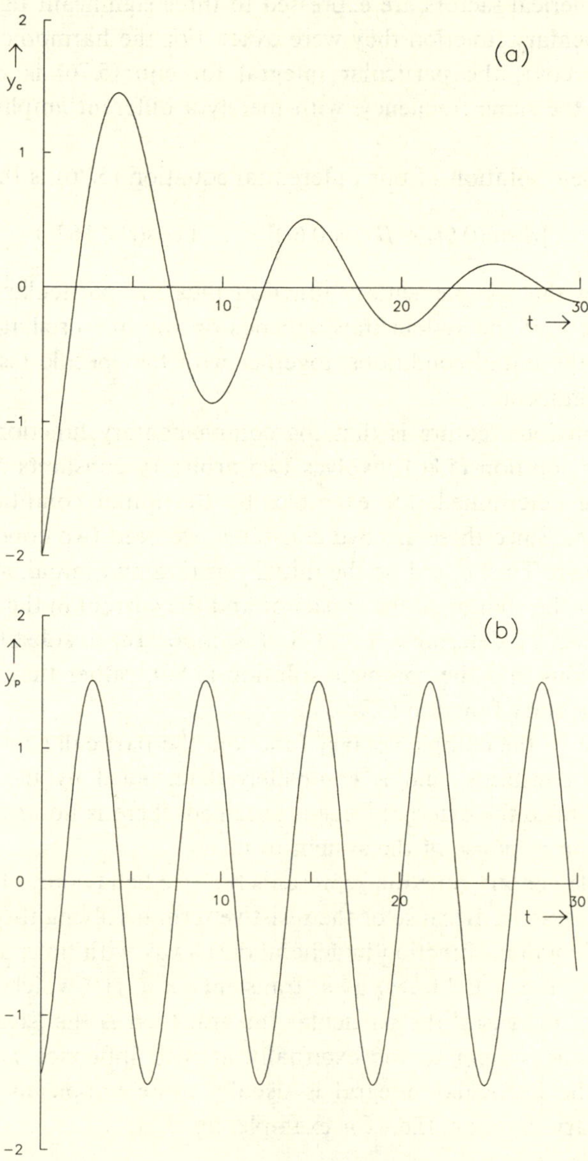

Fig. 5.5 The solution of eqn (5.76) (a) The complementary function y_c, as given by (5.78). (The constants A and B are arbitrary, and have been set to $+1$ and -2 respectively for plotting here.) This is the solution for the free oscillations of the system when no external force is applied. Because of the energy loss caused by the damping term $0.2dy/dt$ in eqn (5.77), these oscillations die away with time. (b) The particular integral y_p (see eqn (5.79)). This is the response of the system to the external force $\cos t$. The system responds at the same frequency $\omega = 1$ (rather than at its natural frequency for free oscillations); and y_p has a constant amplitude. At large times, y_c becomes negligible compared with y_p.

Here the numerical factors are expressed to three significant figures; for the complementary function they were exact. For the harmonic external force $f(t) = \cos t$, the particular integral for eqn (5.76) is thus also harmonic at the same frequency, with merely a different amplitude and phase.

The complete solution of our differential equation (5.76) is thus

$$y = e^{-0.1t}\,[A\sin(0.6t) + B\cos(0.6t)] + 1.51\cos(t - 162^\circ) \qquad (5.80)$$

It consists of the complementary function plus the particular integral. The response y of the system thus depends on the free oscillations that follow from the initial conditions, together with the specific response to the external influence.

Another obvious feature is that the complementary function part of the complete solution (5.80) involves two arbitrary constants A and B. They can be determined, for example, by the initial conditions of a given problem. Since these are two constants, we need two conditions to determine them. They could be the initial position and initial velocity of the spring; or the charge on the capacitor and the current in the electrical circuit, at $t = 0$. To determine A and B, it is important to substitute these given conditions into the complete solution (5.80), rather than just into the complementary function (5.78).

In contrast to the complementary function, the particular integral has no arbitrary constants, and is completely determined by the function $f(t)$. That is, once the external force is specified, there is no arbitrariness concerning the response of the system to it.

The complementary function represents the free behaviour, often oscillatory, of the system. Because of the resistive term involving the constant b, the complementary function in general dies away with time, and hence in the electrical case is known as a 'transient'. For $f(t)$ which does not die away as t increases, the particular integral then is the 'steady state' response of the system to the externally applied influence $f(t)$. Thus at large t, the particular integral is usually more important than the complementary function. (See, for example, fig. 5.5).

5.11 Finding the complementary function

5.11.1 Separate exponentials

We now illustrate the method of finding the complementary function for specific examples of differential equations of the type of eqn (5.73). We start with

$$\frac{d^2y}{dt^2} + 5\frac{dy}{dt} + 4y = t \tag{5.81}$$

As usual, the complementary function is the solution of

$$\frac{d^2y}{dt^2} + 5\frac{dy}{dt} + 4y = 0 \tag{5.81$'$}$$

The standard approach is to guess a functional form for the solution y, containing parameters whose values we find by inserting the trial form into eqn (5.81′). If we make a bad guess, it will be impossible to adjust the parameters to satisfy the equation. The best approach is to know what type of function to use for the trial.

We start, however, with a form that does not work, in order to illustrate how we can tell we have made an unfortunate choice. Our incorrect guess is

$$y = At^n \tag{5.82}$$

Inserting this into (5.81′) yields

$$n(n-1)At^{n-2} + 5nAt^{n-1} + 4At^n = 0 \tag{5.83}$$

Since our differential equation is true for all values of t, we need to make the coefficients of each separate power of t in eqn (5.83) equal to zero (i.e. apart from the uninteresting case of $A = 0$, we require $n(n-1)$, $5n$ and 4(!) all to be zero). Clearly this is impossible, and thus no solution of the form (5.82) exists.

We make more progress with

$$y = Ce^{\lambda t} \tag{5.84}$$

Then eqn (5.81′) becomes

$$C(\lambda^2 + 5\lambda + 4) = 0$$

whence

$$\lambda = -4 \text{ or } -1$$

Thus Ce^{-4t} and Ce^{-t} are solutions (with any value of C being satisfactory). So also is any linear combination of them, because our differential equation is linear, i.e.

$$y = Ae^{-4t} + Be^{-t} \tag{5.85}$$

is our complementary function. We note that it contains two arbitrary constants A and B, as is required for the complementary function of a second order differential equation.

5.11.2 Harmonic solutions

Now we try the same guess (5.84) for the complementary function of

$$\frac{d^2y}{dt^2} + 9y = t \tag{5.86}$$

On evaluating the second derivative, we deduce that (5.84) will be the complementary function provided that

$$\lambda^2 + 9 = 0$$

which has solutions

$$\lambda = \pm 3i \tag{5.87}$$

where i is the square root of -1. (If you are not too familiar with complex numbers, see Chapter 4.)

The complementary function is thus

$$y = Ae^{3i} + Be^{-3i} \tag{5.88}$$

Once again we have two arbitrary constants as expected, but we might be a bit unhappy about imaginary numbers appearing in the answer to a differential equation, which involved purely real (and physical) quantities. However, we can write

$$e^{i\theta} = \cos\theta + i\sin\theta \tag{4.20}$$

and hence

$$y = (A+B)\cos 3t + i(A-B)\sin 3t \tag{5.88$'$}$$

Since A and B are completely arbitrary at this stage, we can choose them as complex numbers such that $(A+B)$ and $i(A-B)$ are both real (In fact all this needs is for A and B to be complex conjugates of each other.) If we call these C and D respectively, we finally obtain

$$y = C\cos 3t + D\sin 3t \tag{5.88$''$}$$

as our complementary function involving only real quantities.

Of course, we could have found the complementary function (5.88″) somewhat faster had we guessed a form

$$y = E\cos\lambda t + F\sin\nu t \tag{5.89}$$

in the first place. The advantage of (5.84) is that it is the correct form for a large variety of cases, provided we do not restrict λ to being real. The small price we have to pay is that the solutions need a bit of interpretation if we want to express them in real terms. With practice, however, we soon learn to recognise that when the exponents are imaginary as in (5.88), the solution can immediately be rewritten as (5.88″).

5.11.3 Decaying oscillations

So much for imaginary exponents. The obvious extension is to think about complex ones. We now have an example in which these arise.

Let us find the complementary function of the differential equation

$$\frac{d^2y}{dt^2} + 2\frac{dy}{dt} + 5y = \cos 2t \tag{5.90}$$

by as usual substituting (5.84) into the differential equation with its right-hand side set to zero. We immediately find that

$$\lambda^2 + 2\lambda + 5 = 0$$

so that

$$\lambda = -1 \pm 2i \tag{5.91}$$

The complementary function is thus

$$y = Ae^{(-1+2i)t} + Be^{(-1-2i)t} \tag{5.92}$$

Again the imaginary parts of the exponents look undesirable, so we factor out the e^{-t}, and then use eqn (4.20) to rewrite the complementary function as

$$y = e^{-t}[(A+B)\cos 2t + i(A-B)\sin 2t] \tag{5.92′}$$

Finally we use the same trick as before of choosing A and B as complex conjugates to obtain

$$y = e^{-t}(C\cos 2t + D\sin 2t) \tag{5.92″}$$

where C and D are real constants. Again with practice, we can immediately write the complementary function as (5.92″) when we find solutions for λ of the form (5.91).

As was explained in the introduction to this chapter and also in Section 4.4, differential equations of the type (5.73) are of very widespread applicability. The complementary function gives the behaviour of the system when it is not subjected to external influences. Thus the solution (5.92″) is of the form of oscillations whose amplitude decays with time, while (5.85) is simply a decaying function. (These are plotted for specific values of the arbitrary constants in fig. 5.6.) These types of behaviour are not surprising, for example, for a swing. If it is initially displaced and released, we would expect it to swing backwards and forwards, with an amplitude which gradually decreases with time (fig. 5.6(b)). If, however, the swing were set up in a way that resulted in considerably more friction (for example, in a large pool filled with honey), then it could well just gradually drop down to the vertical position without oscillating (fig. 5.6(a)).

Which of these types of behaviour applies in any case depends on whether the differential equation (5.73) is such that the exponent λ of our trial complementary function (5.73) is real or complex. This, in turn, depends on the coefficients a, b and c of our differential equation. Thus if

$$b^2 > ac \tag{5.93}$$

the values of λ are real, and no oscillations occur. On the other hand, when

$$b^2 < ac \tag{5.94}$$

λ has complex values, and the system has decaying oscillations for its complementary function.

The difference between (5.93) and (5.94) is sensible, in that in physical systems b represents a resistive or frictional type term, which is responsible for energy loss. Thus a small value of b is required (eqn (5.94)) if we are to have oscillations, while for a large frictional type term, such oscillations do not occur.

In some cases, it is possible to watch the way the nature of the complementary function changes as b is increased. Thus the behaviour of an electric circuit consisting of a resistor, inductor and capacitor can be monitored by an oscilloscope as it responds to an input square wave voltage of low frequency. (This provides a sudden change in voltage, followed by a relatively long period during which the voltage stays constant, and hence the complementary function response of the circuit can be examined.) The constant b in the differential equation is then proportional to the resistance R. For small values of R, the current in the circuit consists of decaying oscillations (fig. 5.6(b)), but as the resistance is increased, a critical value is reached beyond which no oscillations are seen (fig. 5.6(a)).

In almost all physical situations, the constant b is positive. This results in the exponential terms in the complementary functions of the types (5.85) or (5.92″) indeed resulting in solutions that decay (rather than

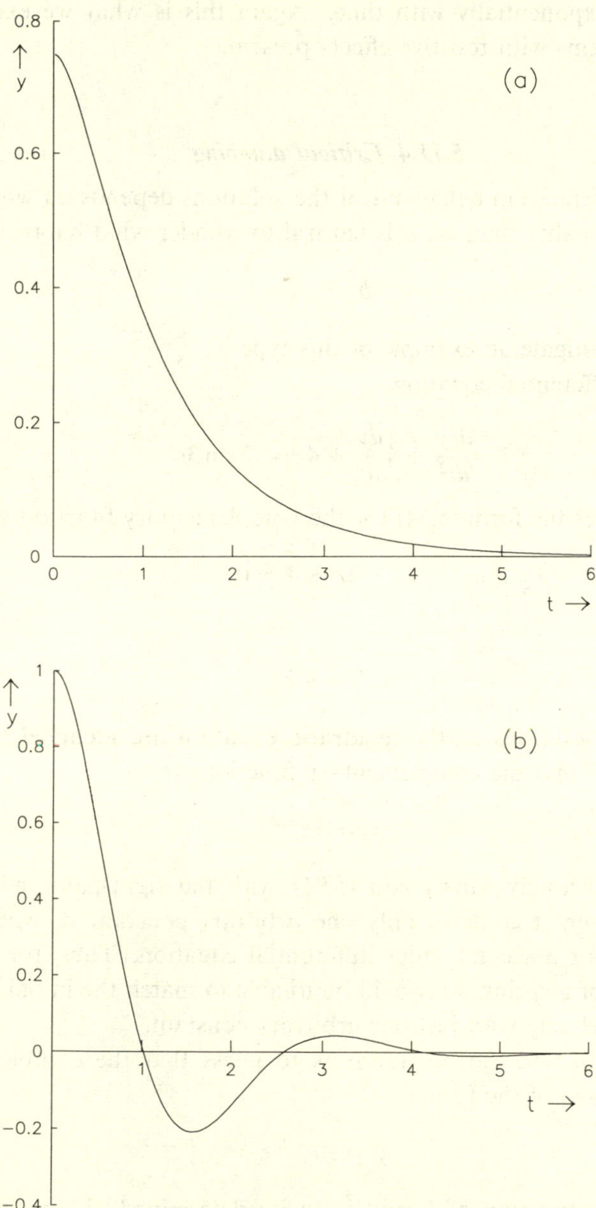

Fig. 5.6 The forms of the complementary function for (a) overdamping, and (b) subcritical damping. These are given by eqns (5.85) and (5.92″) respectively. These both involve arbitrary constants; we have taken $B = 1$ and $C = 1$, and then chosen $A/B = -\frac{1}{4}$ and $C/D = 2$ to give dy_c/dt as zero at $t = 0$. In both cases, the complementary function tends to zero as t becomes large, because of the damping term. The large damping in (a) prevents any oscillations in the complementary function. In (b), an even smaller resistance term b would have resulted in the oscillations being less damped.

increasing) exponentially with time. Again this is what we expect for physical systems with resistive effects present.

5.11.4 Critical damping

Since the difference in behaviour of the solutions depends on whether b^2 is larger or smaller than ac, it is natural to wonder what happens when

$$b^2 = ac \tag{5.95}$$

We now investigate an example of this type.

For the differential equation

$$\frac{d^2y}{dt^2} + 4\frac{dy}{dt} + 4y = 2\sin 3t \tag{5.96}$$

substitution of the form (5.84) for the complementary function yields

$$\lambda^2 + 4\lambda + 4 = 0$$

whence

$$\lambda = -2$$

i.e. the two solutions of the quadratic equation are identical in value. This suggests that the complementary function is

$$y = Ae^{-2t}$$

which does actually satisfy eqn (5.96) with the right-hand side set to zero. However, it contains only one arbitrary constant A, whereas we expect two for a second order differential equation. Thus, for the free oscillations of a spring, we would be unable to match the initial position and initial velocity with just one arbitrary constant.

The way to find the other one is to guess that the complementary function is now of the form

$$y = Ae^{-2t}f \tag{5.97}$$

where f is a function of t, and is to be determined. If this is indeed the complementary function, we substitute it into eqn (5.96) (with its right-hand side set equal to zero), and obtain

$$Ae^{-2t}\left[\left(\frac{d^2f}{dt^2} - 4\frac{df}{dt} + 4f\right) + 4\left(\frac{df}{dt} - 2f\right) + 4f\right] = 0 \tag{5.98}$$

whence

$$\frac{d^2f}{dt^2} = 0$$

This has the solution

$$f = Bt + C \tag{5.99}$$

where B and C are arbitrary constants. Substituting this back into (5.97), we finally obtain for the complementary function

$$y = (A' + Bt)e^{-2t} \tag{5.100}$$

Here $A' = AC$ and is simply an arbitrary constant; we can thus omit the prime from A in (5.100)

Thus when the two roots of the equation for λ are equal, the complementary function still consists of two terms with arbitrary coefficients; the first is the simple exponential term, while the second contains the same exponential multiplied by the independent variable (in this case, t).

This result in fact generalises. Thus if we have an nth order differential equation with constant coefficients, we guess a complementary function of the form (5.84). If the n solutions for λ are all different, the complementary function simply consists of a linear combination of the exponential terms with arbitrary coefficients. If, however, there are $m + 1$ identical values of the root λ_i, the relevant part of the complementary function is

$$e^{\lambda_i t}(A_0 + A_1 t + \ldots A_m t^m)$$

The situation (5.95) is known as 'critical damping', as the value of the resistive term b is such as to divide between the regime of damped oscillations (for smaller b; see Section 5.11.3), and the decaying exponentials of Section 5.11.1 for larger b.

5.12 Finding the particular integral

So far we have concentrated on the complementary function of our second order differential equation (5.73). We now need to find the particular integral, i.e. a solution of the differential equation which takes the term $f(t)$ on the right-hand side into account. As mentioned in Section 5.10, the complementary function is transient in nature (for $b > 0$), and hence from a physical point of view, the particular integral will usually dominate the response of the system at large times.

Rather as with the complementary function, the method of determining the particular integral is to guess a suitable functional form containing arbitrary constants, and then to choose the constants to ensure it is indeed the solution. If our guess is incorrect, then no values of these constants will satisfy the differential equation, and so we have to try a different form. Clearly this procedure could take a long time, so it is a good idea to remember what to try for the common examples of $f(t)$.

We first list these rules, and illustrate some of these with examples later.

5.12.1 The rules

Rule 1 : $f(t)$ = polynomial in t
If $f(t)$ is polynomial in t with highest power t^n, then the trial particular integral is also a polynomial in t, with terms up to the same power. (As in all these rules, the various terms are given arbitrary coefficients, whose values are determined by substituting the trial particular integral into the differential equation – see Section 5.12.3.) The trial particular integral is a power series in t, even if $f(t)$ contains only a single term At^n.

Rule 2 : $f(t) = Ae^{kt}$
Our trial solution is

$$y = Be^{kt}$$

Rule 3 : $f(t) = A\sin lt$
Our trial solution is

$$y = B\sin lt + C\cos lt$$

That is, even though $f(t)$ contains only a sine term, we need both sine and cosine terms for the particular integral.

Rule 4 : $f(t) = A\cos lt$
The same trial solution as in rule 3 is required.

Rule 5 : $f(t) = Ae^{kt}\sin lt$ or $Ae^{kt}\cos lt$
The trial particular integral is

$$y = e^{kt}(B\sin lt + C\cos lt)$$

In fact this rule, as well as rules 3 and 4, can be regarded as extensions of rule 2 for the case where we allow the exponent k to become complex. Indeed, finding the particular integral using a complex number approach is generally much neater and more efficient than the separate use of sines and cosines (see Section 4.4).

Rule 6 : $f(t)$ is a polynomial of order n in t, multiplied by e^{kt}
In this case our trial function is a polynomial in t with coefficients to be determined, multiplied by e^{kt}.

Rule 7 : $f(t)$ is a polynomial of order n in t, multiplied by $\sin lt$
Not surprisingly, we should try

$$y = \sum_{m=0}^{n} (B_m \sin lt + C_m \cos lt) t^m \tag{5.101}$$

The function

$$y = (B \sin lt + C \cos lt) \sum_{m=0}^{n} D_m t^m$$

is unsuitable. It contains only $n+3$ independent coefficients, as compared with $2(n+1)$ in (5.101), and this restricts the relative amounts of the various terms in such a way that in general it is not possible to find a solution.

Rule 0 : $f(t)$ contains a term identical in form to part of the complementary function
We call this 'rule 0', because it is to be taken in conjunction with any of the other rules.

If one of the terms of the complementary function is identical in form to our trial particular integral (as deduced from the rules above) or to part of it, then we need to multiply our trial form by an extra power of t. Here 'identical in form' means that the ratio of their time dependences is a constant. Thus $-2e^{-t}$ and Ae^{-t} are identical in form, but e^{-t} and e^{-2t} are not. The practical application of rule 0 is made clearer by the examples in the next section.

Because of rule 0, it is necessary to find the complementary function before writing down a trial form for the particular integral. Otherwise, we do not know whether to include extra powers of t in our trial form.

5.12.2 Examples of applying the rules

Table 5.1 shows some examples of using the above rules to choose trial forms for the particular integral, for the differential equation

Table 5.1. *Examples of the functional form chosen for the particular integrals for the various $f(t)$ in the differential equation (5.102), whose complementary function is $A + Be^{-t}$.*

$f(t)$	Rule	Trial particular integral
$e^{-t}\sin 2t$	3	$e^{-t}(b\sin 2t + c\cos 2t)$
4	1,0	at
$4 + e^{-2t}$	1,2,0	$at + be^{-2t}$
t^2	1,0	$at + bt^2 + ct^3$
te^{-t}	6,0	$(at + bt^2)e^{-t}$

$$\frac{d^2y}{dt^2} + \frac{dy}{dt} = f(t) \tag{5.102}$$

whose complementary function is

$$y = A + Be^{-t} \tag{5.103}$$

In the first example, $f(t)$ is $e^{-t}\sin 2t$, which is not of the same form as the e^{-t} term in the complementary function. Thus there is no need for extra powers of t, and the trial particular integral follows directly from rule 3.

The second example has $f(t) = 4$. This is of the same functional form as the A term of (5.103), and hence our trial function is at, rather than simply the constant a.

When $f(t) = 4 + e^{-2t}$, only the first term is of the same form as part of the complementary function, and hence only that part of the trial function associated with it is raised by one power of t.

Although $f(t) = t^2$ is not part of the complementary function, our trial solution according to rule 1 is a polynomial in t containing terms up to t^2. The constant term of this is like the A of the complementary function. We thus need a trial solution that extends as far as t^3. This is also illustrated by the last example of the table where $f(t) = te^{-t}$, whose trial particular integral is usually $(a + bt)e^{-t}$, while the second term of the complementary function is Be^{-t}.

As a final example of these rules, we consider the differential equation

$$\frac{d^2y}{dt^2} + 2\frac{dy}{dt} + 1 = e^{-t} \tag{5.104}$$

The complementary function is now

$$y = (A + Bt)e^{-t} \tag{5.105}$$

Thus even if we multiply our first guess of ae^{-t} by t, it is still part of the complementary function. Hence we need to choose

$$y = at^2e^{-t}$$

(See also Problem 5.7.)

5.12.3 Determining the coefficients

Finally we illustrate how we use these trial forms to determine the particular integral completely. We again use the differential equation (5.102), with its complementary function (5.103).

If $f(t) = e^{-3t}$, our trial particular integral is

$$y = ae^{-3t}$$

Substituting this in (5.102) yields

$$9ae^{-3t} - 3ae^{-3t} = e^{-3t}$$

whence $a = 1/6$. The complete solution is thus

$$y = A + Be^{-t} + e^{-3t}/6 \tag{5.106}$$

For $f(t) = 10\sin 2t$, rule 3 tells us to try

$$y = b\sin 2t + c\cos 2t \tag{5.107}$$

On differentiating twice and inserting the derivatives into the differential equation, we obtain

$$(-4b\sin 2t - 4c\cos 2t) + (2b\cos 2t - 2c\sin 2t) = 10\sin 2t \tag{5.108}$$

It is important to remember that this is an equation which must be satisfied *for all values of t*. This can be so only if, when we collect the various $\sin 2t$ and $\cos 2t$ terms, their net coefficients are both zero. We thus rewrite eqn (5.108) as

$$(4b + 2c + 10)\sin 2t + (4c - 2b)\cos 2t = 0 \tag{5.108$'$}$$

and deduce

$$\left.\begin{aligned} b &= -2 \\ c &= -1 \end{aligned}\right\}$$

Our complete solution is then

$$y = A + Be^{-t} - (2\sin 2t + \cos 2t) \tag{5.109}$$

It is interesting to see what happens if we try the plausible but incorrect form for the particular integral

$$y = b\sin 2t$$

When we differentiate and substitute into the differential equation, we obtain

$$-4b\sin 2t + 2b\cos 2t = 10\sin 2t$$

There is no way of choosing b so that this equation is true for all t, and hence our trial solution fails.

A similar example is provided by $f(t) = t^2$ (see Table 5.1). Because of rules 1 and 0, we need to try the form

$$y = at + bt^2 + ct^3 \tag{5.110}$$

The t^3 term is necessary because of the A term in the complementary function. If we omit it, we simply will not be able to satisfy the differential equation for all values of t. Similarly using a trial form dt^3 (on the spurious grounds that $f(t) = t^2$, and we need one higher power of t because of the form of the complementary function) will not work; we need the full polynomial.

When we substitute the correct trial form into (5.102), we obtain

$$(2b + 6ct) + (a + 2bt + 3ct^2) = t^2 \tag{5.111}$$

We require to make the coefficient of each separate power of t vanish, and this is achieved by

$$\left.\begin{aligned} c &= 1/3 \\ b &= -1 \\ a &= 2 \end{aligned}\right\} \tag{5.112}$$

The complete solution is thus

$$y = A + Be^{-t} + 2t - t^2 + \frac{1}{3}t^3 \tag{5.113}$$

After determining the coefficients of our trial solutions as in the examples above, it is a good idea to check that our specific form for the particular integral does indeed satisfy the differential equation. This will ensure that we have made no algebraic errors.

5.13 The D operator method

5.13.1 Introduction

In the first three parts of this section we describe an alternative method, which can be used for finding particular integrals. (Other applications are mentioned in section 5.13.4.) As compared with the method of Section 5.12, it involves less guess-work as to what the form of the solution is, and the constants multiplying the functional forms of the answer are obtained automatically. It does, however, require a fair amount of practice to ensure that you are familiar with how to use it.

The technique involves defining the differential operator

$$D \equiv \frac{d}{dt}$$

On its own, this is fairly meaningless, but we take it always as operating on some function of t, which is written to the right of the D. Thus

$$Dy \equiv \frac{dy}{dt}$$

$$\text{and } D^2y \equiv \frac{d^2y}{dt^2}$$

Thus our differential equation (5.73) becomes

$$aD^2y + 2bDy + cy = f(t)$$

or

$$(aD^2 + 2bD + c)y = f(t) \tag{5.114}$$

If we do not bother too much about what it means, we can immediately write down the solution as

$$y = \frac{1}{aD^2 + 2bD + c} f(t) \tag{5.115}$$

Here we have been careful to write the term with the Ds to the left of $f(t)$, since it is a differential operator which needs to operate on something – in this case, $f(t)$. Another reason is that we derive (5.115) by putting the factor $1/(aD^2 + 2bD + c)$ in front of both sides of eqn (5.114). When factors involve differential operators, their order is important.†

The 'only' trouble with eqn (5.115) is that it contains an expression involving Ds in the denominator. In order to express y in terms of conventional functions, we need to eliminate these. For this, there are several rules to help us.

5.13.2 Rules for D operators

We write as $G(D)$ the particular functional form of the D operator; this is to be distinguished from the functional form on which it operates. Thus in eqn (5.115),

$$G(D) = \frac{1}{aD^2 + 2bD + c}$$

and it operates on $f(t)$. The various rules below apply to different forms of $f(t)$.

Rule (a) : $\quad G(D)e^{at} = G(a)e^{at}$

This simply says that, if we have some function of D operating on an exponential, it is equivalent to multiplying that exponential by the same function of a.

This is readily checked for the case where $G(D)$ can be expressed as terms consisting of D raised to positive or negative integral powers. All we have to remember is that D^n means 'differentiate n times', so it is plausible that D^{-m} requires us to integrate m times. Thus

$$\left(D^2 + 3D + \frac{1}{D}\right) e^{4t} = \left(4^2 + 3 \times 4 + \frac{1}{4}\right) e^{4t}$$

Rule (b) : $\quad G(D)[e^{at}V(t)] = e^{at}G(D+a)[V(t)]$

This states that, when we operate on a product function $e^{at}V(t)$ with some function $G(D)$ of the differential operator D, it is equivalent to operating on $V(t)$ with the same function G, but with argument $D + a$ rather than a, and then multiplying the result by e^{at}. The explanation in

† For example

$$tD \neq Dt$$

as can be seen by checking what happens when they each operate on some function, like e^t. Then

$$t\frac{d}{dt}(e^t) \neq \frac{d}{dt}(te^t)$$

words is probably less comprehensible than the equation, but a simple example should clarify what is meant. Thus

$$D^2[e^{at}t^2] = e^{at}(D+a)^2[t^2]$$

as can be verified by explicit evaluation of both sides of the equation.

Rule (c) : $\quad G(D^2)\sin lt = G(-l^2)\sin lt$
Thus, for example,

$$\frac{1}{D^2}(\sin 3t) = -\frac{1}{9}\sin 3t$$

A similar rule applies if $\sin lt$ is replaced by $\cos lt$ on both sides of the equation.

This rule for sines and cosines follows from rule (a) in the case where the exponent a is imaginary.

The above three rules can to some extent be justified when $G(D)$ can be expressed as the sum of terms involving positive or negative powers of D.

5.13.3 Examples

We now illustrate by a few examples how the rules are used, as well as introducing some useful manipulations which help eliminate the troublesome Ds in expressions in the denominator. We remember that a power series of Ds in the numerator does not worry us, as each term is just an instruction to differentiate the specific function of t the relevant number of times.

Example (i)
For the differential equation

$$\frac{d^2y}{dt^2} + \frac{dy}{dt} = e^{-2t} \tag{5.116}$$

the solution is

$$y = \frac{1}{D^2+D}e^{-2t}$$

Then rule (a) immediately gives

$$y = \frac{1}{2}e^{-2t} \tag{5.117}$$

Example (ii)
Had the right-hand side of eqn (5.116) been e^{-t}, our solution would be

$$y = \frac{1}{D^2 + D} e^{-t} \tag{5.118}$$

Applying rule (a) now produces a difficulty, in that

$$G(a) = \frac{1}{(-1)^2 - 1} = \frac{1}{0}$$

This is meant to give us the message that rule (a) does not work in this case.

We then rewrite (5.118) as

$$y = \frac{1}{D^2 + D}[e^{-t} \cdot 1] \tag{5.118$'$}$$

and use rule (b), with $V(t) = 1$. Then

$$\begin{aligned} y &= e^{-t}\frac{1}{(D-1)^2 + (D-1)}[1] \\ &= e^{-t}\frac{1}{(D-1)D}[1] \end{aligned} \tag{5.119}$$

Now although functions of D in the denominator are a nuisance, a single D on its own is simply an instruction to integrate. So we integrate 1, which gives t, and obtain

$$y = e^{-t}\frac{1}{D-1}[t] \tag{5.120}$$

We are thus making progress as we have less Ds in the expression.

We now have one final D to eliminate from the denominator. We do this by using a binomial expansion† for

$$-\frac{1}{1-D} = -(1 + D + D^2 + \ldots) \tag{5.121}$$

This converts what was previously an obscure function of D into a series of differentiations. This is particularly useful when, as in this case, the series operates on a polynomial of order m in t, since all derivatives beyond the mth then vanish.

† It is not at all obvious that an expression containing D can be manipulated as if it were simply an algebraic quantity. One approach is to regard this expansion (and also the rules given earlier) as providing tentative particular integrals, which we can then check by explicit differentiation. In all simple cases, they are satisfactory.

Thus (5.120) becomes

$$\begin{aligned} y &= -e^{-t}(1 + D + D^2 + \ldots)[t] \\ &= -e^{-t}(t+1) \end{aligned} \tag{5.122}$$

and no further terms are involved.

This then is our particular integral. In fact the complementary function of our differential equation was

$$A + Be^{-t}$$

so the $-e^{-t}$ part of our particular integral can be omitted as it is already contained in Be^{-t}. The complete solution is thus

$$y = A + Be^{-t} - te^{-t} \tag{5.123}$$

We thus see that, although $f(t)$ of our differential equation was e^{-t}, the particular integral is obtained automatically with an extra factor of t. In the previous section, we had to include this extra factor of t explicitly in our trial form for the particular integral.

Two other points are worth noting. The first is that in writing (5.120), we said that the integral of 1 is t, and neglected the constant of integration c. This we did simply because, had we included it, we would have obtained another term ce^{-t} in the answer, which is already part of the complementary function, and hence can be absorbed into it. If in doubt, it is safer to include such constants of integration explicitly, but they should disappear from the final answer, since the particular integral does not involve any arbitrary constants.

The second point refers to the binomial expansion. It would seem equally valid to write

$$\begin{aligned} \frac{1}{D-1} &= \frac{1}{D(1-1/D)} \\ &= \frac{1}{D}\left(1 + \frac{1}{D} + \frac{1}{D^2} + \ldots\right) \end{aligned} \tag{5.121$'$}$$

Then each term is interpreted as an integration. The advantage of (5.121) for our purposes is that, because it involves *differentials* of t, the series quickly terminates. In contrast, (5.121′) operating on t gives an infinite series, so (5.121) is more convenient.

The keen reader may care to contemplate how (5.122) and an infinite series can both be satisfactory for the particular integral of the same differential equation. (See Problem 5.9 for some help with this paradox.)

Example (iii)

The particular integral of the differential equation

$$\frac{d^2y}{dt^2} + \frac{dy}{dt} = t^2 \tag{5.124}$$

is

$$y = \frac{1}{D^2 + D}\left[t^2\right]$$

We could simplify this by the same device used in example (ii), by writing $G(D)$ as $1/(D+1)D$, and then interpreting $1/D$ as an integration. For variety, we use a different trick of writing $G(D)$ in terms of partial fractions:

$$\frac{1}{(D+1)D} = \frac{1}{D} - \frac{1}{1+D} \tag{5.125}$$

Then the first term is simply an integral, while the second we expressed by the binomial theorem as before. Then

$$\begin{aligned} y &= \left\{\frac{1}{D} - (1 - D + D^2 - D^3 + \ldots)\right\} t^2 \\ &= \frac{t^3}{3} - t^2 + 2t - 2 \end{aligned} \tag{5.126}$$

and all further terms are zero. Again the -2 can be absorbed into the A term of the complementary function, and the complete solution is

$$y = A + Be^{-t} + t^3/3 - t^2 + 2t \tag{5.127}$$

(Compare the solution obtained previously in eqn (5.113).) We thus see that the D operator method automatically gives us both the term $2t$ in the particular integral, which is of lower power than the t^2 of $f(t)$; and also the higher one ($t^3/3$) for this special case where the complementary function contains a term (A) of the power series usually required when $f(t) = t^2$.

Example (iv)

For the differential equation

$$a\frac{d^2y}{dt^2} + 2b\frac{dy}{dt} + cy = \sin 2t \tag{5.128}$$

the particular integral is

$$y = \frac{1}{aD^2 + 2bD + c}\sin 2t \tag{5.129}$$

Now our rule (c) for a function of D operating on sines or cosines applies only for $G(D^2)$. Our form is unsatisfactory in that it contains a term in D. We circumvent this problem by a trick similar to the one for converting a complex denominator into a real one, where we multiply the numerator and denominator by the complex conjugate of the denominator (see Section 4.2.4).

Here we use the identity

$$\begin{aligned}\frac{1}{aD^2+2bD+c} &= \frac{aD^2-2bD+c}{aD^2-2bD+c} \times \frac{1}{aD^2+2bD+c} \\ &= \frac{aD^2+c-2bD}{(aD^2+c)^2-4b^2D^2} \qquad (5.130)\end{aligned}$$

What we have achieved by this is that we now have a denominator which includes only D^2 and constant terms, and hence is a function of D^2 only. It is true that there is a D in the numerator, but this is not a problem, since it merely tells us to differentiate. Thus the particular integral is

$$\begin{aligned}y &= (aD^2+c-2bD)\frac{1}{(aD^2+c)^2-4b^2D^2}\sin 2t \\ &= (aD^2+c-2bD)\frac{1}{(-4a+c)^2+16b^2}\sin 2t \\ &= \frac{1}{(c-4a)^2+16b^2}[(-4a+c)\sin 2t - 4b\cos 2t]\end{aligned}$$

We have made use of the fact that if we have a product of functions of differential operators $G(D)H(D)$, we can take them in whichever order is convenient. (This contrasts with the fact that the order of differential operators and a function it operate on is important – see footnote at the end of Section 5.13.1.)

In the second line we have invoked rule (c), and in the last line, we have performed the necessary differentiations as specified by the $(aD^2+c-2bD)$ term.

The rules and some of the useful manipulations with D operators are summarised in Table 5.2.

In conclusion, the D operator method is useful in that it obviates the need for guessing specific forms for particular integrals. Also, if extra factors of t are needed in special cases, the D method provides them automatically. However, it should be clear that it does require a fair amount of experience to discover which rules and manipulations should be applied in any particular problem. Furthermore, for examples like

Table 5.2. *Rules and useful manipulations for D operators*

Rules

(a) $G(D)e^{at} = G(a)e^{at}$

(b) $G(D)[e^{at}V(t)] = e^{at}G(D+a)[V(t)]$

(c) $G(D^2)\sin lt = G(-l^2)\sin lt$

Examples of manipulations

(i) $\dfrac{1}{D^n}$	Integrate n times
(ii) $\dfrac{1}{1+D}$	Expand by binomial theorem
(iii) $\dfrac{1}{(D+a)(D+b)}$	Use partial fractions
(iv) $G(D)e^{at}$	If $G(a) = 1/0$, rewrite as $G(D)[e^{at} \cdot 1]$ and use rule (b)
(v) $\dfrac{1}{aD^2+2bD+c}[\sin lt]$	Multiply numerator and denominator by $aD^2 - 2bD + c$

(iv) above where $f(t)$ involves sines and/or cosines, the use of complex numbers as described in Section 4.4 is strongly recommended.

5.13.4 Other applications of D operators

The use of D operators in solving simultaneous differential equations is described in Section 5.15.1.

The earlier parts of this section explained how D operators are helpful for evaluating particular integrals. Here we describe why they are less useful for determining complementary functions.

The complementary function is the solution of the differential equation

$$(aD^2 + 2bD + c)y = 0 \tag{5.131}$$

Clearly, writing

$$y = \frac{1}{aD^2 + 2bD + c}[0]$$

is going to get us nowhere; we merely discover that the particular integral when $f(t) = 0$ is itself zero. To find the complementary function, we have to resort to guessing a form for y. As in Section 5.11.1, we try

$$y = Ae^{\lambda t} \tag{5.132}$$

Inserting this in (5.131) and using rule (a) for D operators yields

$$a\lambda^2 + 2b\lambda + c = 0 \tag{5.133}$$

which not too surprisingly is equivalent to the equation for λ in Section 5.11.

In fact, rule (a) was rather heavy handed, because $(aD^2 + 2bD + c)e^{\lambda t}$ involves only differentials of $e^{\lambda t}$, which we know how to evaluate directly (and as we indeed did in Section 5.11). The real advantage of the rules for D operators is when the Ds occur in a denominator, as in the particular integral problems of the previous section.

In a similar way, we can use rule (b) for the case of the differential equation.

$$\frac{d^2y}{dt^2} - 2c\frac{dy}{dt} + c^2y = 0$$

which has equal roots of $\lambda = c$. Our guessed complementary function is now

$$y = e^{-ct}V(t)$$

where $V(t)$ is a function of t to be determined (see Problem 5.7). However, as in the previous case, since the D^2 and D are in the numerator of our expression, we can solve the problem directly without the need for rule (b).

5.14 Resonance

As already mentioned, differential equations of the form

$$a\frac{d^2y}{dt^2} + 2b\frac{dy}{dt} + cy = A\cos\omega t \tag{5.134}$$

have very wide application across a whole range of physics situations. Its particular integral can be determined by the methods of Sections 5.11.3, 5.13 or even better Section 4.4. It can be written as

$$y = \frac{A[(c - a\omega^2)\cos\omega t + 2b\omega\sin\omega t]}{(c - \omega^2 a)^2 + 4b^2\omega^2} \tag{5.135}$$

Here we are going to concentrate on the case where ω is close to $\sqrt{c/a}$, and b is small. (As always when some quantity is small, it is sensible to ask with respect to what is this so. In this case, a natural quantity for comparison is b_0, the critical value of b above which there are no free oscillations. This is given by $b_0 = \sqrt{ac}$ (see eqn (5.95)).)

Physically this corresponds to light damping, in which case $\sqrt{c/a}$ is closely equal to the natural frequency of the free oscillations (see Problem 5.14). We are thus examining the behaviour of the system when the frequency of the external influence ($A \cos \omega t$) is near to the natural frequency. In such situations, the response of the system is especially large, and the phenomenon is known as resonance – the external force is 'in resonance' with the natural behaviour of the system.

For example, a swing has its own natural frequency of free oscillations, when left swinging on its own. If we wish to make it have large oscillations of forced motion by pushing it, we should do so at its natural frequency. If we were to push at a different frequency, the oscillations would be much smaller. (Indeed it is hard to imagine this, because we almost invariably push swings at their natural frequency.)

We are going to be interested in the amplitude of the harmonic response of the system, so it is useful to rewrite eqn (5.135) as

$$\left.\begin{aligned} y &= B\cos(\omega t + \phi) \\ \text{with } B^2 &= \frac{A^2}{(c-\omega^2 a)^2 + 4b^2\omega^2} \end{aligned}\right\} \tag{5.136}$$

(see Problem 5.14 and Section 4.4). This gives us the amplitude B as a function of the amplitude and frequency of the external force.

Eqn (5.136) shows that B increases in proportion to A, as expected. But we now wish to examine how B varies with ω, for fixed A (and a, b and c). That is, how does the response of the system vary as the frequency of the external influence is changed, but its magnitude is kept fixed? This is shown in fig. 5.7, for $b = 0.1b_0$ and for $0.03b_0$. As expected we see that the response is largest very close to the natural frequency, and stays large for a range of frequencies that depends on the value of b – small damping gives rise to a narrower resonance. Again not surprisingly, the response at the resonant frequency is larger if the damping is smaller.

In the special case of no damping at all (i.e. b is identically zero), the particular integral becomes

$$y = \frac{A\cos\omega t}{c - a\omega^2} \tag{5.137}$$

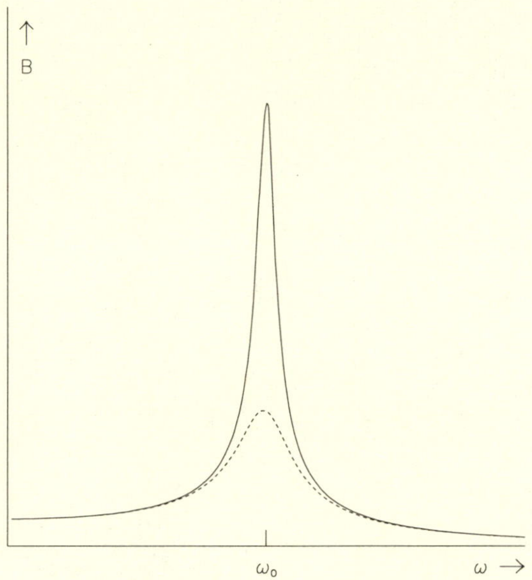

Fig. 5.7 The amplitude of forced oscillations, as the frequency of the driving force is varied (see eqns (5.134) and (5.136)). The response is largest when the frequency is close to the natural frequency of the system in the absence of damping ($\omega_0 = \sqrt{c/a}$). The curves are for a damping constant b of 10% or of 3% of the critical value $b_0 = \sqrt{ac}$ required just to prevent free oscillatons (solid and dashed curves respectively). For even smaller damping, the resonance curve is narrower, and larger.

This is valid unless $c = a\omega^2$, when

$$y = \frac{At \sin \omega t}{2a\omega} \tag{5.138}$$

(Compare Problem 5.13(i).)

This is the example of the undamped system being subjected to an external force exactly at its natural frequency. Because of the factor of t in eqn (5.138), the response consists of oscillations whose amplitude increases with time. In any real situation, this will not go on for ever. Either b is only approximately zero, and for very large oscillations damping eventually becomes important; or the system breaks; etc.

The case of the undamped system forced at its resonant frequency is another example of the particular integral dominating over the complementary function at large times. Usually this occurs because the free oscillations decay exponentially to zero, while the forced oscillations have a time-independent amplitude. Here, because there is no damping,

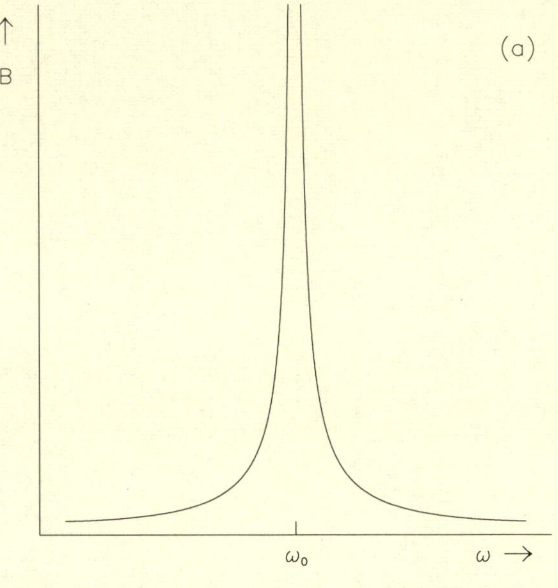

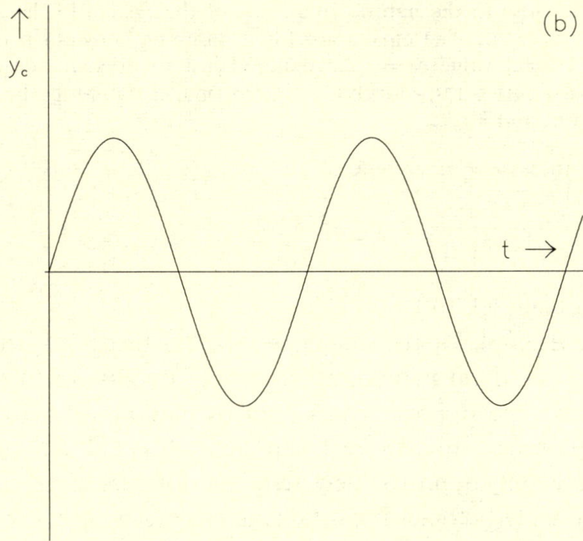

Fig. 5.8 For legend see opposite.

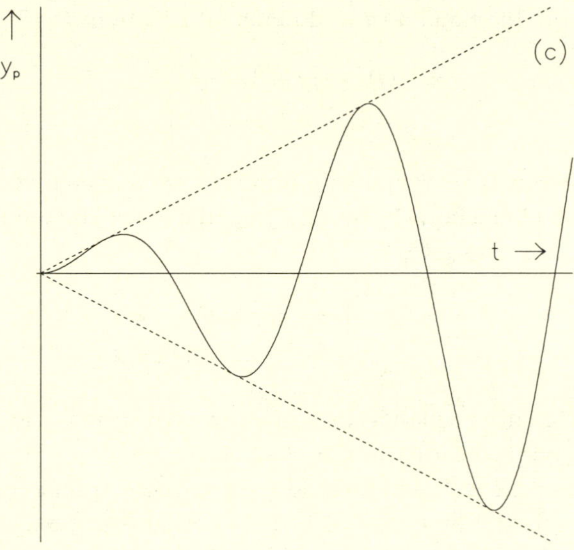

Fig. 5.8 Amplitude B of the response of the system as a function of the frequency of the external force (see eqn (5.137)), when there is no damping. This is the limit of fig. 5.7, as b tends to zero. For the case $\omega = \omega_0$ and $b = 0$, the complementary function $y_c = C \cos \omega t + D \sin \omega t$ is shown in (b) (for the choice $C = 0, D = 1$); and the particular integral y_p of (5.138) in (c). Again the particular integral dominates the complementary function at large times.

it is the complementary function that has constant amplitude, while the particular integral grows (see fig. 5.8).

The subject of resonance is discussed more fully in Sections 4.7 and 4.8, where several physical examples are provided.

5.15 Simultaneous differential equations

Some examples of physical situations giving rise to simultaneous differential equations were already provided in Section 5.1.6 (see also Problem 5.17). Here we consider problems where the differential equations have constant coefficients, and each has up to second order derivatives† in either or both of the dependent variables x and y. (See, for example, eqns (5.17).) The next three sections describe different approaches to solving these equations (i.e. for finding x and y as functions of t).

† The extension to higher order derivatives is straight-forward.

5.15.1 D operator method

We can rewrite the simultaneous differential equations (5.17) as

$$\left.\begin{aligned}(D-1)y+3x&=4\\ y+(D^2+7)x&=e^{-t}\end{aligned}\right\} \tag{5.17$'$}$$

Then we proceed as we would with ordinary simultaneous equations. We decide first to eliminate y, by 'multiplying' the lower equation by $(D-1)$ and subtracting. This gives

$$\begin{aligned}[(D-1)(D^2+7)-3]x&=(D-1)e^{-t}-4\\ &=-2e^{-t}-4\end{aligned} \tag{5.139}$$

This is a differential equation that involves only x and t, and hence we solve it as described earlier in this chapter.

We can then use a similar method to eliminate x from eqns (5.17′), by multiplying the first by (D^2+7), and the second by 3, and subtracting. This will enable us to find y as a function of t, and the problem is completely solved. Alternatively (and much more simply in this case), we can insert our solution for x as a function of t into the second of eqns (5.17′), and obtain y directly.

In this application of converting two simultaneous differential equations into two separate ordinary differential equations each involving only one dependent variable, we are merely using the D operator to tidy the notation, and to make it simpler to eliminate the relevant variable from the two simultaneous equations. We do not make use of the various rules and tricks introduced in Section 5.13.

Of course there is still the need to solve the separate differential equations, but we now no longer have to deal with the problem of having the simultaneous mixture of x and y and their derivatives in each equation.

5.15.2 Complex numbers approach

Consider the differential equations

$$\left.\begin{aligned}\frac{d^2x}{dt^2}+4\frac{dy}{dt}+3x&=e^{-t}\\ \text{and}\quad \frac{d^2y}{dt^2}-4\frac{dx}{dt}+3y&=t^2\end{aligned}\right\} \tag{5.140}$$

If we adopt the surprising idea of multiplying the second equation by i (the square root of -1) and adding, we obtain

$$\frac{d^2(x+iy)}{dt^2}+4\frac{d(y-ix)}{dt}+3(x+iy)=e^{-t}+it^2 \tag{5.141}$$

By defining

$$x+iy=z \tag{5.142}$$

where z is some complex† function of t, we can write (5.141) more neatly as

$$\frac{d^2z}{dt^2}-4i\frac{dz}{dt}+3z=e^{-t}+it^2 \tag{5.141′}$$

Thus we now have a single differential equation for z, which admittedly involves i. We can nevertheless solve (5.141′) using the standard methods for second order differential equations with constant coefficients, and obtain $z(t)$.

Since the functions $x(t)$ and $y(t)$ are both real, we can extract them from $z(t)$ as its real and imaginary parts respectively. In doing this, we must remember that the arbitrary constants in our complementary function can be complex.

Clearly this method works only when there are special relationships between the coefficients of the various terms in the two simultaneous differential equations.

5.15.3 *Use of linear combinations of* x *and* y

A particularly simple case is provided by the example

$$\left.\begin{aligned}\frac{d^2x}{dt^2}+4\frac{dx}{dt}-2\frac{d^2y}{dt}-8\frac{dy}{dt}&=t\\ 2\frac{d^2x}{dt^2}+\frac{d^2y}{dt^2}-4x-2y&=0\end{aligned}\right\} \tag{5.143}$$

All we have to do is to define new variables

$$\left.\begin{aligned}x-2y&=u\\ \text{and } 2x+y&=v,\end{aligned}\right\} \tag{5.144}$$

† In the sense of having a real and an imaginary part, rather than being complicated.

and the equations become

$$\left.\begin{aligned}\frac{d^2u}{dt^2}+4\frac{du}{dt}&=t\\ \frac{d^2v}{dt^2}-2v&=0\end{aligned}\right\}\tag{5.143'}$$

These separate differential equations are solved for u and v, and then the ordinary simultaneous equations (5.144) are inverted to yield x and y.

The above problem was so straight-forward because each of the original simultaneous equations (5.143) became one of the separate equations for u or v. Only a little more complicated is the case

$$a_j\frac{d^2x}{dt^2}+b_j\frac{d^2y}{dt^2}+c_j\frac{dx}{dt}+d_j\frac{dy}{dt}+g_jx+h_jy=f_j(t)\quad (j=1\text{ or }2)\tag{5.145}$$

The special situation relevant here is where the coefficients are such that we can multiply the second equation (i.e. $j=2$) by a suitable constant k and then add the two equations, to achieve a differential equation which involves only one linear combination of x and y, rather than the two variables separately.

We obtain

$$\frac{d^2}{dt^2}[(a_1+ka_2)x+(b_1+kb_2)y]+\frac{d}{dt}[(c_1+kc_2)x+(d_1+kd_2)y]$$
$$+[(g_1+kg_2)x+(h_1+kh_2)y]=f_1(t)+kf_2(t)$$

Then if all three terms in square brackets are proportional to each other, this will be a differential equation for the linear combination

$$W=[(a_1+ka_2)x+(b_1+kb_2)y]$$

which can be solved for W. For this to be so, we require

$$\frac{a_1+ka_2}{b_1+kb_2}=\frac{c_1+kc_2}{d_1+kd_2}=\frac{g_1+kg_2}{h_1+kh_2}\tag{5.146}$$

For the first part of this to be true,

$$(a_1+ka_2)(d_1+kd_2)=(b_1+kb_2)(c_1+kc_2)$$

which is a quadratic equation for k, yielding two solutions. Provided both of these values for k also satisfy the second equality of (5.146),† we have found two different linear combinations W_1 and W_2 satisfying separate uncoupled differential equations, which can be solved individually for the

† In general, g_j and h_j will be such that this is not so, in which case this method does not work.

specified linear combinations of x and y, and from which x and y can be extracted (see, for example, Problem 5.17).

The simplest example of this method is when the multiplicative factors k are ± 1, i.e. the two differential equations (5.145) are to be added, and subtracted, in order to produce the separated differential equations for W_1 and W_2. This may well be worth trying, before attempting the more general analysis.

The complex number method of Section 5.15.2 is a special case of the linear combinations approach, when the multiplicative factor k is imaginary.

5.15.4 Normal modes

We return to the simultaneous differential equations (5.145), but consider the special case of no damping, so that the terms involving dx/dt and dy/dt are absent. If we wish to determine the complementary functions, we have to solve

$$\left.\begin{aligned} a_1\frac{d^2x}{dt^2} + b_1\frac{d^2y}{dt^2} + g_1x + h_1y = 0 \\ \text{and } a_2\frac{d^2x}{dt^2} + b_2\frac{d^2y}{dt^2} + g_2x + h_2y = 0 \end{aligned}\right\} \qquad (5.147)$$

Since the complementary functions describe the free oscillations of the parts of the system (e.g. of two pendula, which are lightly coupled together by being on the same non-rigid support), we may be interested in seeing whether there are solutions for x and y in which the separate parts oscillate at the same frequency, i.e. we look for solutions of the form

$$\left.\begin{aligned} x = A\sin(\omega t + \alpha) \\ y = B\sin(\omega t + \beta) \end{aligned}\right\} \qquad (5.148)$$

This topic, called 'normal modes', is dealt with in detail in Chapter 13 of Volume 2.

5.16 Final comments

These are by no means all the methods that exist for solving differential equations. Here we merely mention four others.

A graphical technique can be used for a first order differential equation where dy/dx is expressed in terms of x and y, e.g.

$$\frac{dy}{dx} = 3.7 \ln y \sin(x/y^2) \tag{5.149}$$

At every point in the x–y plane†, dy/dx is defined, so we can draw very short lines of the relevant gradient at lots of points. By seeing how these can be joined up, we can obtain a visualisation of possible solutions.

Sometimes series solutions are possible, and even work for differential equations where the coefficients of the various derivatives are functions of the independent variable. Thus to solve

$$t^2\frac{d^2y}{dt^2} + t\frac{dy}{dt} + (t^2 - n^2)y = 0 \tag{5.150}$$

we try the series

$$y = t^\rho(a_0 + a_1 t + a_2 t^2 + \ldots) \tag{5.151}$$

where the exponent ρ and the coefficients a_i are to be determined. This is achieved by differentiating the series (5.151) for y, performing the various substitutions into the differential equation (5.150), and then equating the coefficients of each separate power of t to zero (This is to ensure that the differential equation is satisfied for all values of t.) Sometimes the resulting power series can be recognised in terms of well known functions. An example is given in Problem 5.19.

Well chosen changes of variables can often simplify differential equations. Thus the substitution

$$x = e^t$$

reduces the second order differential equation

$$x^2\frac{d^2y}{dx^2} + 3x\frac{dy}{dx} - 3y = 0 \tag{5.152}$$

to

$$\frac{d^2y}{dt^2} + 2\frac{dy}{dt} - 3y = 0 \tag{5.152$'$}$$

Our new second order equation has constant coefficients, and hence is readily solved.

The final method we mention here uses numerical approximations for the derivatives. Thus

$$\frac{dy}{dt} \sim [y(t + \Delta t) - y(t)]/\Delta t \tag{5.153}$$

† Except for $y = 0$.

Table 5.3. *Methods of Solving Differential Equations*

This table should contain a list of the various methods used for solving simple differential equations. It could then be used as a template, with which to compare any differential equation you have to solve.

Because producing such a summary is a useful educational exercise, the main part of the Table has been left blank for the reader to complete.

Type			Example	Method of solution
First order	Separable			
	Single function of x and y			
	Homogeneous			
	Homogeneous but for constant			
	Exact			
	Integrating factor			
	Bernoulli			
Second order, constant coefficients	C.F.	Distinct real roots		
		Identical roots		
		Complex roots		
	PI for $f(t) =$	e^{at}		
		$\sin bt$		
		Polynomial		
		Products of above		
		Contained in C.F.		

where Δt is a suitably small time interval. We could use this in, for example, eqn (5.149) in a sequential manner to obtain $y(t+\Delta t)$, and then $y(t+2\Delta t)$, $y(t+3\Delta t)$, etc. in terms of a starting value $y(t)$. In a similar manner we can rewrite a second order differential equation like (5.152) as

$$\left.\begin{aligned} \frac{dv}{dt} &= \frac{3y}{x^2} - \frac{3v}{x} \\ \text{where} \qquad v &= \frac{dy}{dt} \end{aligned}\right\} \tag{5.152''}$$

Then, given y and v at some starting time t, we can use eqn (5.153), and its analogue with y replaced by v, in order to calculate y and v at subsequent times $t+n\Delta t$, where $n = 1, 2, 3 \ldots$. Clearly a computer will make this procedure very much simpler. For such problems, the choice of a suitable step size Δt can be important if we want an accurate solution.

The message is thus that there are a host of more advanced methods for solving differential equations, and that there are many vastly more complicated equations than have been described here, many of which have no analytic solution in terms of simple functions. Even if you can now solve all the problems at the end of this chapter, it is a mistake to think that you have mastered the whole subject.

Nevertheless a typical examination syllabus is very limited in terms of the types of question examiners are allowed to set, so it is well worth while becoming very familiar with the techniques discussed earlier in this chapter (see also Table 5.3).

Problems

5.1 What is a linear differential equation? And what does homogeneous mean?

The differential equation

$$\frac{d^3y}{dt^3} + f(t)\frac{d^2y}{dt^2} + g(t)\frac{dy}{dt} + h(t)y = 0$$

has solutions $y_1(t), y_2(t)$ and $y_3(t)$. ($f(t), g(t)$ and $h(t)$ are arbitrary functions of t, but are independent of y.) Show that any linear combination of them is also a solution.

Investigate whether the corresponding property holds for the solutions $y_a(t)$ and $y_b(t)$ of

(i) $a\frac{d^2y}{dt^2} + 2b\frac{dy}{dt} + cy = 1$

(ii) $a\frac{d^2y}{dt^2} + 2by\frac{dy}{dt} + cy = 0$

5.2 Check that, when $A = 0$, the differential equation

$$y \sin y + (1 + xy \cos y + x \sin y)\frac{dy}{dx} = Ax^2$$

is exact, and hence solve it.

When $A \neq 0$, solve the differential equation (i) by integrating each side separately with respect to x; and (ii) by taking the Ax^2 term to the left-hand side, showing that the resulting equation is exact, and hence finding the solution.

5.3 Derive a function $f(x)$ such that when the differential equation

$$\frac{dy}{dx} + g(x)y = h(x) \qquad [1]$$

is multiplied by it, the left-hand side of the equation will be an exact derivative.

Show that the equation

$$\frac{dy}{dx} + k(x)y = l(x)y^m \qquad [2]$$

can be reduced to the form of eqn [1] by the substitution $z = y^\alpha$, provided that α is suitably chosen.

Solve the differential equation

$$\frac{dy}{dx} + 2y \tan x = y^2 \tan^2 x$$

5.4 Solve the following differential equations:

(i) $\dfrac{dy}{dx} = -\dfrac{2xy + 3x^2}{2 + x^2}$

(ii) $\dfrac{dy}{dt} = \dfrac{2t + y + 3}{4t + 2y}$

(iii) $\dfrac{dx}{dt} = \dfrac{2t + x + 3}{t + 2x}$

(iv) $z\dfrac{dy}{dz} = y + \sqrt{z^2 - y^2}$

(v) $\dfrac{dy}{dx} = e^{-x^2} - 2xy$

(vi) $\sin 2x\dfrac{dy}{dx} = y - 1/y$

5.5 Use the D operator rules of Table 5.2 to evaluate the following. Check your answers by explicit differentiation or integration.

(i) $\sum_n b_n D^n e^{ax}$

(ii) $\dfrac{1}{D^n} e^{ax}$

(iii) $\dfrac{1}{(D+k)^n}e^{ax}$

(To check your answer, first multiply both sides of your solution by the relevant function of D)

(iv) $D^2(x^n e^{ax})$

(v) $D^m(x^3 e^{ax})$

(Use Leibnitz's Theorem)

(vi) $\dfrac{1}{(D-a)^2}[xe^{ax}]$

(vii) $D^2 \cos^2 x$

(To apply the relevant rule, first express $\cos^2 x$ in terms of $\cos 2x$)

5.6 Verify that rule (a) for D operators applied to $D^n[e^{ax}]$ gives the same answer as rule (b) applied to $D^n[f(x)e^{ax}]$, where $f(x) = 1$.

5.7 The complementary function of the differential equation

$$(D+a)^3 y = \frac{d^3y}{dt^3} + 3a\frac{d^2y}{dt^2} + 3a^2\frac{dy}{dt} + a^3 y = te^{-at}$$

can be written as

$$y = e^{-at}g(t)$$

where $g(t)$ is a function to be determined. Use the rules for D operators (see Table 5.2) to determine $g(t)$.

Find also the particular integral.

5.8 Evaluate the following, where D stands for the operator $\frac{d}{dt}$:

(i) $\dfrac{1}{D-3}[2e^{bt}]$, where $b \neq 3$.

(ii) $\dfrac{1}{D-3}[2e^{bt}]$, where $b = 3$.

(iii) $\dfrac{1}{1+D}[t^3]$

(iv) $\dfrac{1}{D^2+D}[te^{-t}]$

(v) $\dfrac{1}{D^2+3D+2}[e^{-3t}]$

(vi) $\dfrac{1}{D^2+3D+2}[\sin x]$

5.9 The particular integral

$$y = \frac{1}{D+1}[t]$$

can be evaluated by using the binomial theorem to give

$$\begin{aligned} y &= (1+D)^{-1}t \\ &= (1 - D + D^2 - \ldots)t \\ &= t \end{aligned}$$

Alternatively, we could write

$$\begin{aligned} y &= D^{-1}(1+1/D)^{-1}t \\ &= \frac{1}{D}\left(1 - \frac{1}{D} + \frac{1}{D^2} - \ldots\right)t \\ &= \frac{t^2}{2!} - \frac{t^3}{3!} + \frac{t^4}{4!} - \ldots \end{aligned}$$

We thus appear to have two alternative answers. Comment.

(Try to resolve this paradox yourself. If you cannot, a hint is given later in these problems.)

5.10 (i) An ellipse may be defined as the locus of points whose distance from the origin is a constant $e(<1)$ times the distance from a fixed line of gradient $\tan\beta$ and which does not pass through the origin. Show that the equation of the ellipse can be written in polar coordinates as

$$l/r = 1 + e\cos(\theta - \theta_0)$$

where $\theta_0 = \beta - 90°$ and l is a constant.

(ii) A particle is at a position r and moves in a plane with variable angular velocity $d\vartheta/dt$ about the origin. Show that its acceleration has radial and tangential components of $d^2r/dt - r(d\theta/dt)^2$ and $rd^2\theta/dt^2 + 2(dr/dt)d\vartheta/dt$ respectively.

(iii) Show that for a central attractive force

$$\mathbf{F} = -\frac{a}{r^3}\mathbf{r}$$

the particle's angular momentum $mr^2d\vartheta/dt$ is constant. By writing $u = 1/r$ and deriving a differential equation for u in terms of θ, show that the particle may describe an elliptic orbit.

(iv) In a universe with n spatial dimensions, the gravitational force would presumably vary with distance r like $r^{-(n-1)}$. Rederive the differential equation relating θ and u in n dimensions.

It is highly desirable that planetary orbits around the sun are stable against small disturbances. Investigate whether this is so for circular orbits as follows. Consider an orbit with

$u = u_0 + \varepsilon(\theta)$, where u_0 is a constant and $\varepsilon(\theta)$ is a small perturbation. Derive an approximate differential equation for $\varepsilon(\theta)$, and comment on the form of the solutions for $n = 3$, $n = 4$, and $n = 5$.

5.11 Find the solutions of the differential equations below, which satisfy the initial conditions that at $t = 0$, $y = 0$ and $dy/dt = V$:

(i) $\dfrac{d^2y}{dt^2} + 25\dfrac{dy}{dt} + 100y = 0$

(ii) $\dfrac{d^2y}{dt^2} + 20\dfrac{dy}{dt} + 100y = 0$

(iii) $\dfrac{d^2y}{dt^2} + 16\dfrac{dy}{dt} + 100y = 0$

5.12 How many arbitrary constants do you expect there to be in the complementary function for the differential equation

$$\frac{d^4y}{dt^4} - 2\frac{d^2y}{dt^2} + y = 0$$

Show by direct substitution that the following satisfy the differential equation:

(i) $y = e^t$
(ii) $y = e^{-t}$
(iii) $y = \cosh t$
(iv) $y = \sinh t$

Do you expect that there are any other functional forms for the complementary function, apart from those listed above?

Write down the full complementary function, containing the correct number of arbitrary constants.

5.13 Solve the following differential equations:-

(i) $\dfrac{d^2y}{dt^2} + 9y = f(t)$, where
(a) $f(t) = \sin t$
(b) $f(t) = \sin 3t$

(ii) $\dfrac{d^2y}{dt^2} + 3\dfrac{dy}{dt} = g(t)$, where
(a) $g(t) = t$
(b) $g(t) = e^{-3t}$
(c) $g(t) = e^{-3t} \sin t$
(d) $g(t) = \sin 3t$
(e) $g(t) = t \sin 3t$

(iii) $\dfrac{d^2y}{dt^2} + 6\dfrac{dy}{dt} + 9y = h(t)$, where

(a) $h(t) = e^{-t}$

(b) $h(t) = e^{-3t}$

(c) $h(t) = te^{-3t}$

(iv) $\dfrac{d^4y}{dt^4} - 2\dfrac{d^3y}{dt^3} + 2\dfrac{dy}{dt} - y = e^t$

5.14 Verify that the particular integral of eqn (5.134) is given by (5.135), and that the amplitude of the forced oscillations is as stated in eqns (5.136). Show that the maximum value of the amplitude is obtained when $\omega^2 = \omega_{\max}^2 = c/a - 2(b/a)^2$.

5.15 A system is described by the differential equation

$$a\frac{d^2x}{dt^2} + 2b\frac{dx}{dt} + cx = A\cos\omega t$$

Show that, in the absence of an external force, the system may exhibit damped oscillations, provided b is smaller than some critical value b_0. Express b_0 in terms of the constants of the differential equation.

The value of b_0 is now set at $0.01b_0$. Calculate approximately how many oscillations (of the complementary function) take place before their amplitude is reduced by a factor of e, compared with that of the first maximum near $t = 0$.

For times that are large enough so that the complementary function can be neglected, find an expression for the velocity dx/dt of the system, in response to the external force $A\cos\omega t$. If ω is varied but A is kept fixed, what value ω_R of ω will result in the amplitude of the dx/dt response being a maximum?

The frequency is now changed from ω_R to $\omega_R + \delta\omega$, such that the amplitude of the dx/dt oscillations is smaller by a factor of $\sqrt{2}$, as compared with those at $\omega = \omega_R$. Show that $\delta\omega \sim \omega_R/100$. What would the approximate value of $\delta\omega/\omega_R$ have been if the reduction factor had been $\sqrt{5}$, rather than $\sqrt{2}$?

5.16 Calculate numerically the fractional difference between $\omega_{\max}$ of Problem 5.14 for the case where $b = 0.01b_0$; and ω_R of Problem 5.15. (These two frequencies correspond to the maximum amplitude for the oscillations, and for their speed respectively.) Compare this with the numerical value for the fractional width of the resonance $\delta\omega/\omega_R$ of Problem 5.15.

This should confirm that, for small damping, the differences in the frequencies corresponding to different definitions of the resonance are small compared with the width of the resonance response curve.

5.17 Solve the simultaneous differential equations

$$\left.\begin{aligned} \frac{d^2x}{dt^2} + 2\frac{dy}{dt} + 3x = \cos 2t \\ \frac{d^2y}{dt^2} - 2\frac{dx}{dt} + 3y = \sin 2t \end{aligned}\right\}$$

by the three methods of Sections 5.15.1, 5.15.2 and 5.15.3. Check that they give the same answers.

5.18 Use the substitution $x = e^z$ to reduce the differential equation

$$x^2\frac{d^2y}{dx^2} + 2x\frac{dy}{dx} - 2y = 0$$

to one relating y and z, and which is linear. Hence solve it.

5.19 Solve the differential equation

$$4x\frac{d^2y}{dx^2} + 2\frac{dy}{dx} + y = 0$$

by looking for a series solution

$$y = x^\alpha[a_0 + a_1x + a_2x^2 + \ldots + a_nx^n + \ldots]$$

as follows:

(i) Differentiate the series term by term, and substitute into the differential equation.

(ii) Collect terms of the same power in x, and equate them to zero (because we want the differential equation to be true for all values of x). Do this explicitly for the term of lowest power in x. This should provide a quadratic equation for α, with solutions $\alpha = 0$ or $1/2$.

(iii) Next set the net coefficient of the term in $x^{m+\alpha}$ to zero. This should give the relationship

$$\frac{a_{m+1}}{a_m} = -\frac{1}{(2\alpha + 2m + 1)(2\alpha + 2m + 2)}$$

between consecutive coefficients in the series. Thus if we choose a_0 as A when $\alpha = 0$, and as B when $\alpha = \frac{1}{2}$, in each case all subsequent terms in the series are exactly specified.

(iv) Write out the two series (for $\alpha = 0$ and for $\alpha = \frac{1}{2}$) explicitly. Try to recognise them as series expansions of well known functions.

Now solve the differential equation by an entirely different method. Use the substitution

$$z = \sqrt{x}$$

to obtain a differential equation relating y and z. Solve it, and resubstitute for x. Compare your answer with that from the series method.

(The second method is quicker but requires us to find the correct substitution. The series method requires no special assumptions.)

Hint for Problem 5.9 : Reconstruct the differential equation for which we are trying to find the particular integral. What is its complementary function?

6
Partial derivatives

The final draft of a mathematics book contained the sentence: '$\partial f/\partial x$ means the ratio at constant y of δf and δx, where δf and δx are vanishingly small.' When the author received the publisher's proofs, this appeared as '$\partial f/\partial x$ means the ratio at constant y of . and ..' On closer examination with a powerful magnifying glass, however, it turned out that the first two full stops were in fact the smallest δf and δx that the publisher was able to produce.

Many branches of science involve partial derivatives. The aim of this chapter is to make you understand what they are, and become so fluent at manipulating them that this sort of operation becomes as familiar and as accepted as the arithmetic operations with ordinary numbers. This will then enable you to concentrate on the basic principles of your science problem, rather than battling with the mathematics of the partial derivatives involved.

6.1 Introduction

We are often interested in calculating the derivatives df/dx, d^2f/dx^2, etc for a function $f(x)$ of a single variable x. Similarly, for functions of more than one variable $f(x, y, \ldots)$, we may well also want the derivatives. These are written as, for example, $\partial f/\partial x$, which means 'the rate of change of the function f with respect to small changes in x, assuming that all the other independent variables are kept constant'. That is, $\partial f/\partial x$ is the limit as δx tends to zero of $\delta f/\delta x$, where δf is the difference in the function f for a small change δx in just the x variable, i.e.

$$\delta f = f(x + \delta x, y, \ldots) - f(x, y, \ldots)$$

Apart from the restriction that all the other variables must be kept constant, this is very much the same sort of way that derivatives of a single variable are defined. The curly ∂s in $\partial f/\partial x$ are used to denote the partial derivative. Unless it is completely obvious from the context, we should really write the partial derivative as $(\partial f/\partial x)_{y,\ldots}$, to denote that $y,\ldots$ are being kept constant.

Thus, for example, for

$$f = x^2 \sin x \tag{6.1}$$

$$\left.\begin{aligned} \frac{df}{dx} &= x^2 \cos x + 2x \sin x \\ \text{and} \quad \frac{d^2 f}{dx^2} &= -x^2 \sin x + 4x \cos x + 2 \sin x \end{aligned}\right\} \tag{6.2}$$

In contrast, for

$$f = x^2 \sin y \tag{6.3}$$

$$\frac{\partial f}{\partial x} = \left(\frac{\partial f}{\partial x}\right)_y = 2x \sin y \tag{6.4}$$

because $\sin y$ is simply a constant if y is held fixed. Similarly

$$\frac{\partial f}{\partial y} = \left(\frac{\partial f}{\partial y}\right)_x = x^2 \cos y \tag{6.5}$$

Because $\partial f/\partial x$ and $\partial f/\partial y$ are both functions of x and y, we can differentiate them with respect to x or y. Thus

$$\left.\begin{aligned} \frac{\partial}{\partial x}\left(\frac{\partial f}{\partial x}\right) &= \frac{\partial^2 f}{\partial x^2} = 2 \sin y \\ \frac{\partial}{\partial y}\left(\frac{\partial f}{\partial x}\right) &= \frac{\partial^2 f}{\partial y \partial x} = 2x \cos y \\ \frac{\partial}{\partial y}\left(\frac{\partial f}{\partial y}\right) &= \frac{\partial^2 f}{\partial y^2} = -x^2 \sin y \\ \text{and} \quad \frac{\partial}{\partial x}\left(\frac{\partial f}{\partial y}\right) &= \frac{\partial^2 f}{\partial x \partial y} = 2x \cos y \end{aligned}\right\} \tag{6.6}$$

In eqns (6.6) we have written the second derivatives in their more compact conventional form, just as $(d/dx)(df/dx)$ in a one-dimensional problem is written as d^2f/dx^2. Specifying the variables that are kept constant at each stage of the partial differentiation is cumbersome; they are usually taken to be understood.

One feature we immediately see from eqns (6.6) is that $\partial^2 f/\partial y \partial x = \partial^2 f/\partial x \partial y$. The fact that it does not matter in which order we perform the differentiations is a general property; we derive this for second derivatives in eqn (7.45).

Another point to realise is that the mechanism of performing partial differentiation for several variables is as straight-forward as ordinary differentiation with only one independent variable. Any difficulties arise in understanding what it is we are doing, and in appreciating the richer variety of results. It is to these that we now turn our attention.

6.2 What does it mean?

Just in case you have gained the impression that all this is too simple, we now pose two paradoxes for you to resolve.

Paradox 1

The second derivative $\partial^2 f/\partial x \partial y$ is actually $(\partial/\partial x)[(\partial f/\partial y)_x]_y$. That is, in the first stage denoted by the inner brackets, we vary y at constant x, while in the second we vary x while keeping y fixed. Given that x was set to a constant in the first stage, how can we then differentiate with respect to it in the second?

Paradox 2

Consider the functions

$$\left.\begin{aligned} f_1 &= x + 2y \\ \text{and} \qquad f_2 &= u + y \end{aligned}\right\} \tag{6.7}$$

These are identical to each other if

$$u = x + y \tag{6.8}$$

But

$$\frac{\partial f_1}{\partial y} = 2 \tag{6.9}$$

and this is not equal to

$$\frac{\partial f_2}{\partial y} = 1 \tag{6.9$'$}$$

How can two identical functions have different derivatives?

Clearly you will benefit from solving these problems yourself, so think about them before reading on.

Those of you who tried reading on immediately to find out the answers without attempting to discover them yourselves will be disappointed because we next have a short diversion, to explain in a slightly different way what partial derivatives mean. So please go back and think about the above paradoxes now.

The price of a second-hand car depends on several factors. These include the original price, the car's age, the number of owners, its mileage, whether or not its general condition is average for its age, and the degree of opportunism of the seller. Thus we can write

$$p = p(v_1, v_2, v_3, v_4, v_5, v_6) \tag{6.10}$$

where p is the price of the second-hand car and v_1–v_6 are the six variables mentioned above (and where we have assigned some numerical scale to general condition and to opportunism). Then $\partial p/\partial v_2$ gives the rate of change of the price with respect to its age, assuming that all the other variables are kept constant. Except for vintage cars, it will be negative, and it could well depend on all six variables. Thus the older the car is, the smaller its annual depreciation, while expensive cars lose more value in absolute terms each year then do cheaper cars.

For example, if a car which originally cost £10 000 depreciated in value from £6000 to £5000 between years two and three, then $\partial p/\partial v_2 = -£1000/\text{year}$ after $\sim 2\frac{1}{2}$ years. On the other hand, if the car had many extra features and its price when new was £12 000, its second-hand value may be £6900 or £5700 after two or three years respectively. In that case

$$\frac{\partial p}{\partial v_2} = -£1200/\text{year} \tag{6.11}$$

Now if we compare these two cases, we see that the difference in $\partial p/\partial v_2$, the rate at which the second-hand value changes with age, is $-£200/\text{year}$, for a difference in initial value v_1 of £2000. Thus

$$\frac{\partial}{\partial v_1}\left(\frac{\partial p}{\partial v_2}\right) \equiv \frac{\partial^2 p}{\partial v_1 \partial v_2} = \frac{-£200/\text{year}}{£2000} \tag{6.12}$$

(These numbers are set out in Table 6.1.) Of course, we are assuming that at each stage of the comparison, all other relevant variables are kept constant. These are v_1 and v_3–v_6 in the first stage; and v_2–v_6 in the second.

An alternative analysis of the data of Table 6.1 is that for two-year-old cars of these types, $\partial p/\partial v_1$ is +0.45, while for three-year-old ones it is +0.35. (The latter number means that, after three years, the more expensive car maintains only 35% of its extra initial cost.) Comparison of these two values shows that the change in $\partial p/\partial v_1$, the rate of change

Table 6.1. *Price p of second-hand cars*

$$p = p(v_1, v_2, v_3, v_4, v_5, v_6)$$

v_1 original price; v_2 age; v_3 number of owners; v_4 mileage; v_5 condition compared with average for age; v_6 degree of opportunism of seller.

	Car 1	Car 2	$\frac{\partial p}{\partial v_1}$	$\frac{\partial^2 p}{\partial v_2 \partial v_1}$
Original price (v_1)	£10000	£12000		
Price at age $v_2 = 2$ years	£6000	£6900	+0.45	$\frac{-0.1}{\text{year}}$
Price at age $v_2 = 3$ years	£5000	£5700	+0.35	
$\frac{\partial p}{\partial v_2}$	−£1000/year	−£1200/year		
$\frac{\partial^2 p}{\partial v_1 \partial v_2}$	$\frac{-£200/\text{year}}{£2000}$			

(We assume that all variables other than the one explicitly varied are kept constant, e.g. the condition of all cars is average for their age.)

The table provides an example of the fact that it is irrelevant in which order the derivatives of $\partial^2 p/\partial v_1 \partial v_2$ are performed.

of second-hand price with respect to the original price, is −0.1 between years 2 and 3, i.e.

$$\frac{\partial}{\partial v_2}\left(\frac{\partial p}{\partial v_1}\right) \equiv \frac{\partial^2 p}{\partial v_2 \partial v_1} = -0.1/\text{year} \tag{6.13}$$

Once again we see that the derivatives $\partial^2 p/\partial v_1 \partial v_2$ and $\partial^2 p/\partial v_2 \partial v_1$ (eqns (6.12) and (6.13)) are equal.

We now return to the paradoxes at the beginning of this section. We first asked how it could be meaningful to differentiate $(\partial f/\partial y)_x$ with respect to x. In fact our discussion about car prices turns out to be relevant after all. There we calculated $\partial p/\partial v_2$, the rate of change of second-hand price with respect to the age of the car. This was done assuming that we kept constant all the other variables, including v_1, the price of a new car. However, we could, and indeed did, calculate $\partial p/\partial v_2$ for more than one constant value of v_1; we used £10 000 and £12 000. We found that $\partial p/\partial v_2$ at constant v_1 depended on the v_1 value, and hence it is meaningful to ask what is the rate of change of $(\partial p/\partial v_2)_{v_1}$ with respect to v_1; this is $\partial^2 p/\partial v_1 \partial v_2$.

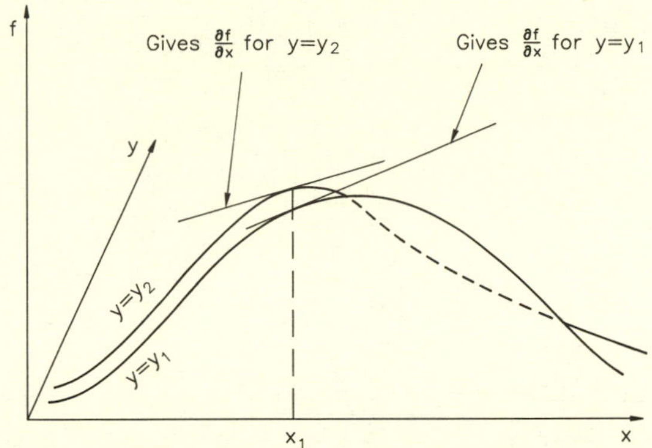

Fig. 6.1 Demonstration of the meaning of $(\partial/\partial y)[(\partial f/\partial x)_y]$. Two vertical slices through a hill are shown; both are parallel to the x axis. The height of the hill is given by a function $f(x, y)$. The first slice is at $y = y_1$, while the second is at $y = y_2 = y_1 + \delta y$. The gradient of the hill as we change x is $(\partial f/\partial x)_y$. At x_1 the gradients for the two separate slices are not quite equal. Then $(\partial/\partial y)[(\partial f/\partial x)_y]$ tells us how quickly the gradient in the x direction changes as we alter the slice. In the example shown, this second derivative at (x_1, y_1) is negative, since $\partial f/\partial x$ decreases as y increases.

As a further example, we can think of a function $h(x, y)$ giving the height h of a hill in terms of coordinates x and y. Then $(\partial h/\partial x)_y$ gives the gradient of the hill in the x direction as we keep y fixed. That is, we imagine slicing downwards through the hill with a cut parallel to the x axis at $y = y_1$, looking at the profile, and measuring the gradient $(\partial h/\partial x)_y$ (see fig. 6.1). This gradient will in general depend on x, and if we take a second cut at $y = y_2$, we will be able to see whether it depends on y as well. If we choose y_2 close to y_1, then

$$\frac{\left(\dfrac{\partial h}{\partial x}\right)_{y=y_2} - \left(\dfrac{\partial h}{\partial x}\right)_{y=y_1}}{y_2 - y_1} \tag{6.14}$$

will give us a good approximation to $[(\partial/\partial y)(\partial h/\partial x)_y]_x$, provided that the hill is smooth enough. Stated in a more formal way, $(\partial h/\partial x)_y$ is simply a function $g(x, y)$ of x and y, which we can differentiate partially with respect to x or with respect to y. In the latter case, we obtain $\partial g/\partial y$, which is $\partial^2 h/\partial y \partial x$.

In writing the second derivative as $\partial^2 h/\partial y \partial x$, we have suppressed the subscripts which tell us which variable(s) remain constant at each stage of the differentiation. This is reasonable provided there is no ambiguity.

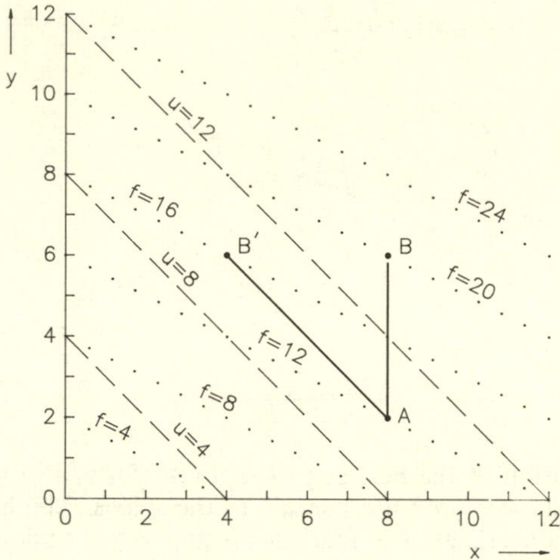

Fig. 6.2 Comparison of the derivatives $(\partial f/\partial y)_x$ and $(\partial f/\partial y)_u$, for $f = x + 2y$ and $u = x + y$. The dotted lines are contours of f, while the dashed lines show constant u. The derivative $(\partial f/\partial y)_x$ can be calculated from the differences in f and in y for the points B and A; for $(\partial f/\partial y)_u$, we use B′ and A. The y differences are the same, but f is larger at B than at B′, so $(\partial f/\partial y)_x$ is larger than $(\partial f/\partial y)_u$. (Since f is a linear function of x and y, we do not have to restrict ourselves to small changes in order to evaluate the derivatives.)

Failure to be explicit about the 'constant variables' can, however, cause problems. This was the source of the problem in paradox 2 earlier.

Thus $\partial f_1/\partial y$ of eqn (6.9) implicitly assumes that it is x (the other variable in the definition of f_1) which is kept constant, while for $\partial f_2/\partial y$ of (6.9′), y is varied at fixed u. Thus

$$\left(\frac{\partial f_1}{\partial y}\right)_x = 2 \quad \text{and} \quad \left(\frac{\partial f_2}{\partial y}\right)_u = 1 \tag{6.15}$$

Even though f_1 and f_2 are identical, the fact that different variables are constant in the two cases explains why the two partial derivatives are not equal.

The above is perfectly correct, but a less formal explanation provides more insight. In fig. 6.2, we show contour lines in the x–y plane of the function f_1 of eqn (6.7). To visualise the meaning of the derivative $(\partial f_1/\partial y)_x$, we imagine starting at some point A, and moving to another point B *which is at the same x value.* For the example shown

$$\left(\frac{\partial f_1}{\partial y}\right)_x = \frac{f_B - f_A}{y_B - y_A} = \frac{20 - 12}{6 - 2} = 2 \tag{6.16}$$

Now since, for the definition of u in eqn (6.8), f_2 is identical to f_1, the same contour lines in the x–y plane apply to f_2. The derivative $(\partial f_2/\partial y)_u$, however, requires us to move from A to B′, *which is at the same u value* (see fig. 6.2). Since $u = x+y$, B' must be on a line of gradient -1 through A. Then

$$\left(\frac{\partial f_2}{\partial y}\right)_u = \frac{f_{B'} - f_A}{y_{B'} - y_A} = \frac{16-12}{6-2} = 1 \qquad (6.17)$$

Since the points B′ and B are physically different, we are not at all surprised that $f_{B'}$ and f_B are not equal, and hence $(\partial f/\partial y)_x$ and $(\partial f/\partial y)_u$ can differ.

Thus we can think of $(\partial f/\partial y)_x$ and $(\partial f/\partial y)_u$ as giving us the gradients of the hill whose contours are shown in fig. 6.2, but in different directions, the former being for the direction AB (constant x) and the latter for AB′ (constant u). Anyone who has ever stood on a hillside knows that the steepness of a climb will depend on the direction in which you choose to walk.

An aside about the above discussion is that, because f_1 and f_2 are linear in their variables, their first derivatives are constants (i.e. independent of x and y, or equivalently of u and y). Thus in this particular case, we would obtain the identical values for the derivatives for any choice of the point A. It also follows that the points B and B′ can be any distances from A along the lines of constant x and constant u respectively. In general, to evaluate derivatives numerically, the second point must be close to the first one. Otherwise eqn (6.16) would be only approximate, corresponding to the first term in the Taylor series for $f_B - f_A$ (compare Chapter 7).

Yet another way of resolving this paradox is to use the result

$$\left(\frac{\partial f}{\partial y}\right)_u = \left(\frac{\partial f}{\partial y}\right)_x + \left(\frac{\partial f}{\partial x}\right)_y \left(\frac{\partial x}{\partial y}\right)_u \qquad (6.29')$$

which we derive later in Section 6.3.3. Now from eqn (6.7) we obtain $(\partial f/\partial x)_y = 1$, while (6.8) gives $(\partial x/\partial y)_u = -1$. Thus

$$\left(\frac{\partial f}{\partial y}\right)_u = \left(\frac{\partial f}{\partial y}\right)_x - 1 \qquad (6.18)$$

which agrees satisfactorily with (6.9) and (6.9′).

Although this explanation seems rather formal, in essence the derivation of (6.29′) is equivalent to staring hard at fig. 6.2 and thinking a bit, which was what we did in our second approach.

In the language of second-hand car prices, we could consider $(\partial p/\partial v_2)_{v_4}$, which is the rate of change of price with age at constant mileage (and all the other variables except v_2 too); or alternatively $(\partial p/\partial v_2)_{v'_4}$, where v'_4 is a new variable which is simply the average yearly mileage (i.e. $v'_4 = v_4/v_2$), and is kept constant. Thus $(\partial p/\partial v_2)_{v'_4}$ assumes that in the ageing process the car drove an extra $\sim$10 000 miles per year. On the other hand $(\partial p/\partial v_2)_{v_4}$ gives the rate of depreciation, assuming the car did not drive any significant distance during the period δv_2. Because wear and tear affect the value of a car, $(\partial p/\partial v_2)_{v_4}$ is presumably less negative than $(\partial p/\partial v_2)_{v'_4}$.

A more physical example is provided by the specific heats of gases. Thus c_v, the specific heat at constant volume, and c_p, the specific heat at constant pressure, both involve the ratio of the heat supplied to unit mass of gas and the resultant temperature rise, but the conditions under which the gas is heated are different in the two cases. Since gases expand when heated at constant pressure and work has to be done against the atmosphere, c_p is larger than c_v. This can be written as

$$\left(\frac{\partial S}{\partial T}\right)_p > \left(\frac{\partial S}{\partial T}\right)_v \tag{6.19}$$

where S is in fact the entropy of the gas (although this is irrelevant for appreciating that the partial derivatives with different quantities held constant differ from each other).

6.3 Relationships among partial derivatives

Partial derivatives are much used in theoretical aspects of physical sciences. For example, thermodynamics sometimes seems to involve more manipulation of partial derivatives than consideration of the basic physics. Now that we understand what partial derivatives mean, we next need to become confident about the rules and relations between them. For example, under what circumstances is

$$\frac{\partial z}{\partial x}\frac{\partial x}{\partial y} = \frac{\partial z}{\partial y} \tag{6.20}$$

and when instead is

$$\frac{\partial z}{\partial x}\frac{\partial x}{\partial y} = -\frac{\partial z}{\partial y}\ ? \tag{6.21}$$

Clearly signs are important for both mathematics and for science, and so we need to know which of (6.20) or (6.21) is applicable in any given problem. This will be made clear later in this section.

Since there are lots of formulae that occur in this subject, and since many of them have a rather similar appearance, it is very easy to become confused if you try to remember them all by heart. For example, it is not particularly easy to recall whether it is $(\partial f/\partial y)_u$ or $(\partial f/\partial y)_x$ which occurs on the right-hand side of eqn (6.29′). Our strategy instead is to provide fast derivations of the useful formulae, so that there is no need to remember them.

The starting point for all these derivations is the relationship

$$\delta f \sim \left(\frac{\partial f}{\partial x}\right)_y \delta x + \left(\frac{\partial f}{\partial y}\right)_x \delta y \tag{6.22}$$

which gives the small change in the function $f(x, y)$ of the two variables x and y, when each of them undergoes a small change. We can regard eqn (6.22) as the first order term of the Taylor series expansion for $f(x, y)$ – this is discussed extensively in Section 7.5.

Alternatively (6.22) follows almost immediately from the definition of the partial derivatives. Thus if only x changes by an infinitesimal amount, the corresponding change in f is given by the first term on the right-hand side of eqn (6.22); similarly the second term corresponds to a change in y only. Thus if both x and y are free to vary, the net change in f is the sum of these two terms, as given in (6.22).

Eqn (6.22) is just about the only formula we need to remember; all else follows from it.

Our basic procedure for deriving a useful formula from eqn (6.22) consists in first dividing it by some small quantity (e.g. δx, δy, $\delta t \ldots$), and then going to the limit where all the small quantities go to zero in some specified way (e.g. u constant, f constant, everything varying, etc). This will become clearer with the specific examples which follow immediately.

6.3.1 $\dfrac{df}{dt} = \dfrac{\partial f}{\partial x}\dfrac{dx}{dt} + \dfrac{\partial f}{\partial y}\dfrac{dy}{dt}$

If we assume that x and y in eqn (6.22) are each separate functions of t, we can divide by δt to obtain

$$\frac{\delta f}{\delta t} = \left(\frac{\partial f}{\partial x}\right)_y \frac{\delta x}{\delta t} + \left(\frac{\partial f}{\partial y}\right)_x \frac{\delta y}{\delta t} \tag{6.23}$$

When we go to the limit where δf, and hence δx and δy, and consequently δt, go to zero, then the ratios $\delta x/\delta t$ and $\delta y/\delta t$ become dx/dt and dy/dt respectively. Note that these are ordinary derivatives, rather than partial

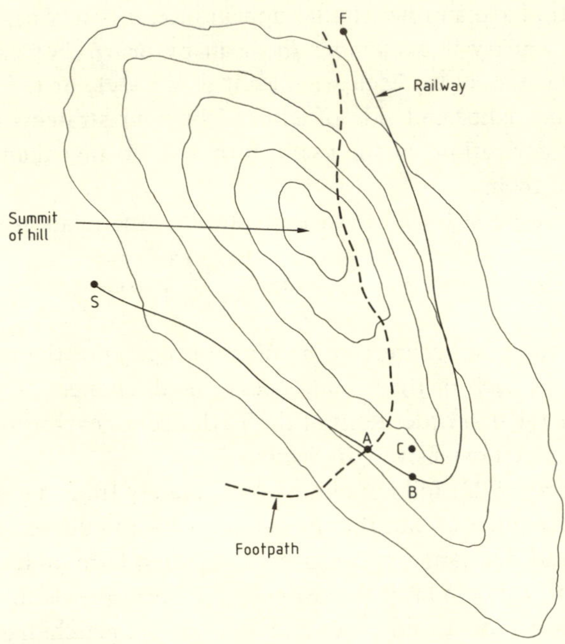

Fig. 6.3 A train travels along the track SAF over a hill, whose contours are shown. AB is a short segment of the track. C is a point on the hill (but not on the track) whose x value is a bit larger than that at A, and whose y value equals A's. Similarly the x values of C and B are the same.

A footpath, defined by $u(x, y) = 0$ and which passes through A, is also shown.

ones, since x depends only on the single variable t (and similarly for y). Although f is a function of the two variables x and y, it can as well be expressed in terms of t alone (because each of x and y can be written in terms of t), and hence $\delta f/\delta t$ also tends to the ordinary derivative df/dt. Thus our formula (6.23) becomes

$$\frac{df}{dt} = \frac{\partial f}{\partial x}\frac{dx}{dt} + \frac{\partial f}{\partial y}\frac{dy}{dt} \tag{6.24}$$

We can understand what this means by considering a specific example. We imagine a train moving along a railway line which winds its way over a hill (see fig. 6.3). Here f gives the height of the hill as a function of x and y, and df/dt is the rate of change of height with time as the train moves along the track. According to eqn (6.24), this is made up of two components. The first is the product of $\partial f/\partial x$, the rate of change of height with x when y is kept constant (i.e. when we move to the right along AC in fig. 6.3), and dx/dt, which is the x component of the velocity of the train as it moves along the track. The second term is a similar one with x replaced by y. The rate of change of height df/dt is a total derivative, not a partial one (for example, at constant x or y), since it indeed tells us how the height varies with time as the train moves along the track, for which x and y (and of course t) are varying. Similarly, dx/dt refers to the way x changes with t along the actual path taken, with y varying as well. On the other hand $\partial f/\partial x$ is a partial derivative, since we are expressing the change in height as a sum of two terms, as if someone had got off the train and run along the path ACB rather than remaining on the train as it travelled along the track from A to B.

Formula (6.24) is almost obvious when we think of the man running along ACB to rejoin the train at B.

If we attempt to write down eqn (6.24) from memory, we could make the mistake of writing down only the first product on the right-hand side, in some sort of analogy with the one-dimensional problem. This is clearly incorrect since the problem is symmetric in x and y. Furthermore since $\partial f/\partial x$ refers to the rate of change of height with x at constant y, the first product corresponds only to the part AC in fig. 6.3. Since in this diagram C is higher than B, the answer would be too big. The second product, applied to the path CB, is negative since near A, $\partial f/\partial y$ is positive but dy/dt is negative; this reduces the answer to the correct value.

As is apparent from the above discussion, eqn (6.24) applies at any particular point. If we are going to approximate the derivatives by the ratio of small changes over finite paths like AC, AB or CB in fig. 6.3, we must ensure that they are small enough so that the approximation is reasonable.

The summary of the above is that we obtained eqn (6.24) by starting with eqn (6.22), dividing by δt, and going to the suitable limit. This derivation was straight-forward and fast; what does take some time, at least initially, is to understand what the formula means, and to what situations it applies.

6.3.2 $\dfrac{df}{dx} = \dfrac{\partial f}{\partial x} + \dfrac{\partial f}{\partial y}\dfrac{dy}{dx}$

Now we will derive a slightly different formula. Rather than use the same procedure of commencing with eqn (6.22), we will instead make a short cut of starting with eqn (6.24), and simply changing the variable t to x. (The reader is advised to go back to (6.22), and to derive our new formula from there.) We obtain

$$\frac{df}{dx} = \left(\frac{\partial f}{\partial x}\right)_y + \left(\frac{\partial f}{\partial y}\right)_x \frac{dy}{dx} \tag{6.25}$$

This is a formula which relates the total derivative df/dx to the partial one $\partial f/\partial x$. Again the question is to find a situation to which eqn (6.25) applies.

Fig. 6.3 turns out to be useful here too. As the train moves from A to B, df/dx gives the rate of change of height with respect to small changes in x, *for the actual path of the train.* This contrasts with $\partial f/\partial x$, which corresponds as before to the theoretical path AC *at constant* y. Clearly the two are different, and the final term in (6.25) allows for this. The factor dy/dx is just the gradient of the train's track at A, and is negative; this makes df/dx smaller than $\partial f/\partial x$ in our problem.

Another example of eqn (6.25) can be obtained by changing the variable x to t. Then

$$\begin{aligned}\frac{df}{dt} &= \frac{\partial f}{\partial t} + \frac{\partial f}{\partial y}\frac{dy}{dt} \\ &= \frac{\partial f}{\partial t} + v\frac{\partial f}{\partial y}\end{aligned} \tag{6.26}$$

where we have written the speed v for dy/dt. Eqn (6.26) applies to problems with just one space dimension (y) and relates the total rate of change of some property f to it partial rate of change. For example, we can imagine that f is the temperature, and that we set off in the morning and drive southwards. The total derivative df/dt gives the rate of change of temperature that we experience as we drive. In contrast $\partial f/\partial t$ is the rate of temperature change at the place we started at. These, in general, will be different since the temperature may well depend on where we are. As we drive south, we could be moving into a region of higher temperatures. The term $v\ \partial f/\partial y$ gives us the contribution from this effect to our total derivative df/dt.

6.3.3 $\left(\frac{\partial f}{\partial x}\right)_u = \left(\frac{\partial f}{\partial x}\right)_y + \left(\frac{\partial f}{\partial y}\right)_x \left(\frac{\partial y}{\partial x}\right)_u$

If we return to our starting point (6.22), we can divide by δx to obtain

$$\frac{\delta f}{\delta x} = \left(\frac{\partial f}{\partial x}\right)_y + \left(\frac{\partial f}{\partial y}\right)_x \frac{\delta y}{\delta x} \tag{6.27}$$

This is exactly how we would have started our previous example. However, instead of going to the limit where we consider total changes (i.e. df/dx and dy/dx), here we assume that the change in x is performed such that some other variable u is kept constant. Then

$$\left.\begin{aligned} \frac{\delta f}{\delta x} &\to \left(\frac{\partial f}{\partial x}\right)_u \\ \text{and } \frac{\delta y}{\delta x} &\to \left(\frac{\partial y}{\partial x}\right)_u \end{aligned}\right\} \tag{6.28}$$

so that eqn (6.27) becomes

$$\left(\frac{\partial f}{\partial x}\right)_u = \left(\frac{\partial f}{\partial x}\right)_y + \left(\frac{\partial f}{\partial y}\right)_x \left(\frac{\partial y}{\partial x}\right)_u \tag{6.29}$$

This formula relates the partial derivatives giving the rate of change of f with respect to x, but with different variables held constant.

We can, of course, obtain an equivalent formula by interchanging x and y in eqn (6.29). This yields

$$\left(\frac{\partial f}{\partial y}\right)_u = \left(\frac{\partial f}{\partial y}\right)_x + \left(\frac{\partial f}{\partial x}\right)_y \left(\frac{\partial x}{\partial y}\right)_u \tag{6.29'}$$

which we used in Section 6.2 to resolve paradox 2.

For our contour line example of fig. 6.3, we can define some function $u(x, y) = 0$, which specifies a footpath passing through A. As we walk along this path, $(\partial f/\partial x)_u$ gives the change in height with respect to x at constant u (i.e. as we stay on the footpath). Very much as in the previous example, this is given in terms of $(\partial f/\partial x)_y$ – the corresponding rate of change had we stayed at constant y – plus an extra term which allows for the fact that the footpath is not at constant y. For the example shown, at A $(\partial f/\partial x)_u$ is larger than $(\partial f/\partial x)_y$.

6.3.4 Changes of variable

Yet another valid equation can be obtained by our short cut method of starting with eqn (6.24), and changing the 'd's of the total derivatives to partial '∂'s. Then

$$\frac{\partial f}{\partial t} = \frac{\partial f}{\partial x}\frac{\partial x}{\partial t} + \frac{\partial f}{\partial y}\frac{\partial y}{\partial t} \tag{6.30}$$

The question once again is 'For what situation is this relevant?'

As we have mentioned before, we can omit the subscript on a partial derivative that specifies which variable is constant provided this is obvious. Is this so for eqn (6.30)?

The derivatives $\partial f/\partial x$ and $\partial f/\partial y$, given the fact that they appear here by virtue of eqn (6.22), are clearly to be evaluated at constant y and constant x respectively. But what about the partial derivatives with respect to t?

The answer is that we are free to choose what we want, depending on what limiting process we select as δt tends to zero. However, all three partial derivatives must then have the same variable constant. There are two useful choices. The second we shall come back to later in Section 6.3.5. The one we use here is to assume that some new variable u is constant. Then

$$\left(\frac{\partial f}{\partial t}\right)_u = \left(\frac{\partial f}{\partial x}\right)_y \left(\frac{\partial x}{\partial t}\right)_u + \left(\frac{\partial f}{\partial y}\right)_x \left(\frac{\partial y}{\partial t}\right)_u \tag{6.30$'$}$$

From the fact that we are writing $(\partial x/\partial t)_u$, it is clear that x is a function of both u and t (and similarly for y), while f on the right-hand side of the equation is as usual a function of x and y. In contrast on the left-hand side, f is expressed in terms of the variables u and t.

This equation thus applies to the situation where

$$f(t,u) = f[x(t,u), y(t,u)] \tag{6.31}$$

That is, f was originally expressed as a function of x and y, and we then transform them to other variables t and u, in order to obtain f in terms of these new variables. Then eqn (6.30′) gives the derivative of f with respect to one of the new variables in terms of its derivatives with respect to the old ones and in terms of the derivatives of the transformation equations $x = x(t,u)$ and $y = y(t,u)$.

An example will make this clearer. Our function f is

$$f = x^2 + y^2 \tag{6.32}$$

with the consequence that

$$\left.\begin{aligned} \frac{\partial f}{\partial x} &= 2x \\ \text{and} \quad \frac{\partial f}{\partial y} &= 2y \end{aligned}\right\} \tag{6.33}$$

We then transform to polar coordinates r and θ, in terms of which x and

y are given by

$$\left.\begin{aligned} x &= r\cos\theta \\ y &= r\sin\theta \end{aligned}\right\} \tag{6.34}$$

We want $(\partial f/\partial r)_\theta$. In fact we know that

$$f = r^2 \tag{6.32$'$}$$

and so

$$\left(\frac{\partial f}{\partial r}\right)_\theta = 2r \tag{6.35}$$

Our formula (6.30′) for this problem becomes

$$\left(\frac{\partial f}{\partial r}\right)_\theta = \left(\frac{\partial f}{\partial x}\right)_y \left(\frac{\partial x}{\partial r}\right)_\theta + \left(\frac{\partial f}{\partial y}\right)_x \left(\frac{\partial y}{\partial r}\right)_\theta \tag{6.36}$$

On substituting for $\partial f/\partial x$ and $\partial f/\partial y$ on the right-hand side, and using (6.34) to obtain $\partial x/\partial r$ and $\partial y/\partial r$, we find

$$\begin{aligned} \left(\frac{\partial f}{\partial r}\right)_\theta &= 2x\cos\theta + 2y\sin\theta \\ &= (2r\cos\theta)\cos\theta + (2r\sin\theta)\sin\theta \\ &= 2r \end{aligned} \tag{6.37}$$

This agrees with our previous result (6.35).

By this point, you should have been asking yourself why we have considered only $(\partial f/\partial t)_u$ when in fact f is a function of both t and u. In fact, because of the symmetry in t and u, we of course have an exactly similar equation to (6.30′) with t and u interchanged. Thus we have a pair of equations relating the new derivatives $\partial f/\partial t$ and $\partial f/\partial u$ to the old ones $\partial f/\partial x$ and $\partial f/\partial y$, *viz*

$$\left.\begin{aligned} \frac{\partial f}{\partial t} &= \frac{\partial f}{\partial x}\frac{\partial x}{\partial t} + \frac{\partial f}{\partial y}\frac{\partial y}{\partial t} \\ \text{and}\quad \frac{\partial f}{\partial u} &= \frac{\partial f}{\partial x}\frac{\partial x}{\partial u} + \frac{\partial f}{\partial y}\frac{\partial y}{\partial u} \end{aligned}\right\} \tag{6.38}$$

where we have omitted the constants on each of the partial derivatives in the hope that they are now indeed clear.

We can check that the second of the eqns (6.38) also gives the correct answer for our simple function (6.32) or (6.32′). We have

$$\begin{aligned}\left(\frac{\partial f}{\partial \theta}\right)_r &= \left(\frac{\partial f}{\partial x}\right)_y \left(\frac{\partial x}{\partial \theta}\right)_r + \left(\frac{\partial f}{\partial y}\right)_x \left(\frac{\partial y}{\partial \theta}\right)_r \\ &= (2x)(-r\sin\theta) + (2y)(r\cos\theta) \\ &= -2r^2\cos\theta\sin\theta + 2r^2\sin\theta\cos\theta \\ &= 0 \end{aligned} \tag{6.39}$$

which is correct since, from eqn (6.32′), f is independent of θ.

Although it is clear that eqns (6.38) must contain x and y in a symmetric way, it is a common mistake to forget to include the second term on the right-hand side of each equation. This is because the product $(\partial f/\partial x)(\partial x/\partial t)$ indeed has the same dimensions as $\partial f/\partial t$, and by analogy with the one-dimensional chain rule, it looks as if they may be equal.† It is important to remember to add on the second terms as specified in (6.38) if we want to obtain the correct answer.

(Following this specific example, the earlier discussion of this section should be more meaningful. It is highly recommended that you now go back and read again from the beginning of 6.3.4.)

It is straight-forward to extend eqns (6.38) to problems with more than two variables. Thus if the original function $f(x_1, x_2, \ldots, x_n)$ is expressed in terms of n variables x_i, which are then transformed to n new variables y_j via equations

$$y_j = y_j(x_1, x_2, \ldots, x_n), \qquad j = 1, 2, \ldots, n$$

then the derivatives of $f(y_1, y_2, \ldots, y_n)$ are given by

$$\frac{\partial f}{\partial y_j} = \sum_{i=1}^{n} \frac{\partial f}{\partial x_i}\frac{\partial x_i}{\partial y_j}, \qquad j = 1, 2, \ldots, n \tag{6.40}$$

The formulae (6.38) or (6.40) have wide application in problems involving changes of variables. For example, partial differential equations, which may be difficult to solve directly when expressed in one set of variables, may become much simpler for another choice. The way the derivatives in the equation are transformed as the variables are changed is given by eqns (6.38) or (6.40). If the partial differential equation involves

† Indeed we shall soon find an equation that is equivalent to

$$\frac{\partial f}{\partial t} = \frac{\partial f}{\partial x}\frac{\partial x}{\partial t}$$

(see Section 6.3.5), but this is with the constant of the partial differentiation process being different from those specified here.

second order derivatives or higher, it will be necessary to apply these equations more than once (see Problem 6.1 and Chapter 11 of Volume 2).

In this section, we wrote down the transformation equations (6.38) by the mnemonic of changing each 'd' in eqn (6.24) to a '∂'. We leave it as a short exercise for the reader to derive these properly from eqn (6.22), including the specification of what is kept constant for each of the partial derivatives that appears in the equations.

6.3.5 $\dfrac{\partial f}{\partial x} = \pm \dfrac{\partial f}{\partial y}\dfrac{\partial y}{\partial x}$

In this section we will derive the formulae

$$\frac{\partial f}{\partial x} = -\frac{\partial f}{\partial y}\frac{\partial y}{\partial x} \tag{6.41}$$

and

$$\frac{\partial f}{\partial x} = +\frac{\partial f}{\partial y}\frac{\partial y}{\partial x} \tag{6.42}$$

These appear to be inconsistent with each other. Obviously in any particular problem you have to make a decision about which is the correct one to use. It is thus crucial to realise the different situations to which they apply. If you have met these equations before, you should at this stage clear up this point in your mind; if they are new to you, you should do so later in this section.

We first derive (6.41), starting as usual from eqn (6.22) for the function $f(x, y)$. We divide through by δx to obtain

$$\frac{\delta f}{\delta x} = \frac{\partial f}{\partial x} + \frac{\partial f}{\partial y}\frac{\delta y}{\delta x} \tag{6.43}$$

and choose to go to the limit of infinitesimal changes in such a way that δx and δy are adjusted in order to make δf zero. This results in the ratio on the left-hand side of (6.43) vanishing, while $\delta y/\delta x$ becomes $(\partial y/\partial x)_f$. Thus we can rewrite eqn (6.43) as

$$0 = \left(\frac{\partial f}{\partial x}\right)_y + \left(\frac{\partial f}{\partial y}\right)_x \left(\frac{\partial y}{\partial x}\right)_f \tag{6.44}$$

Although we have set $\delta f = 0$, neither $\partial f/\partial x$ nor $\partial f/\partial y$ necessarily vanishes. This is because δf is zero only because of our special way of selecting δx and δy together; for arbitrary δx and δy, δf will not vanish. In particular, δf is in general non-zero for $\delta x = 0, \delta y \neq 0$, which implies

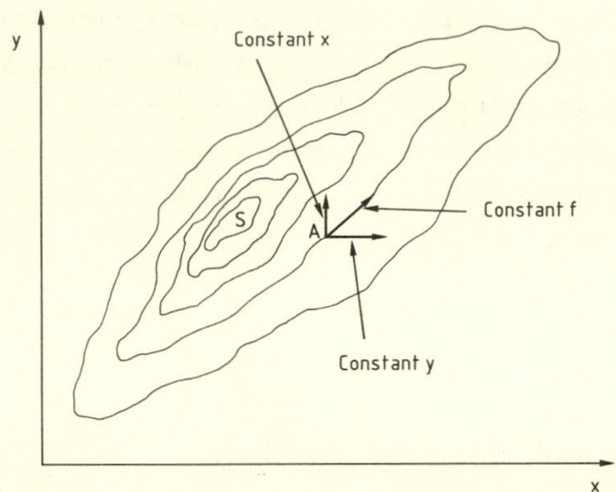

Fig. 6.4 Contour lines of a hill whose summit is at S. The derivatives $\partial f/\partial x$ and $\partial f/\partial y$ give the gradients of the hill as we travel in two perpendicular directions, parallel to the x and y axes respectively. For example, at A $\partial f/\partial x$ is negative – increasing x corresponds to going downhill – and $\partial f/\partial y$ is positive. $(\partial y/\partial x)_f$ is the tangent of the direction in the x–y plane in which we keep at constant height; at A this is positive. Thus the product of all three partial derivatives at A is negative.

that $(\partial f/\partial y)_x$ is non-zero (and similarly for $(\partial f/\partial x)_y$). This will become clearer from the example illustrated in fig. 6.4.

Now all we have to do is to rearrange (6.44) as

$$\left(\frac{\partial f}{\partial x}\right)_y = -\left(\frac{\partial f}{\partial y}\right)_x \left(\frac{\partial y}{\partial x}\right)_f \tag{6.44$'$}$$

and this is our formula (6.41).

Sometimes this is written in a slightly different form by multiplying through by $(\partial x/\partial f)_y$, to obtain

$$\left(\frac{\partial f}{\partial y}\right)_x \left(\frac{\partial y}{\partial x}\right)_f \left(\frac{\partial x}{\partial f}\right)_y = -1 \tag{6.45}$$

Here we have made use of the relationship

$$\left(\frac{\partial f}{\partial x}\right)_y \left(\frac{\partial x}{\partial f}\right)_y = 1 \tag{6.46}$$

This is clearly true if we backtrack one stage before going to the limit where the small quantities become infinitesimal. Then the left-hand side is

$$\frac{\delta f}{\delta x} \cdot \frac{\delta x}{\delta f} \tag{6.46$'$}$$

where each ratio is understood to be calculated at constant y. Since the conditions are the same for each factor of (6.46′), the δxs are the same, and so are the δfs. They thus cancel, and the product is unity, as expressed in (6.46).

Now that we have proved eqn (6.44′), we need to understand it, and in particular its minus sign.

The simplest test is just to try some simple function. For example,

$$f = 2x + y \tag{6.47}$$

Then

$$\left.\begin{aligned} \frac{\partial f}{\partial x} &= 2 \\ \text{and } \frac{\partial f}{\partial y} &= 1 \end{aligned}\right\} \tag{6.48}$$

To evaluate $(\partial y/\partial x)_f$, it is useful to rewrite (6.47) as

$$y = f - 2x \tag{6.47′}$$

and thus

$$\frac{\partial y}{\partial x} = -2 \tag{6.48′}$$

The derivatives (6.48) and (6.48′) clearly satisfy (6.44′), and we see that the minus sign in eqn (6.44′) is needed.

A more physical picture can be obtained by considering a case where $f(x, y)$ gives the height of a hill. In fig. 6.4, we draw some of the hill's contour lines, and show some arbitrary point A on it. For the situation as depicted in the diagram, $\partial f/\partial x$ evaluated at A is negative (i.e. we go downhill as x increases while y is kept constant); in contrast $\partial f/\partial y$ is positive. Now $(\partial y/\partial x)_f$ is the gradient in x–y plane of the contour line through A, since by definition contour lines are at constant height f. At A, this gradient is positive. Thus of the three derivatives occurring in eqn (6.44′), two are positive and one is negative. This is consistent with the minus sign in the formula.

Now one simple function and one point on a hill do not prove that eqn (6.44′) is valid. It does, nevertheless, give us some feeling that the negative sign is not as unreasonable as we might initially have thought.

At this stage, it is a good idea for you to test out a few other simple functions, and to draw another set of contour lines and investigate a few points on it.

It is also an appropriate time to think about when you should use

(6.42) instead of (6.41). Indeed (6.42) has a rather obvious and sensible look about it, and seems a logical extension of the one-dimensional chain rule. Also if we look back at eqn (6.30), we see that the first term on the right-hand side indeed does have the same structure as eqn (6.42); if we can find some way of removing the second term, we obtain

$$\frac{\partial f}{\partial t} = +\frac{\partial f}{\partial x}\frac{\partial x}{\partial t}$$

with its positive sign, rather than eqn (6.41).

The way we eliminate the second term is as follows. When we go to the limit of infinitesimal changes, we choose to do so with δy always zero, so that $\partial y/\partial t$ vanishes. Then eqn (6.30) becomes

$$\left(\frac{\partial f}{\partial t}\right)_y = \left(\frac{\partial f}{\partial x}\right)_y \left(\frac{\partial x}{\partial t}\right)_y \tag{6.49}$$

where $\partial f/\partial x$ is evaluated at constant y because eqn (6.22) tells us to do so, while the other two partial derivatives have y constant because we chose it to be so.

We see that eqn (6.49) agrees with eqn (6.42), with its positive sign. Now the resolution of the paradox at the beginning of this section is clear. Once again, confusion can be created by not specifying clearly what is being kept constant for each partial derivative. We need the plus sign of eqn (6.42) when some fourth variable (i.e. not f, x or y) is kept constant for each of the three partial derivatives. In contrast the minus sign of eqn (6.41) is for a situation involving only three variables, and the one kept constant for each of the derivatives is that which does not appear in the top or the bottom of that partial derivative; thus the constant quantity is different for each of the derivatives, and is as specified explicitly in eqn (6.44′).

Earlier we used the simple function (6.47) to illustrate formula (6.41). Now we use it as an example for formula (6.42) with t kept constant for each partial derivative. We assume that x and y are defined in terms of some other variables t and u. For example

$$\left.\begin{aligned} x &= t + u \\ \text{and } y &= t - u \end{aligned}\right\} \tag{6.50}$$

Since according to eqn (6.42) we need to evaluate $(\partial f/dx)_t$, $(\partial f/\partial y)_t$ and $(\partial y/\partial x)_t$, we must first express x in terms of t and y. We then need f in terms of these variables, and also as a function of t and x.

From eqns (6.50),

$$x = 2t - y \tag{6.50$'$}$$

and then

$$\begin{aligned} f &= 2x + y && (6.47)\\ &= 4t - y && (6.51)\\ &= x + 2t && (6.51') \end{aligned}$$

Thus

$$\left.\begin{aligned} \left(\frac{\partial f}{\partial y}\right)_t &= -1 \text{ from (6.51)}\\ \left(\frac{\partial f}{\partial x}\right)_t &= +1 \text{ from (6.51}'\text{)}\\ \text{and } \left(\frac{\partial y}{\partial x}\right)_t &= -1 \text{ from (6.50}'\text{)} \end{aligned}\right\} \tag{6.52}$$

These derivatives verify that we indeed need the positive sign to relate the three partial derivatives when the same fourth quantity (in this case t) is kept constant.

We set up this problem as one involving a change of variable from the old pair x and y to the new ones t and u, as given by eqns (6.50). It is more useful, however, to regard the transformation as being from x and y to t and y, where y is unchanged and only x is redefined in terms of t and y via eqn (6.50′). Since we are altering only one of the variables, we are not too surprised to obtain the relationship (6.42), which with its positive sign looks very much like the chain rule

$$\frac{df}{dx} = \frac{df}{dy}\frac{dy}{dx}$$

for changing variables in problems dealing with functions of only one variable.

We summarise this section by rewriting formulae (6.41) and (6.42) explicitly with the constant quantities as

$$\left(\frac{\partial f}{\partial x}\right)_y = -\left(\frac{\partial f}{\partial y}\right)_x \left(\frac{\partial y}{\partial x}\right)_f \tag{6.41$'$}$$

and

$$\left(\frac{\partial f}{\partial x}\right)_t = +\left(\frac{\partial f}{\partial y}\right)_t \left(\frac{\partial y}{\partial x}\right)_t \tag{6.42$'$}$$

If you keep in mind precisely which variables are held constant in any particular problem, you will have no difficulty in deciding which of the formulae is to be used.

The conclusion for the whole of Section 6.3 is as follows. There are several formulae relating partial derivatives which are very useful for mathematics and/or scientific applications. These have a rather similar look about them, and anyone trying to learn them off by heart is liable to confuse one with another. The formulae, however, follow so readily from (6.22) that it is far better to become efficient at performing these derivations of the formulae as needed, and to recognise the situation to which each of the formulae applies. It is, however, especially important to remember and understand the different conditions under which (6.41′) and (6.42′) are relevant.

6.4 Jacobians

We begin with a paradox. We want to check the relationship of the form (6.46). We are going to use the transformation between polar and rectangular coordinates (6.34) and their inverse

$$\left.\begin{aligned} \tan\theta &= y/x \\ \text{and } r^2 &= x^2 + y^2 \end{aligned}\right\} \tag{6.34′}$$

From the above, we have for example that

$$\frac{\partial r}{\partial x} = \frac{x}{r} \tag{6.53}$$

while from the first equation of (6.34)

$$\frac{\partial x}{\partial r} = \cos\theta \tag{6.54}$$

Thus

$$\begin{aligned} \frac{\partial r}{\partial x}\frac{\partial x}{\partial r} &= \frac{x\cos\theta}{r} \\ &= \cos^2\theta \end{aligned} \tag{6.55}$$

Unless θ happens to be zero, this is not equal to unity, and hence seems to contradict (6.46).

While you are resolving this, we backtrack to remind ourselves about what happens when we change variables in one-dimensional problems. Changing variables is often helpful for evaluating integrals. Thus if we want to evaluate

$$\int_a^b f(y)dy \tag{6.56}$$

we might decide to change to a new variable x which is related to y by

$$y = y(x) \tag{6.57}$$

Then the integral becomes

$$\int_{a'}^{b'} f[y(x)]\frac{dy}{dx}dx \tag{6.58}$$

Here $f[y(x)]$ is simply $f(y)$ reexpressed in terms of the variable x; the limits a' and b' are those values of x for which y equals the old limits a and b respectively; and dy/dx must be included because we have changed the variable of integration.

A trivial example (where in fact the change in variable does not make the integral much easier) is

$$\int_1^2 y^3 dy \tag{6.59}$$

with the change of variable

$$y = \sqrt{x} \tag{6.60}$$

Thus from (6.58), the integral in terms of x is

$$\int_1^4 x^{\frac{3}{2}}\left(\frac{1}{2}\frac{1}{\sqrt{x}}\right)dx = \frac{1}{2}\int_1^4 xdx$$

$$= \frac{1}{4}[x^2]_1^4 = \frac{15}{4} \tag{6.61}$$

This is what we would have found by performing the integral (6.59) directly. The factor in brackets on the top line of eqn (6.61) is simply dy/dx, and is clearly an essential ingredient if we are to get the correct answer.

We thus write our formula for transforming integrals as

$$\int_a^b f(y)dy \to \int_{a'}^{b'} f[y(x)]T dx \tag{6.62}$$

where T is a factor to do with the transformation from y to x (and is simply dy/dx).

Now we consider the corresponding situation in two (or more) dimensions.† We will first deal with a specific example of changing between rectangular and polar coordinates.

† Multiple integrals are discussed in Chapter 9 of Volume 2.

We assume we have to perform an integral

$$\int f(x,y)dxdy \tag{6.63}$$

over some specific region in the x–y plane. If, for example, this was a circle of radius R, and the function $f(x,y)$ was simply unity, the integral would equal the area of the circle, πR^2.

To convert to polar coordinates, we utilise the transformation (6.34). The function f is now expressed in terms of r and θ, thus:

$$f(x,y) \to f[x(r,\theta), y(r,\theta)] \tag{6.64}$$

Of course, if $f(x,y) = 1$, it remains unity when we change variables. In general we shall write the right-hand side of the above equation as $f(r,\theta)$. It is the function that $f(x,y)$ transforms into when we change the variables.

It is tempting, but wrong, to rewrite the integral (6.63) as

$$\int f(r,\theta)drd\theta.$$

One way we can see that this is invalid is that it has the wrong dimensions. The function f has the same meaning and keeps the same dimensions when the variables are changed. The infinitesimal integration region $dxdy$ was an area, but $drd\theta$ is only a length. This cannot be right.

Another hint is provided by the one-dimensional analogue eqn (6.62). There we had to include the extra factor

$$T = \frac{dy}{dx} \tag{6.65}$$

when we changed the integration variables. It seems most plausible that we need something similar here.

Indeed this is correct. In Chapter 9 of Volume 2, we shall show that this extra factor is $|J|$, where the modulus sign implies that we take the numerical value of J, which is known as the Jacobian of the transformation. It is conventional to write J in the form

$$J = \frac{\partial(x,y)}{\partial(r,\theta)} \tag{6.66}$$

which does not tell us what J actually is, but does have a sensible look about it as a factor for converting $drd\theta$ to be equivalent to $dxdy$.

The definition of J is

$$J = \frac{\partial(x, y)}{\partial(r, \theta)} = \begin{vmatrix} \dfrac{\partial x}{\partial r} & \dfrac{\partial x}{\partial \theta} \\ \dfrac{\partial y}{\partial r} & \dfrac{\partial y}{\partial \theta} \end{vmatrix} \tag{6.67}$$

where the vertical lines imply the determinant. If we compare this with the corresponding one-dimensional conversion factor (6.65), we see that (6.67) seems plausible. We expect it to have partial derivatives rather than total ones, and the new variables r and θ have to occur in the denominator, with x and y in the numerator. This does not give us too many other sensible possibilities for J.

One alternative could have been

$$J = \begin{vmatrix} \dfrac{\partial x}{\partial r} & \dfrac{\partial y}{\partial r} \\ \dfrac{\partial x}{\partial \theta} & \dfrac{\partial y}{\partial \theta} \end{vmatrix} \tag{6.67'}$$

The properties of determinants, however, are such that this is identical to (6.67), and so this too is correct. Although the above discussion is far from a proof, it does mean that, provided you remember that J is a determinant of partial derivatives, it is quite hard to write it down incorrectly.

Thus the transformation of the integral is

$$\int f(x, y) dx\, dy \to \int f(r, \theta)|J| dr d\theta \tag{6.68}$$

We must, of course, change the limits of integration (see Chapter 9 of Volume 2).

With the relationship between the old and the new variables as given in eqn (6.34), we can evaluate the partial derivatives, to find from eqn (6.67) that

$$J = \begin{vmatrix} \cos\theta & -r\sin\theta \\ \sin\theta & r\cos\theta \end{vmatrix} = r \tag{6.69}$$

Thus the integral becomes

$$\int f(r, \theta) r dr d\theta$$

which is the well known result for polar coordinates.

The general transformation from variables $x_1, x_2, \ldots, x_n$ to new ones

$y_1, y_2, \ldots, y_n$ results in the integral changing thus

$$\int f(x_1, \ldots, x_n) dx_1 \ldots dx_n \to \int f(y_1, \ldots, y_n) |J| dy_1 \ldots dy_n \tag{6.70}$$

where

$$J = \frac{\partial(x_1, \ldots, x_n)}{\partial(y_1, \ldots, y_n)} \tag{6.71}$$

and is a determinant whose element in the (i, j)th position is $\partial x_i / \partial y_j$ (alternatively they can instead all be $\partial x_j / \partial y_i$ – compare eqns (6.67) or (6.67′) respectively). This then provides us with a simple recipe for what happens to integrals on changing variables. Further details and more insights can be found in Chapter 9 of Volume 2.

The Jacobian is also to be found in equations relating derivatives. Again we choose the 'Cartesian-to-polar' case as a specific example of a transformation of variables. We have already considered the derivatives

$$\left(\frac{\partial x}{\partial r}\right)_\theta, \left(\frac{\partial y}{\partial r}\right)_\theta, \left(\frac{\partial x}{\partial \theta}\right)_r \text{ and } \left(\frac{\partial y}{\partial \theta}\right)_r \tag{6.72}$$

We are now going to see how these are related to the corresponding set when we perform the reverse transformation, *viz*

$$\left(\frac{\partial r}{\partial x}\right)_y, \left(\frac{\partial \theta}{\partial x}\right)_y, \left(\frac{\partial r}{\partial y}\right)_x \text{ and } \left(\frac{\partial \theta}{\partial y}\right)_x \tag{6.73}$$

As usual our starting point is the equivalent of eqn (6.22), which enables us to write

$$\left.\begin{aligned} \delta x &= \frac{\partial x}{\partial r}\delta r + \frac{\partial x}{\partial \theta}\delta\theta \\ \text{and } \delta y &= \frac{\partial y}{\partial r}\delta r + \frac{\partial y}{\partial \theta}\delta\theta \end{aligned}\right\} \tag{6.74}$$

These already contain the partial derivatives (6.72) in which we are interested. We then solve these simultaneous equations for δr and $\delta\theta$, and remember that these solutions can be expressed as

$$\left.\begin{aligned} \delta r &= \frac{\partial r}{\partial x}\delta x + \frac{\partial r}{\partial y}\delta y \\ \text{and } \delta\theta &= \frac{\partial \theta}{\partial x}\delta x + \frac{\partial \theta}{\partial y}\delta y \end{aligned}\right\} \tag{6.75}$$

which contain the other partial derivatives (6.73).

Trivial algebra gives the solutions of (6.74) as

$$\left.\begin{aligned}\delta r &= \left(\frac{\partial y}{\partial \theta}\delta x - \frac{\partial x}{\partial \theta}\delta y\right) \Big/ \left(\frac{\partial x}{\partial r}\frac{\partial y}{\partial \theta} - \frac{\partial x}{\partial \theta}\frac{\partial y}{\partial r}\right)\\ \delta \theta &= \left(-\frac{\partial y}{\partial r}\delta x + \frac{\partial x}{\partial r}\delta y\right) \Big/ \left(\frac{\partial x}{\partial r}\frac{\partial y}{\partial \theta} - \frac{\partial x}{\partial \theta}\frac{\partial y}{\partial r}\right)\end{aligned}\right\} \tag{6.76}$$

We notice that the denominator of both these expressions is simply the Jacobian $\partial(x, y)/\partial(r, \theta)$ of the transformation.

Identification of the corresponding independent terms of (6.75) and (6.76) then yields

$$\left.\begin{aligned}\frac{\partial r}{\partial x} &= \frac{\partial y}{\partial \theta}/J\\ \frac{\partial r}{\partial y} &= -\frac{\partial x}{\partial \theta}/J\\ \frac{\partial \theta}{\partial x} &= -\frac{\partial y}{\partial r}/J\\ \text{and}\quad \frac{\partial \theta}{\partial y} &= \frac{\partial x}{\partial r}/J\end{aligned}\right\} \tag{6.77}$$

These then are the required equations relating the two sets of partial derivatives. So far they are, in fact, completely general for changes of pairs of variables, i.e. we have in no way made use of the fact that at the back of the our minds we were thinking about rectangular and polar coordinates.

At first sight, eqns (6.77) seem to have a somewhat random look about them. However, we see that each numerator on the right-hand side contains the pair of variables that do not appear on the other side. Of course, whereas r and θ appear in the upper part of the derivatives on the left, x and y do so on the right; this is necessary since we are relating the set (6.72) to the inverse ones (6.73). Since J is $\partial(x, y)/\partial(r, \theta)$ and appears in the denominator, the dimensions of each of the equations (6.77) is correct, whatever the physical significance of the variable. Finally, even the signs are not haphazard. The second and third equations have minus signs in front of the derivatives $\partial x/\partial \theta$ and $\partial y/\partial r$, which appear in the negative term of the determinant J (see eqn (6.67)).

If for a moment we consider our particular case of rectangular and polar co-ordinates, we simply substitute r for J in eqns (6.77) (see eqn (6.69)). We can now return to our paradox at the beginning of this section. From eqns (6.77), we have

$$\left.\begin{aligned} \left(\frac{\partial r}{\partial x}\right)_y &= \frac{1}{r}\left(\frac{\partial y}{\partial \theta}\right)_r \\ \text{and} \quad \left(\frac{\partial \theta}{\partial y}\right)_x &= \frac{1}{r}\left(\frac{\partial x}{\partial r}\right)_\theta \end{aligned}\right\} \tag{6.78}$$

There is thus no reason to expect $(\partial r/\partial x)_y$ and $(\partial x/\partial r)_\theta$ to be reciprocals of each other. Once again, the paradox has been produced by neglecting to be careful with what is kept constant in the partial derivatives. Indeed we do expect $(\partial r/\partial x)(\partial x/\partial r)$ to equal unity, provided the omitted subscripts are identical. What we evaluated, however, when we produced the paradox were the derivatives as specified in eqn (6.78). Indeed in the absence of explicit subscripts, these are the natural choices of the constants. Finally to convince ourselves that all is completely satisfactory, we can check that the derivatives (6.72) and (6.73) for the 'Cartesian-to-polar' case do satisfy eqns (6.77).

We discover a further propitious aspect of eqns (6.77) when we calculate the Jacobian J' of the reverse transformation from the variables (r, θ) to (x, y). By analogy with (6.67),

$$J' = \frac{\partial(r, \theta)}{\partial(x, y)} = \begin{vmatrix} \dfrac{\partial r}{\partial x} & \dfrac{\partial r}{\partial y} \\ \dfrac{\partial \theta}{\partial x} & \dfrac{\partial \theta}{\partial y} \end{vmatrix} \tag{6.79}$$

Substitution from eqn (6.77) gives

$$\begin{aligned} J' &= \left(\frac{\partial y}{\partial \theta}\frac{\partial x}{\partial r} - \frac{\partial x}{\partial \theta}\frac{\partial y}{\partial r}\right) \Big/ J^2 \\ &= 1/J \end{aligned} \tag{6.80}$$

Thus, for example, when we transform an integral from (r, θ) to (x, y), the Jacobian that we need is simply the reciprocal of the original one for (x, y) to (r, θ), i.e.

$$\begin{aligned} \int g(r, \theta)drd\theta \rightarrow & \int g(x, y)J'dxdy \\ &= \int \frac{g(x, y)}{J}dxdy \end{aligned} \tag{6.81}$$

This implies that if we transform form (x, y) to (r, θ) and then back again, the two Jacobians J and J' cancel, and our integral returns identically to its original form.

Another way of looking at this result is that J in eqn (6.68) or (6.81) is not some magical function which ensures that the two integrals in the different coordinates are equal, but is rather a factor which converts each infinitesimal area $\delta x \delta y$ into the corresponding one in the new system, i.e.

$$\delta x \delta y \rightarrow J \delta r \delta \theta \tag{6.82}$$

It is thus hardly surprising that

$$\delta r \delta \theta \rightarrow J' \delta x \delta y = \delta x \delta y / J \tag{6.83}$$

Another satisfactory feature arises from the notation of eqn (6.67) for J. Then eqn (6.80) can be rewritten

$$\frac{\partial(r, \theta)}{\partial(x, y)} = 1 \bigg/ \frac{\partial(x, y)}{\partial(r, \theta)} \tag{6.84}$$

which has a rather plausible look about it.

Again the results from (6.79) to (6.84) are general, and are not restricted to our specific choices for the two pairs of variables. Indeed, with suitable generalisations, they also apply to changes of variables in multi-dimensional situations.

Finally we can contemplate the following. In transformations of one-dimensional integrals, the factor T occurs (see eqn (6.62)), in analogy with the Jacobian in the multi-dimensional integrals (6.68). The Jacobian also occurs in the relations (6.77) among the partial derivatives, whereas the one-dimensional case has the simpler formula

$$\frac{dy}{dx} = 1 \bigg/ \frac{dx}{dy} \tag{6.85}$$

Why?

In fact (6.77) and (6.85) are not inconsistent, in that the former reduces to the latter (with a suitable change in notation) when we have transformations between one single variable and another. An alternative way of looking at this is that if we group the formulae (6.77) together to construct the inverse Jacobian J', we obtain the relationship (6.80), which is indeed similar to the one-dimensional (6.85). Thus the simple example appears as a special case of the general formulae for multi-dimensional problems.

We conclude that the Jacobian is a particular combination of partial derivatives that is especially relevant for problems involving changes of variables.

Problems

6.1 Laplace's Equation in two dimensions is

$$\frac{\partial^2 V}{\partial x^2} + \frac{\partial^2 V}{\partial y^2} = 0.$$

Show that in plane polar coordinates given by

$$x = r\cos\theta$$

and

$$y = r\sin\theta,$$

the equation becomes

$$r^2\frac{\partial^2 V}{\partial r^2} + r\frac{\partial V}{\partial r} + \frac{\partial^2 V}{\partial \theta^2} = 0.$$

6.2 (i) Three sets of coordinates (x, y), (u, v) and (s, t) are related by the transformations

$$\left.\begin{aligned} u &= u(x, y) \\ v &= v(x, y) \end{aligned}\right\} \qquad [1]$$

and

$$\left.\begin{aligned} s &= s(u, v) \\ t &= t(u, v) \end{aligned}\right\} \qquad [2]$$

Show that the Jacobian of the transformation from variables (x, y) to (s, t) is simply the product of those for the transformations [1] and [2]. i.e.

$$\frac{\partial(s,t)}{\partial(x,y)} = \frac{\partial(s,t)}{\partial(u,v)}\frac{\partial(u,v)}{\partial(x,y)}$$

(ii) By explicitly evaluating the relevant Jacobians, show that the Jacobian for the two-dimensional transformation from rectangular co-ordinates (x, y) to polar coordinates (r, θ) is the reciprocal of that for the inverse transformation.

6.3 Evaluate the Jacobian $\partial(x, y)/\partial(r, \theta)$ for the transformation

$$\left.\begin{aligned} x &= r\cosh\theta \\ y &= r\sinh\theta \end{aligned}\right\}$$

Hence evaluate the area in the x–y plane enclosed by the curves

$$\left.\begin{aligned} x^2 - y^2 &= 9 \\ x^2 - y^2 &= 4 \\ y &= 4x \\ \text{and } y &= x. \end{aligned}\right\}$$

7
Taylor series

The grand old Duke of York
He had ten thousand men
He marched them up to the top of the hill
And he marched them down again.
And when they were up they were at a maximum,
And when they were down they were at a minimum,
And when they were at a saddle point they were neither entirely up nor entirely down.

(Based on nursery rhyme)

7.1 Taylor series in one dimension

In fig. 7.1, we show a graph of a fairly arbitrary function $f(x)$ of a single variable x. We assume that we know the value of $f(x)$ at a particular x position ($x = a$), as well as a few other things concerning $f(x)$ at this point. (Exactly what these 'other things' are will become clear soon.) What we are interested in is the value of $f(x)$ at some position $x = a+\delta x$, which is reasonably close to the original point $x = a$.

If we want only a crude approximation, we would write

$$f(a+\delta x) \approx f(a) \tag{7.1}$$

An example of this is the non-meteorological approach to weather forecasting. If we want to make a good guess at what the weather will be like tomorrow, we make a forecast which is simply today's weather. In most countries, the success rate of this recipe is around 70–80%, and, particularly if you do not disclose the basis of the method, the results can be impressive.

We know, of course, that eqn (7.1) is not exact, and it is fairly easy to produce a somewhat better estimate. In eqn (7.1), we are in effect

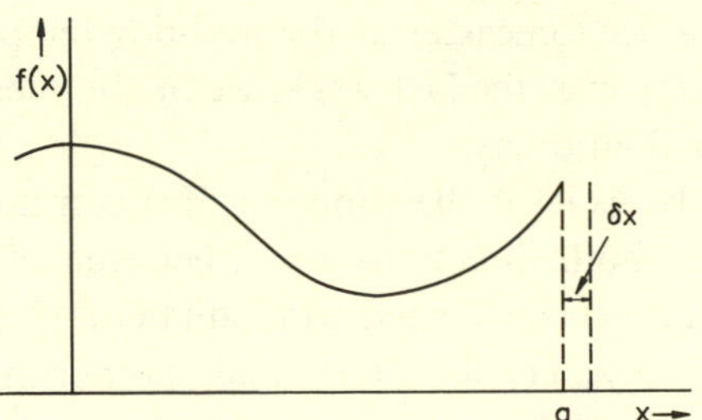

Fig. 7.1 $f(x)$ is a function of x. The value of $f(x)$ and its derivatives are known at $x = a$. We are interested in the value $f(a+\delta x)$ of the function at some nearby point.

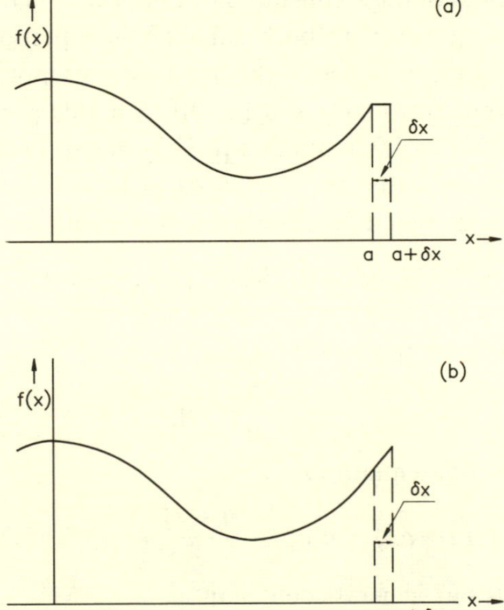

Fig. 7.2 Approximations to $f(a+\delta x)$. In (a), $f(a+\delta x)$ is simply set equal to $f(a)$, while in (b) a correction is added to allow for the gradient of $f(x)$ at $x = a$ (eqn (7.2)).

assuming that our function continues horizontally between $x = a$ and $x = a+\delta x$. It is generally more reasonable to consider that $f(x)$ continues in a straight line with the same gradient as the function had at $x = a$ (see fig. 7.2(b)). Then a trivial calculation gives us our next approximation as

$$f(a+\delta x) \approx f(a) + f'(a)\delta x \tag{7.2}$$

where $f'(a)$ means the value of the derivative df/dx, evaluated at $x = a$. In terms of our weather forecaster, if the mid-day temperature has been rising by 0.8° each day over the last week, we might guess that tomorrow will be 0.8° warmer than today.

Eqn (7.2) would be exact if the function $f(x)$ continued as a straight line for evermore in both directions on either side of $x = a$. This in general is not so, and hence we need to modify eqn (7.2). Now since the right-hand side contains a constant term and another that is linear in δx, it is reasonable to guess that a power series in δx would provide a better approximation. We thus write

$$f(a + \delta x) = f(a) + f'(a)\delta x + c_2(\delta x)^2 + c_3(\delta x)^3 + \ldots \tag{7.3}$$

where $c_2, c_3 \ldots$ are constants to be determined.

We find the values of these constants by adopting a general prescription of 'doing something clever' to both sides of the equation, in order that all the terms except one on the right-hand side disappear (compare the discussion between eqns (3.49) and (3.50)). For this particular case, the 'something clever' is to differentiate eqn (7.3) the appropriate number of times with respect to x, and then to set $\delta x = 0$.

For example, after two differentiations, we obtain

$$f''(a+\delta x) = 2c_2 + 3 \times 2c_3\delta x + 4 \times 3c_4(\delta x)^2 + \ldots + n(n-1)c_n(\delta x)^{(n-2)} + \ldots \tag{7.4}$$

On putting $\delta x = 0$, we determine

$$c_2 = \frac{1}{2}f''(a) \tag{7.5}$$

Similarly, after n differentiations

$$f^n(a + \delta x) = n!c_n + \frac{(n+1)!}{2}c_{n+1}\delta x + \ldots \tag{7.6}$$

which then yields our general coefficient as

$$c_n = \frac{1}{n!}f^n(a) \tag{7.7}$$

Here, of course, $f^n(a + \delta x)$ means the nth derivative of the function f with respect to x, evaluated at $x = a + \delta x$.

When we substitute our values of the constants c_n into the expression (7.3), we obtain the Taylor series

$$\begin{aligned} f(a + \delta x) = f(a) + f'(a)\delta x + \frac{1}{2!}f''(a)(\delta x)^2 + \frac{1}{3!}f'''(a)(\delta x)^3 + \ldots \\ + \frac{1}{n!}f^n(a)(\delta x)^n + \ldots \end{aligned} \tag{7.8}$$

We note in passing that we could have obtained the first two terms of this series (i.e. $f(a)$ and $f'(a)\delta x$) in exactly the same way as we derived the others, by writing our power series (7.3) instead as

$$f(a+\delta x) = c_0 + c_1\delta x + c_2(\delta x)^2 + \ldots \qquad (7.3')$$

With $\delta x = 0$, we at once obtain

$$c_0 = f(a) \qquad (7.9)$$

while differentiating once and setting $\delta x = 0$ yields

$$c_1 = f'(a) \qquad (7.10)$$

This gives us the first two terms of the series exactly as we deduced them in eqn (7.2)

There is another way of understanding why the Taylor series (7.8) contains an infinite number of terms, rather than simply the two of eqn (7.2). As we have already remarked, just as it was not exact to assume that our function continued horizontally between $x = a$ and $x = a + \delta x$ (cf eqn (7.1)), neither is it, in general, completely correct to assume that we can extrapolate simply by a straight line of the same gradient that $f(x)$ has at $x = a$. In particular this would be incorrect if the gradient is changing in the region around $x = a$. Since that would imply that d^2f/dx^2 is non-zero, we might guess that we need a further correction term that involves the second derivative of f. And since d^2f/dx^2 itself may not be constant, we would not be surprised to find a whole series of correction terms involving the higher derivatives of f with respect to x. Now if we want to add terms to the right-hand side of eqn (7.2), and since it is essential for them to have the correct dimensions, a derivative like d^2f/dx^2 must be multiplied by a distance squared (e.g. $(\delta x)^2$). Thus the various terms of eqn (7.8) have a sensible structure. The only bits we have to make an effort to remember are the $1/n!$ factors in front of each term. In Section 7.3, we will provide a further justification of the Taylor series.

Sometimes Taylor series are written in ways that look a little different from eqn (7.8), but in reality are completely equivalent. Thus we could have

$$\begin{aligned} f(x_2) = f(x_1) + (x_2 - x_1)f'(x_1) + \frac{1}{2!}(x_2 - x_1)^2 f''(x_1) \\ + \ldots \frac{1}{n!}(x_2 - x_1)^n f^n(x_1) + \ldots \end{aligned} \qquad (7.8')$$

where the value of the function at $x = x_2$ is given in terms of the function and its derivatives at $x = x_1$.

A special case of the Taylor series arises when the value of a in eqn (7.8) is zero. Then

$$f(\delta x) = f(0) + f'(0)\delta x + \frac{1}{2!}f''(0)(\delta x)^2 + \dots \quad (7.11)$$

This is usually written with x instead of δx, i.e.

$$f(x) = f(0) + xf'(0) + \frac{1}{2!}x^2 f''(0) + \dots \frac{1}{n!}x^n f^n(0) + \dots \quad (7.11')$$

Because this is such a common usage, it goes by the separate name of the Maclaurin series. In fact the Taylor series is so useful and important that it is well worth remembering eqn (7.8) off by heart (and, of course, understanding it); the Maclaurin series is so readily derived from it that it is not necessary to remember it separately.

What the Taylor series thus does is to enable us to find an approximation to a function at a given position in terms of its value and its derivatives somewhere else. It is most important to remember that all the derivatives and also the function itself in the series expansion are to be evaluated at the same place (i.e. at $x = a$ in eqn (7.8), or at $x = x_1$ in (7.8′)).

For all but a few curious functions, the infinite series will converge to the exact value of the function required. The main use of the Taylor series, however, is that, provided the two relevant x values are close together, it will usually be sufficient to evaluate the sum of only a couple of terms on the right-hand side in order to obtain a good approximation. This we can see from the Taylor series itself. Provided the derivatives $f^n(a)$ don't increase too rapidly in magnitude as n increases, the higher order terms in the expansion each contain an extra factor of δx (as well as a $1/n$), and hence for small δx become less significant.

7.2 Examples of Taylor and Maclaurin series

7.2.1 An accelerating body

We all remember the formula

$$s = ut + \frac{1}{2}at^2 \quad (7.12)$$

giving the position s after a time t of a body starting from the origin with an initial speed u and having a constant acceleration a. This is simply a truncated version of the Maclaurin series (7.11′) with the variable x replaced by t and with $f(t)$ written as s. Since the body starts from the origin, $f(0)$ is zero. The derivatives $f'(0)$ and $f''(0)$ are u and a respectively, while higher terms vanish since the derivatives d^nf/dt^n are zero for n larger than 2 because the acceleration d^2f/dt^2 is constant. Thus

$$\underset{s}{\underset{\uparrow}{f(t)}} = \underset{zero}{\underset{\uparrow}{f(0)}} + \underset{u}{\underset{\uparrow}{f'(0)}}t + \tfrac{1}{2!}\underset{a}{\underset{\uparrow}{f''}}(0)t^2 + \underbrace{\tfrac{1}{3!}f'''(0)t^3 + \ldots}_{zero} \tag{7.12'}$$

The factor of $\frac{1}{2}$ in eqn (7.12) is a direct consequence of the $\frac{1}{2!}$ in the third term of the Maclaurin series, and hence again confirms the need for the factorials.

7.2.2 Exponentials

We know that the value of e is 2.7183..., but we want to estimate $e^{1.1}$. We write our Taylor series for $f(x) = e^x$ as

$$f(1.1) = f(1) + 0.1f'(1) + \frac{1}{2!}(0.1)^2f''(1) + \frac{1}{3!}(0.1)^3f'''(1) + \ldots \tag{7.13}$$

Now since the derivatives of e^x are all equal to e^x, the factors $f^n(1)$ are each simply e. Thus

$$f(1.1) = 2.7183\ldots(1+0.1+\frac{1}{2}\times 0.01+\frac{1}{6}\times 0.001+\ldots) = 3.00415\ldots \tag{7.14}$$

This is just over 0.00001 smaller than the correct value.

Apart from being pleased at how well the series worked out, we can see that the Maclaurin series is converging, and is doing so rather quickly (i.e. we do not have to take too many terms to get a goodish approximation to the answer). As explained at the end of Section 7.1, it is the factorial in the denominator of each term and the extra δx in each numerator which produces this effect. Of course, had we required a larger δx (e.g. in order to evaluate $e^{1.5}$), we would have needed a larger number of terms to achieve similar accuracy. That is, the larger extrapolation we are performing, the more sophisticated our procedure needs to be.

7.2.3 sin x

If we use $f(x) = \sin x$ in the Maclaurin series, we obtain

$$\sin x = f(x) = f(0) + f'(0) + \frac{1}{2!}f''(0)x^2 + \frac{1}{3!}f'''(0)x^3 + \ldots \tag{7.15}$$

Now

$$\left.\begin{aligned} f'(x) &= \cos x \\ f''(x) &= -\sin x \\ f'''(x) &= -\cos x \\ &\text{etc.} \end{aligned}\right\} \tag{7.16}$$

and so

$$\begin{aligned} \sin x &= \sin(0) + x\cos(0) - \tfrac{1}{2!}x^2\sin(0) - \tfrac{1}{3!}x^3\cos(0) + \ldots \\ &= x - x^3/3! + \ldots \end{aligned} \tag{7.17}$$

This, of course, is the well known series for $\sin x$.

In a similar manner we could use the Taylor series to deduce $\sin 59°$, in terms of $\sin x$ at $60°$ and its derivatives there. In both these examples, it is absolutely vital to remember that the derivatives of $\sin x$ are as given in (7.16) provided that x is expressed in radians. If we had incorrectly expressed x in degrees, we would have deduced from eqn (7.17) that $\sin 2° \sim 1.95$. In this case, we are fortunate in that the answer is immediately recognisable as being stupid, and so we are persuaded to look for what we have done wrong. In other cases we may not be so lucky.

7.2.4 A trivial example

If $f(x)$ is given by a power series in x

$$f(x) = b_0 + b_1x + b_2x^2 + b_3x^3 + \ldots \tag{7.18}$$

we can use the Maclaurin series to determine the value of the function at an arbitrary position x in terms of the values of the function and its derivatives at the origin. It is easy to confirm that

$$\left.\begin{aligned} f(0) &= b_0 \\ f'(0) &= b_1 \\ f''(0) &= 2b_2 \\ f^n(0) &= n!b_n \end{aligned}\right\} \tag{7.19}$$

When we substitute these into the Maclaurin series, we find

$$\begin{array}{rcccc} f(x) = & f(0) + & f'(0)x + & \tfrac{1}{2!}f''(0)x^2 + \ldots & \tfrac{1}{n!}f''(0)x^n + \ldots \\ & \uparrow & \uparrow & \uparrow & \uparrow \\ & b_0 & b_1 & 2b_2 & n!b_n \\ = & b_0 + & b_1x + & b_2x^2 + \ldots & b_nx^n + \ldots \end{array} \tag{7.20}$$

This is, of course, identical with the functional form we started with in eqn (7.18). Thus by using the Maclaurin series for a function $f(x)$ that is already expressed as a power series, we do not discover anything new about the function. What we do convince ourselves, however, is that at least it gives us back the correct answer. In particular it provides experimental justification of the fact that we really do need those factorials in front of each term of the Taylor and Maclaurin series.

7.2.5 *tan x, and difficulties*

We can derive the Maclaurin series for $\tan x$ in exactly the same manner as we did for $\sin x$. It is a little more tedious since the derivatives of $\tan x$ are somewhat messier. They are

$$\left.\begin{aligned} f'(x) &= \sec^2 x \\ f''(x) &= 2\sec^3 x \sin x \\ f'''(x) &= 2\sec^2 x + 6\sec^4 x \sin x \end{aligned}\right\} \tag{7.21}$$

which give

$$\left.\begin{aligned} f'(0) &= 1 \\ f''(0) &= 0 \\ f'''(0) &= 2 \end{aligned}\right\} \tag{7.21$'$}$$

Then, on setting $x = 0$, we obtain

$$\tan x = x + x^3/3 + \ldots \tag{7.22}$$

First we note that this is an odd series in x, as it should be. Also it has the correct behaviour at small x, being $\sim x$ at very small x, and then being larger than x as x increases somewhat. Thirdly, it agrees with the answer we obtain by writing

$$\begin{aligned} \tan x &= \sin x / \cos x \\ &= (x - x^3/3! + \ldots)(1 - x^2/2! + \ldots)^{-1}, \end{aligned} \tag{7.23}$$

expanding the second term by the binomial theorem, and multiplying out.

An interesting feature emerges if we try to use the Taylor series for the somewhat artificial problem of obtaining $\tan 91°$ in terms of $\tan x$ and its derivatives at 89°. We appear to have

$$\begin{aligned}\tan 91^\circ &= \tan 89^\circ + (\tan x)'_{89}\delta\theta + \frac{1}{2!}(\tan x)''_{89}\delta\theta^2 + \frac{1}{3!}(\tan x)'''_{89}\delta\theta^3 + \dots \\ &= 57.3 + 3283 \times 0.035 + \frac{1}{2} \times 3.76 \times 10^5 \times 1.22 \times 10^{-3} \\ &\quad + \frac{1}{6} \times 6.47 \times 10^7 \times 4.25 \times 10^{-5} + \dots \end{aligned} \tag{7.24}$$

where of course we have expressed $\delta\theta$ in radians. Thus

$$\tan 91^\circ = 57 + 114 + 229 + 458 + \dots \tag{7.25}$$

Thus, because the derivatives are increasing so quickly, we appear to have a problem with the series' convergence. Also if we check on our calculator, we find that $\tan 91^\circ = -57.3$; it is hard to believe that our series is going to give the correct answer.

A glance at a graph of $\tan x$ in the relevant region (see fig. 7.3(a)) should tell us why our Taylor series is having trouble. We are using information about the function and its derivatives at A in order to try to deduce its value at B. Between A and B, however, we have a discontinuity. This prevents us from obtaining a sensible extrapolation. Thus, for example, had we considered instead the function $|\tan x|$ (see fig. 7.3(b)), we would have found identical derivatives at A, but the function at B is now +57.3. This illustrates the impossibility of using the Taylor or Maclaurin series across a discontinuity.

In fact the situation is even more restrictive than this. We can immediately see from fig. 7.3(c) that we again will not obtain the correct value of the function at B from its properties at A. Here the function itself has no discontinuities, but its gradient does. Indeed, in order to use the Taylor or Maclaurin series we need to have all the derivatives free from discontinuities in the relevant range.†

Most functions for which we write down an algebraic or trigonometric expression will satisfy these conditions, provided that we do not divide by zero, or go to the limit of the region where the function is defined (e.g. log(0), $\cos^{-1}(1)$, etc.). Our example of $\tan x$ was of the former type, since $\tan x = \sin x/\cos x$, and $\cos x$ is zero in the range between 89° and

† Or at least all the derivatives we are going to include in our series should be discontinuity-free. Thus if we are going to consider terms up to the one containing d^2f/dx^2, it does not matter if any higher derivatives have discontinuities.

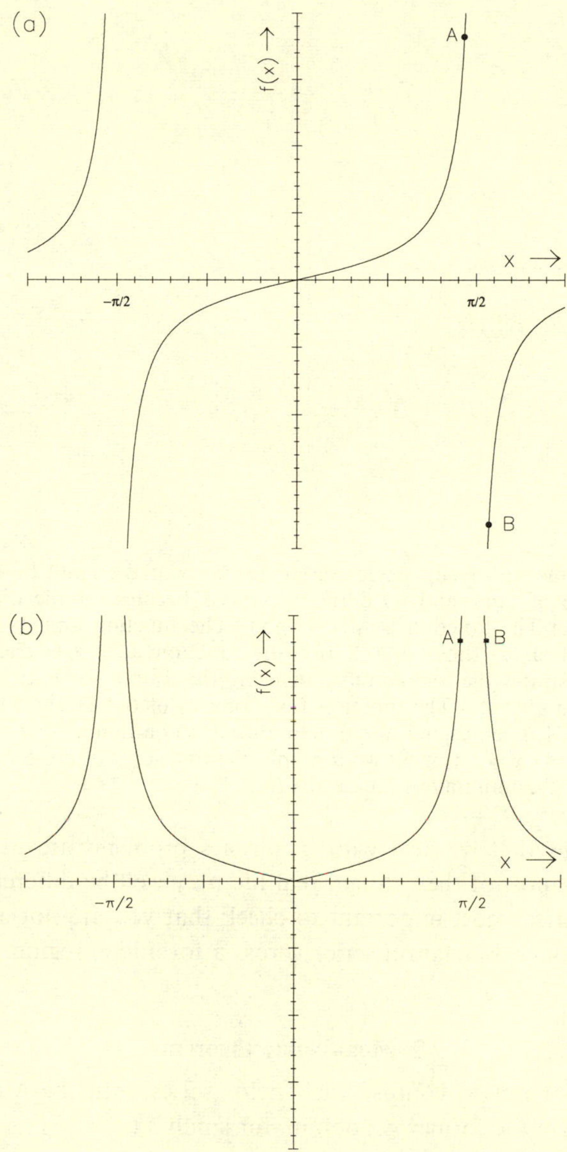

Fig. 7.3 For legend see next page.

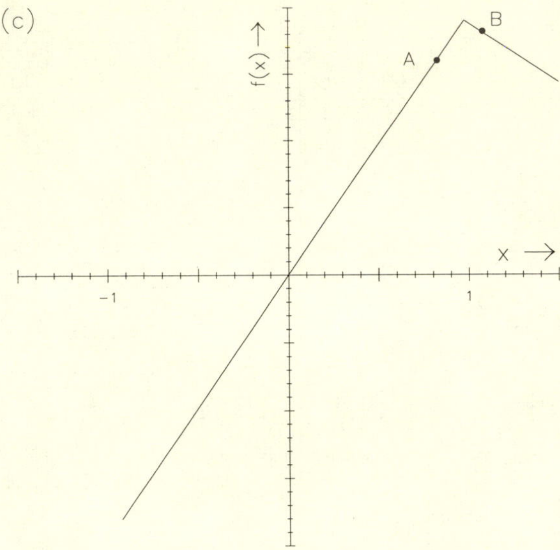

Fig. 7.3 Problems with using Taylor series. (a) $\tan x$ at B cannot be determined from the value of $\tan x$ and its derivatives at A because of the discontinuity at $x = \pi/2$. (b) The function is now $|\tan x|$. The function and its derivatives at A are identical to those in (a), but the function at B has changed sign. This again illustrates the impossibility of using the Taylor series to extrapolate across discontinuities. (c) The function $f(x)$ consists of two straight lines joined together at $x = 1$. Even though the function itself is continuous there, we cannot extrapolate across $x = 1$ (unless we use only the first term of the series). This is because of the discontinuity in the derivative.

91°. Functions that we deal with in physics problems are usually well behaved over most of their range; real life may well be different.†

It is, of course, most important to check that you are not attempting to use a Taylor or Maclaurin series across a forbidden region.

7.3 Mean value theorem

It is important not to confuse the Taylor series with the Mean Value Theorem. From the former we obtain, for small δx

$$f(a + \delta x) \sim f(a) + f'(a)\delta x \tag{7.26}$$

while the Mean Value Theorem (see fig. 7.4) states that

$$f(a + \delta x) = f(a) + f'\delta x \tag{7.27}$$

† A careful study of a friend's behaviour at the present time, including its time derivatives, is not guaranteed to predict correctly his future behaviour.

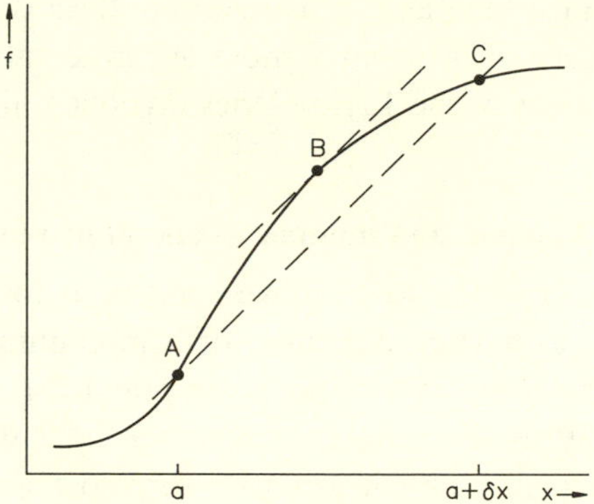

Fig. 7.4 Illustration of the Mean Value Theorem. The difference in the values of f at C and at A is given by δx times the gradient df/dx at some point B between A and C. This is true for any magnitude of δx, provided f and df/dx are continuous between A and C. The point B is such that the gradient there is equal to that of the line joining A and C.

which admittedly does look rather like (7.26). However, eqn (7.27) is exact, even for large δx, provided that the derivative f' is evaluated at some particular point in the range between $x = a$ and $x = a + \delta x$.† Since this point is unspecified, it is not too useful in evaluating $f(a+\delta x)$. In contrast the Taylor or Maclaurin series in principle require all the derivatives, but they are all evaluated at the same point at the beginning of the range.

We now return to the question of why the factor $\frac{1}{2}$ occurs in front of the $f''(a)\delta x^2$ term in the Taylor series of eqn (7.8). The Mean Value Theorem eqn (7.27) is exact for $f'(x)$ evaluated somewhere in the relevant range of x. Now if $f'(x)$ is not varying too rapidly in this region, it is usually a reasonable approximation to substitute its value at the middle of the range (or more or less equivalently, its average value). In order to determine $f'(a+\delta x/2)$, we apply the Taylor series to $f'(x)$, rather than to $f(x)$ as previously. Using only the first two terms on the right-hand side, we obtain

$$f'(a+\delta x/2) = f'(a) + f''(a)\delta x/2 + \ldots \tag{7.28}$$

Then from eqn (7.27), we obtain

$$\begin{aligned} f(a+\delta x) &\sim f(a) + [f'(a) + f''(a)\delta x/2]\delta x \\ &= f(a) + f'(a)\delta x + \frac{1}{2!}f''(a)\delta x^2 \end{aligned} \tag{7.29}$$

† Again it relies on the function and its derivative being continuous over this range.

Thus we have obtained the factor $\frac{1}{2}$ as required. Because of the approximations we have made, if we want a more accurate value for $f(a+\delta x)$ we require further terms of the Taylor series involving higher derivatives.

7.4 Maxima and minima in one dimension

In this section, we derive the well known conditions for a function $f(x)$ of a single variable to have a maximum or a minimum. There are two reasons for doing this. First, it provides more practice at using the Taylor series; and secondly, it is useful to remember how we approach the simpler one-dimensional problem when we turn to the more interesting case of the stationary values of a function of two variables (see Section 7.6).

In order to consider the behaviour of a function in the neighbourhood of a point $x = a$, we rewrite the Taylor series as

$$\begin{aligned}\delta f &= f(a+\delta x) - f(a) \\ &= f'(a)\delta x + \frac{1}{2!}f''(a)\delta x^2 + \frac{1}{3!}f'''(a)\delta x^3 + \ldots \qquad (7.30)\end{aligned}$$

This gives us the difference between the value of the function at a, and at a nearby point $x = a + \delta x$. For small δx, we will, in general, expect the successive terms of the series to decrease in magnitude. Thus in the immediate neighbourhood of $x = a$, the behaviour of δf will be dominated by the first term in the series, i.e.

$$\delta f \approx f'(a)\delta x \qquad (7.31)$$

Now if the function has a stationary value at $x = a$, then for small (but of course non-zero) δx, δf must be zero, since this is what we mean by a stationary value. From eqn (7.31), this then implies

$$f'(a) = 0 \qquad (7.32)$$

This condition ensures that the function will be flat in the immediate vicinity of $x = a$.

What happens slightly further away now depends on the next term of the series (7.30). Since $f'(a)$ is zero, this becomes

$$\delta f = \frac{1}{2!}f''(a)\delta x^2 + \ldots \qquad (7.33)$$

Thus if $f''(a)$ is positive, then so is δf for both positive and negative δx. Since the function increases as we go away from $x = a$ in either direction, we clearly have a minimum at $x = a$ (see fig. 7.5(a)). In contrast, if $f''(a)$

is negative, the function decreases in both directions, and we have a maximum.

Thus the conditions for maxima or minima are:

$$\left.\begin{array}{ll}\text{Maximum} & : f'(a)=0 \quad f''(a)<0 \\ \text{Minimum} & : f'(a)=0 \quad f''(a)>0\end{array}\right\} \tag{7.34}$$

The only other possibility is that $f''(a)$ is also zero. Then in analogy with the previous argument, the properties of the function near $x = a$ are determined by the third term of the Taylor series. Thus

$$\delta f \sim \frac{1}{3!} f'''(a) \delta x^3 \tag{7.35}$$

The presence of the δx^3 factor implies that, whatever the sign of $f'''(a)$, on one side of $x = a$ the function increases and on the other it decreases, even though in the region very close to $x = a$ it is flat (see fig. 7.5 (b)); an example of such a function is $(x - a)^3$. Because d^2f/dx^2 is zero at $x = a$, the stationary point there is a point of inflection.

In general, if the first few derivatives of $f(x)$ at $x = a$ are zero, the nature of the stationary value is determined by the first non-vanishing derivative. If this is an odd derivative, the function will go up in one direction and down in the other. If it is even, it will go up in both directions if this derivative is positive; or down in both directions if it is negative. These last possibilities look like minima or maxima respectively, although from a conventional mathematical viewpoint they are not since they do not satisfy the conditions (7.34).

As a final point on functions of a single variable, we consider

$$f(\cos\vartheta) = (\cos\vartheta - 0.5)^2 \tag{7.36}$$

which has a minimum at $\cos\vartheta = \frac{1}{2}$, but does not have a maximum. It does, however, have its largest value of $\frac{9}{4}$ at $\cos\vartheta = -1$, which is the boundary of the physically sensible region (see fig. 7.6). In many applications, we may be interested in finding the largest (or smallest) value of $f(\cos\vartheta)$. It is worth remembering that we might not discover this by looking where $f'(\cos\vartheta)$ is zero. In such cases, we have to investigate separately the values of the function on the boundaries of the physical region.

7.5 Two-dimensional Taylor series

We are now going to derive the two-dimensional Taylor series, using as our starting point the one-dimensional series with which we are already

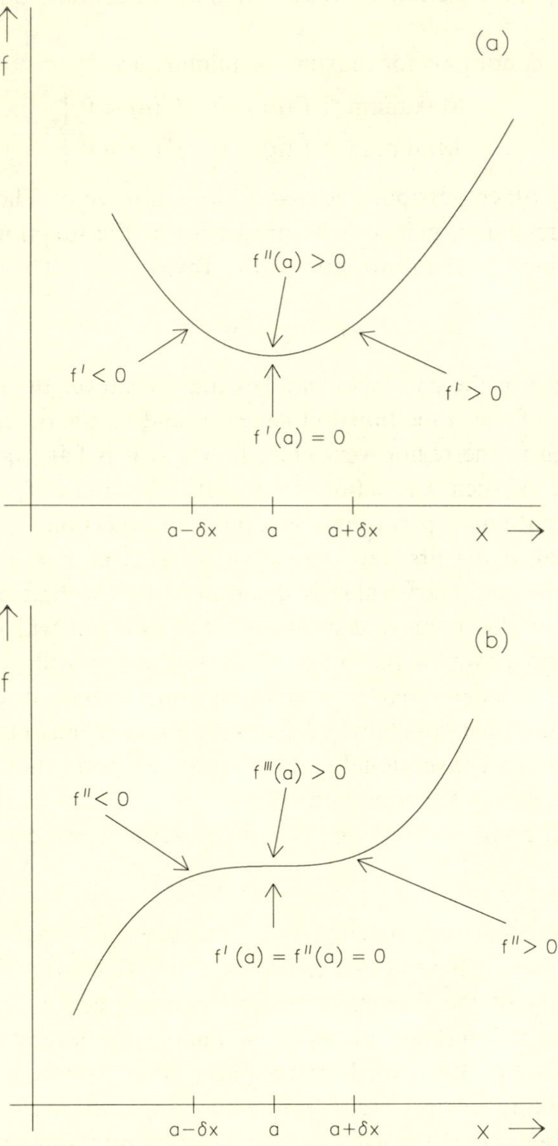

Fig. 7.5 (a) The function f has a minimum at $x = a$. At that point, the first derivative $f' = 0$. For somewhat smaller x, f' is negative, while at larger x it is positive. Thus $f''(a)$ is positive. (b) If f' and f'' are both zero at $x = a$, the behaviour of f in the neighbourhood of $x = a$ is determined by the term $f'''(a)\delta x^3/3!$ in the Taylor series. The illustration here is for the case where $f'''(a)$ is positive.

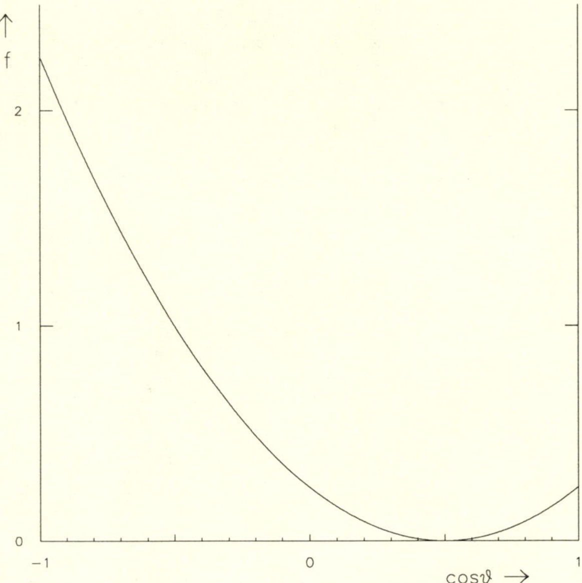

Fig. 7.6 The function $f = (\cos\vartheta - 0.5)^2$. The largest value of f is at $\cos\vartheta = -1$. The derivative $df/d\cos\vartheta$ is not equal to zero there.

familiar. We will perform the derivation in two slightly different ways. Both of them make use of partial derivatives. If you are unsure of what they mean, you should at this stage go back to Chapter 6.

By analogy with the one-dimensional problem, the two-dimensional version of the Taylor series will give us the value of a function $f(x, y)$ of two variables at a point $(x = a + \delta x, y = b + \delta y)$ in terms of the function and its derivatives at another point $(x = a, y = b)$. In general we expect δx and δy to be not too large (which is why we have written them with 'δ's) in order that we will obtain a good approximation without having to use too many terms of the series.

An example of such a function would be $\theta(d, t)$, where θ is the temperature, which is a function of the distance d from the sea, and of the time of day t. Alternatively $h(x, y)$ could give the height of a hill as a function of distances x and y respectively east and north of some origin.

Fig. 7.7 shows the point B at which we want to determine our function, and A at which it and its derivatives are known. Since both coordinates of B differ from those of A, this is very much the two-dimensional problem that we want to study. As up till now we have discovered how to deal with only one-dimensional situations, we split the problem

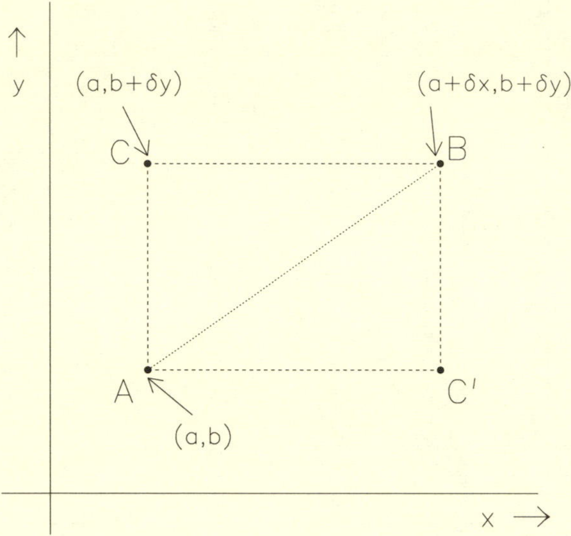

Fig. 7.7 We wish to specify the value of the function $f(x,y)$ at the point B in terms of its value and its derivatives at A. In order to make use of the one-dimensional Taylor series, we do this in two steps, first relating $f(x,y)$ at B to quantities at C (i.e. only x changes), and then these in turn are expressed in terms of the corresponding quantities at A (only y changes). This enables us to derive the two-dimensional Taylor series (7.42). An alternative to the path BCA is BC′A. A comparison of these gives the relation (7.45) between second order partial derivatives. A diagonal path from A to B can (almost) be used for deriving the two-dimensional Taylor series more directly. For rapid convergence of the Taylor series, ∂x and ∂y should be small; they are drawn large here merely for clarity.

into two one-dimensional parts. Thus first we relate $f(a+\delta x, b+\delta y)$ to the function and its derivatives at C; this involves changing only the x coordinate while keeping y constant, and hence is a typical one-dimensional situation. Then we convert all the required quantities at C to those at the original starting point A; here only y changes, so again we know how to do this. The net result is that we express the function at B in terms of its known behaviour at A. This provides us with our required two-dimensional Taylor series. We now implement this procedure.

We first need the extension of the Taylor series (7.8) to the case of a function of two variables, one of which remains constant. It does not require too much imagination to convince oneself that this is given by

$$f(a+\delta x, b) = f(a,b) + \frac{\partial f}{\partial x}\delta x + \frac{1}{2!}\frac{\partial^2 f}{\partial x^2}\delta x^2 + \ldots \qquad (7.37)$$

where the derivatives are all partial ones with respect to x, since this is the variable that is changing, while y is held constant; and they are all evaluated at our starting point $(x = a, y = b)$. We will not write explicitly any more terms in our series, since in general we will be interested in the series up to and including second order terms in the small distances δx and δy.

We need only a slight modification of eqn (7.37) for relating f at B $(x = a + \delta x, y = b + \delta y)$ to that at C $(x = a, y = b + \delta y)$. This is

$$f(a+\delta x, b+\delta y) = f(a, b+\delta y) + \frac{\partial f}{\partial x}(a, b+\delta y)\delta x + \frac{1}{2!}\frac{\partial^2 f}{\partial x^2}(a, b+\delta y)\delta x^2 + \ldots \tag{7.37$'$}$$

where we have this time written explicitly where we evaluate the derivatives.

Eqn (7.37′) involves f, $\partial f/\partial x$ and $\partial^2 f/\partial x^2$ (plus all the higher order derivatives we have not bothered to write down), evaluated at C. We need now to express each of these in terms of the properties of the function at A.

First we deal with the term $f(a, b + \delta y)$. In analogy with eqn (7.37), we relate f at C to its properties at A by

$$f(a, b + \delta y) = f(a, b) + \frac{\partial f}{\partial y}(a, b)\delta y + \frac{1}{2!}\frac{\partial^2 f}{\partial y^2}(a, b)\delta y^2 + \ldots \tag{7.38}$$

Now all the derivatives on the right-hand side of this equation are evaluated at $(x = a, y = b)$, which is where we want them. They can thus form part of our answer.

Now we turn our attention to the next term in eqn (7.37′). This involves $(\partial f/\partial x)$ evaluated at $(a, b + \delta y)$; this too needs to be expressed in terms of what is happening at $(x = a, y = b)$, just as we did for f in the previous paragraph. The easiest way to do this is to define

$$g(x, y) = \frac{\partial f}{\partial x}(x, y) \tag{7.39}$$

and then to write the Taylor series for $g(a, b + \delta y)$ in complete analogy with that for f in eqn (7.38). We obtain

$$\begin{aligned}\frac{\partial f}{\partial x}(a, b + \delta y) &= g(a, b + \delta y) \\ &= g(a, b) + \frac{\partial g}{\partial y}(a, b)\delta y + \frac{1}{2!}\frac{\partial^2 g}{\partial y^2}(a, b)\delta y^2 + \ldots \\ &= \frac{\partial f}{\partial x}(a, b) + \frac{\partial^2 f}{\partial y \partial x}(a, b)\delta y + \ldots \end{aligned} \tag{7.40}$$

where in the last line we have resubstituted $\partial f/\partial x$ for g. Again we note that, as required for our answer, the derivatives are evaluated at A.

Because we are interested in obtaining the Taylor series up to second order terms in δx and δy, we have written only two terms in the last line of eqn (7.40). This is because $(\partial f/\partial x)(a, b+\delta y)$ is to be multiplied by δx to obtain our answer (see eqn (7.37′)), and so the next term would already be third order.

What we did for $f(a, b+\delta y)$ and $(\partial f/\partial y)(a, b+\delta y)$, we must now do for $(\partial^2 f/\partial y^2)(a, b+\delta y)$. This is even simpler, since our requirement of keeping terms up to second order in the answer means that this time we need only the first one in our Taylor series. Thus without too much further thought we can write

$$\frac{\partial^2 f}{\partial y^2}(a, b+\delta y) = \frac{\partial^2 f}{\partial y^2}(a, b) + \ldots \tag{7.41}$$

At last we are ready to substitute eqns (7.38), (7.40) and (7.41) into (7.37′). After a little rearrangement, we obtain

$$\begin{aligned} f(a+\delta x, b+\delta y) = {} & f(a, b) \\ & + (\frac{\partial f}{\partial x}\delta x + \frac{\partial f}{\partial y}\delta y) \\ & + \frac{1}{2!}\left(\frac{\partial^2 f}{\partial x^2}\delta x^2 + 2\frac{\partial^2 f}{\partial y \partial x}\delta x \delta y + \frac{\partial^2 f}{\partial y^2}\delta y^2\right) \\ & + \ldots \end{aligned} \tag{7.42}$$

This is our desired result for the Taylor series in two dimensions, up to and including second order terms. To make it neat, we have omitted to write down explicitly in the formula that all the derivatives on the right-hand side are to be evaluated at $(x = a, y = b)$. The formula is so important that it should be learnt by heart. It is, not surprisingly, a fairly straight-forward extension of the one-dimensional series. The only term that is in danger of being forgotten is the one involving $\partial^2 f/\partial y \partial x$.

The first term on the right-hand side of eqn (7.42) gives the crude approximation that the function at the point $(a+\delta x, b+\delta y)$ has more or less the same value as at (a, b). If we include the first order terms on the second line, we have the same approximation as was used, for example, in eqn (6.22). Finally the bottom line gives the second order corrections.

The series can readily be extended to include higher order terms, although these are rarely needed. The next term is

$$\frac{1}{3!}\left(\frac{\partial^3 f}{\partial x^3}\delta x^3 + 3\frac{\partial^3 f}{\partial y \partial x^2}\delta x^2 \delta y + 3\frac{\partial^3 f}{\partial y^2 \partial x}\delta x \delta y^2 + \frac{\partial^3 f}{\partial y^3}\delta y^3\right) \tag{7.43}$$

where the constants within the brackets are the binomial coefficients. The nature of even higher terms should now be obvious.

If we look back at fig. 7.7, we see that we derived our Taylor series (7.42) by moving back from B to A via C. We did this because we wanted to change only one variable at a time. Another route which shares this property is to travel via C′ (i.e. first reduce y, and then x, instead of vice versa). To discover whether this would have made any difference for our Taylor series, we can either repeat the whole discussion of this section for the alterntive route; or else use an argument based on the symmetry in our problem between x and y. In either case, we obtain

$$\begin{aligned} f(a+\delta x, b+\delta y) = f(a,b) & \\ & + \left(\frac{\partial f}{\partial x}\delta x + \frac{\partial f}{\partial y}\delta y\right) \\ & + \frac{1}{2!}\left(\frac{\partial^2 f}{\partial x^2}\delta x^2 + 2\frac{\partial^2 f}{\partial x \partial y}\delta x \delta y + \frac{\partial^2 f}{\partial y^2}\delta y^2\right) \\ & + \dots \end{aligned} \tag{7.44}$$

This is the same as the original Taylor series (7.42), except for the fact that the derivative $\partial^2 f/\partial y \partial x$ there has been replaced by $\partial^2 f/\partial x \partial y$ here. Since we usually expect to obtain the same answer for $f(a+\delta x, b+\delta y)$ independent of the route between A and B, this requires that generally

$$\frac{\partial^2 f}{\partial x \partial y} = \frac{\partial^2 f}{\partial y \partial x} \tag{7.45}$$

i.e. the order in which we perform the partial differentiations is unimportant. It is worth trying out a few random functions $f(x,y)$ of your own choice, to verify that they do indeed satisfy eqn (7.45).

The relationship (7.45) will not be true if $f(x,y)$ is such that, when we start from (a,b), the value at $(a+\delta x, b+\delta y)$ depends on the route we take. For example, if we were on a spiral staircase, our change in vertical position after completing a half turn of the staircase depends on whether we are going up or down. Alternatively we could imagine a sea front dominated by a vertical cliff whose height varies. If we start at a point where the cliff's height has just become zero, a short walk takes us to a point at the top of the cliff, while another could be along the sea shore; these could end at two separate points, one vertically above the other. Again we have an example where $h(x,y) - h(0,0)$ depends on the path taken. (h is the height as a function of the horizontal coordinates x and y.) These examples depend on the fact that the function f is multi-valued.

Almost all functions that we can write down reasonably simply will satisfy eqn (7.45).

Now we turn to our second method of obtaining eqn (7.42), which is something between a mnemonic and a derivation. We change our notation slightly, and we try to find $f(a+\alpha, b+\beta)$ in terms of $f(a,b)$ and its derivatives. We again do this by using only the one-dimensional Taylor series, but this time we choose a route from A to B directly (see fig. 7.7). Points along this diagonal direction are given by

$$(x = a + k\alpha, y = b + k\beta) \tag{7.46}$$

where k varies; at B, it has the value 1. Now along this line, the function $f(a+k\alpha, b+k\beta)$ depends on only a single variable k, since a, b, α and β are to be thought of as constants. Then

$$f(a+k\alpha, b+k\beta) = f(a,b) + \frac{df}{dk}(a,b)k + \frac{1}{2!}\frac{d^2f}{dk^2}(a,b)k^2 + \ldots \tag{7.47}$$

where the derivatives are to be evaluated at (a,b), i.e. at $k=0$. Now a little thought tells us that the reason f changes when k varies is because both x and y change (by $(dx/dk)\delta k$ and $(dy/dk)\delta k$ respectively). These give contributions of $(\partial f/\partial x)\delta x$ and $(\partial f/\partial y)\delta y$ to the change in f. Thus

$$\frac{df}{dk} = \frac{\partial f}{\partial x}\frac{dx}{dk} + \frac{\partial f}{\partial y}\frac{dy}{dk} \tag{7.48}$$

$$= \frac{\partial f}{\partial x}\alpha + \frac{\partial f}{\partial y}\beta \tag{7.48$'$}$$

where we have used (7.46) to obtain dx/dk and dy/dk.

Substituting this into eqn (7.47) directly for df/dk, and as an operator for d^2f/dk^2, and then putting $k=1$, we obtain

$$\begin{aligned} f(a+\alpha, b+\beta) &= f(a,b) + \left(\frac{\partial f}{\partial x}\alpha + \frac{\partial f}{\partial y}\beta\right) + \frac{1}{2!}\left(\frac{\partial}{\partial x}\alpha + \frac{\partial}{\partial y}\beta\right)^2 f + \ldots \\ &= f(a,b) + \left(\frac{\partial f}{\partial x}\alpha + \frac{\partial f}{\partial y}\beta\right) \\ &\quad + \frac{1}{2!}\left(\frac{\partial^2 f}{\partial x^2}\alpha^2 + 2\frac{\partial^2 f}{\partial x \partial y}\alpha\beta + \frac{\partial^2 f}{\partial y^2}\beta^2\right) \\ &\quad + \ldots \end{aligned} \tag{7.49}$$

This is once again our Taylor series. As usual all the derivatives are to be evaluated at A. We have also made use of the identity (7.45) in order to simplify the second order term.

The main reason that this is not a complete proof is because we have not really justified eqn (7.48). To do so, we can either use the Taylor series as in Section 6.3.1 (but then we cannot use it in order to derive the Taylor series); or else we can essentially repeat the argument used in our first derivation (in which case the second derivation is not much simpler than the first).

Finally we give a simple mnemonic for the Taylor series in any number of dimensions, and up to terms of any required order. We merely write

$$f(a+\alpha, b+\beta, \ldots) = f(a, b, \ldots) + Df + \frac{1}{2!}D^2 f + \frac{1}{3!}D^3 f + \ldots \qquad (7.50)$$

where the differential operator

$$D = \alpha\frac{\partial}{\partial x} + \beta\frac{\partial}{\partial y} + \ldots \qquad (7.51)$$

and the derivatives are all to be evaluated at $(a, b \ldots)$. In two dimensions this reduces to eqn (7.49), while in one dimension, since the derivatives can now be written as total rather than partial ones, it has exactly the same form as eqn (7.8).

We summarise this section by writing down the various terms of the two-dimensional Taylor series. To zeroth order,

$$f(a+\alpha, b+\beta) \approx f(a, b) \qquad (7.52)$$

This says that, for example, tomorrow's weather in Oxford will be more or less like today's in London.

The next approximation consists of adding the first order terms

$$\frac{\partial f}{\partial x}\alpha + \frac{\partial f}{\partial y}\beta \qquad (7.53)$$

This corrects for the way the temperature is changing with position along the London-to-Oxford road and with time.

If we want to allow for the non-linear dependence of the temperature on time and position, we include also the second order terms

$$\frac{1}{2!}\left(\frac{\partial^2 f}{\partial x^2}\alpha^2 + 2\frac{\partial^2 f}{\partial x \partial y}\alpha\beta + \frac{\partial^2 f}{\partial y^2}\beta^2\right) \qquad (7.54)$$

For even better accuracy, we can, in principle, add on as many higher order terms as we like.

In the next section we go on to study the way these various terms determine the behaviour of a function in the vicinity of a specified point.

As an aside, we note that the Taylor series (7.42) can be used to write down the tangent plane and the direction of the normal at the point

(x_0, y_0) to a surface $z(x, y)$ (i.e. we have replaced the function f in (7.42) by the third coordinate z). For small deviations from (x_0, y_0), the z coordinate of a point in the surface is given by

$$z = z_0 + \frac{\partial z}{\partial x}(x - x_0) + \frac{\partial z}{\partial y}(y - y_0)$$

where z_0 is the value of z at the point (x_0, y_0). For larger deviations, the above equation simply gives the tangent plane at the required point. The normal to the surface at (x_0, y_0) is in the direction of the normal to the tangent plane, and has direction cosine ratios $(\partial z/\partial x, \partial z/\partial y, -1)$, as can be read off directly from the equation of the plane (see Section 2.4).

7.6 Maxima, minima and saddle points

In Section 7.4, we dealt with the conditions for a function $f(x)$ of a single variable to have a maximum or a minimum. Here we are going to extend the discussion to a function $f(x, y)$ of two variables. If you are not completely confident about the discussion of the single variable problem, go back and reread the earlier section.

Once again, we shall consider the behaviour of the function in the neighbourhood of a point, which this time we specify as (a, b). In analogy with eqn (7.30), we rewrite the two-dimensional Taylor series as

$$\begin{aligned}\delta f &= f(a + \alpha, b + \beta) - f(a, b) \\ &= \left(\frac{\partial f}{\partial x}\alpha + \frac{\partial f}{\partial y}\beta\right) + \frac{1}{2!}\left(\frac{\partial^2 f}{\partial x^2}\alpha^2 + 2\frac{\partial^2 f}{\partial x \partial y}\alpha\beta + \frac{\partial^2 f}{\partial y^2}\beta^2\right) + \dots \quad (7.55)\end{aligned}$$

Since we are interested in stationary values, we require that, in the immediate vicinity of our point, the function is flat. This implies that the first order term vanishes, i.e.

$$\frac{\partial f}{\partial x}\alpha + \frac{\partial f}{\partial y}\beta = 0 \quad (7.56)$$

Now most of us in fact know that a necessary condition for a stationary value is

$$\frac{\partial f}{\partial x} = \frac{\partial f}{\partial y} = 0 \quad (7.57)$$

The way we derive (7.57) from (7.56) is to require that the latter must be true for *all* small values of α and β. This follows from the fact that in the neighbourhood very close to the stationary point, $\delta f = 0$ in *every* possible direction. Then since we can, for instance, choose β to be zero, $\partial f/\partial x$ must vanish; and similarly for $\partial f/\partial y$. Hence eqns (7.57) follow.

Of course eqn (7.56) will be satisfied at any general position that is not a stationary value for some *specific* choice of α/β, since at any given point $\partial f/\partial x$ and $\partial f/\partial y$ are simply constants. (Thus if $\partial f/\partial x$ is 3 and $\partial f/\partial y$ is -1, we merely choose $\alpha/\beta = 1/3$ in order to obtain $\delta f = 0$, to first order.) This corresponds to the fact that, if we are standing on the side of a smooth hill, then in some directions the slope is upwards, in others it is downwards, while along what is the boundary between these regions, we could walk along a flat path (at least for a short distance, until the higher order terms in the Taylor series become important). This is the direction of the contour line through the point at which we are standing, and contemplating the beauty of the Taylor series. As we emphasised in the last paragraph, the important feature that distinguishes a stationary value is that the function is constant to first order for *all* choices of α and β; or equivalently, when we are at the top of our hill, we can step a short distance in *any* direction, without climbing or descending significantly.

As in the one-dimensional problem of Section 7.4, it is the second order terms that determine the nature of the stationary value. This is because, since the first order derivatives vanish,

$$\delta f \sim \frac{1}{2!}\left(\frac{\partial^2 f}{\partial x^2}\alpha^2 + 2\frac{\partial^2 f}{\partial x \partial y}\alpha\beta + \frac{\partial^2 f}{\partial y^2}\beta^2\right) \tag{7.58}$$

Thus if this is positive for all choices of α and β, then $f(x, y)$ increases in all directions from $(x = a, y = b)$, and this point is a minimum. Similarly, if (7.58) is always negative, then our point is a maximum.

At this stage, a new feature arises as compared with the one-dimensional problem. This is the existence of a new type of stationary value called a saddle point. It is so called because a saddle provides an example of this type of behaviour; in its normal orientation, there is a stationary value at the place where you sit, in that the seat is locally flat. However, if you consider slightly larger displacements, it slopes downwards on both sides of the horse, while it goes up towards the horse's head or its tail. This is thus neither a maximum nor a minimum.

Another example is a train going from one side of an uneven mountain range to the other, with the track chosen to pass over the top of the range at its easiest (i.e. lowest) place on the ridge. As the train travels up towards the range, it is not at a stationary value at all, but once it reaches its highest point, it is. This is because the gradient at that point is zero, irrespective of whichever direction you choose to consider. Thus this is true with respect to the direction of the train's track, since it is at the highest point on the track; and the gradient is also zero in the

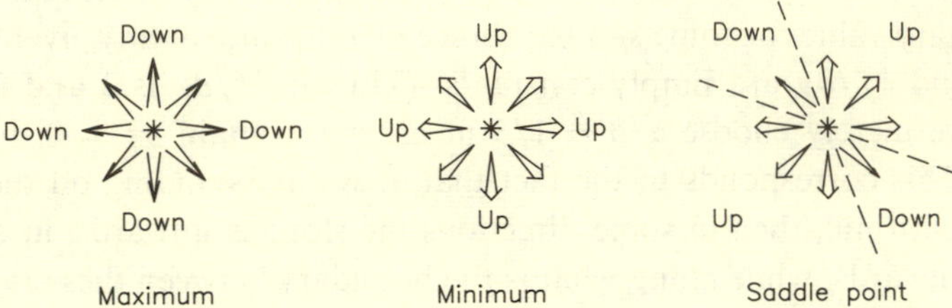

Fig. 7.8 Schematic diagrams illustrating a maximum, a minimum and a saddle point, as seen from above. The stars denote the positions of the stationary points. For a maximum, all the paths away from the star lead downwards, while for a minimum they all go up. At a saddle point, some directions go up and others go down; the dashed lines divide between these two types of directions.

transverse direction, since the train passes over at the position of the minimum height on the ridge. Now if the train stopped at this point, and we looked around, we would see the mountain going downwards in the forward and backward directions, but upwards as we looked out of either side window. Assuming that the mountain was reasonably smooth (as our mathematical functions are, even if real mountains are not), this would imply that there were a couple of in-between directions in which the ground was level (at least until the third order derivatives began to become important at larger distances).

The different types of behaviour of maxima, minima and saddle points are shown schematically in fig. 7.8. In one way, the saddle point resembles a stationary value with a point of inflection in the one-dimensional problem, in that the function increases in one direction, and decreases in another. For the latter, however, the behaviour was due to the third order term, when the first and second derivatives vanished; the saddle point's properties, however, are determined by the second order terms.

In order to derive the mathematical conditions that distinguish maxima, minima and saddle points, we rewrite eqn (7.58) as

$$\delta f \sim \frac{\beta^2}{2!}\left[\frac{\partial^2 f}{\partial x^2}\left(\frac{\alpha}{\beta}\right)^2 + 2\frac{\partial^2 f}{\partial x \partial y}\left(\frac{\alpha}{\beta}\right) + \frac{\partial^2 f}{\partial y^2}\right] \tag{7.59}$$

Here the ratio α/β determines the direction from (a, b) to $(a + \alpha, b + \beta)$ (the gradient of this line in the x–y plane is simply β/α), while β is the difference in the y co-ordinates of the points. As already mentioned, we are interested in how δf varies as a function of α/β. From fig. 7.8, we have:

$$\left.\begin{array}{rl} \text{Maximum:} & \text{always negative} \\ \text{Minimum:} & \text{always positive} \\ \text{Saddle point:} & \text{positive for some } \alpha/\beta, \text{ negative} \\ & \text{for others, and zero for specific } \alpha/\beta \\ & \text{in between.} \end{array}\right\} \tag{7.60}$$

Thus if a point is a saddle point, we must have

$$\frac{\partial^2 f}{\partial x^2}\left(\frac{\alpha}{\beta}\right)^2 + 2\frac{\partial^2 f}{\partial x \partial y}\left(\frac{\alpha}{\beta}\right) + \frac{\partial^2 f}{\partial y^2} = 0 \tag{7.61}$$

for the specific values of α/β that divide between the 'up' and the 'down' regions. Now since for a given function f the partial derivatives at any particular point are constants, eqn (7.61) is simply a quadratic equation for α/β. This is the equation which determines the two directions in which at a saddle point the function stays constant (up to second order terms).

Now the solutions of eqn (7.61) for α/β are

$$\alpha/\beta = \left(-\frac{\partial^2 f}{\partial x \partial y} \pm \sqrt{\left(\frac{\partial^2 f}{\partial x \partial y}\right)^2 - \frac{\partial^2 f}{\partial x^2}\frac{\partial^2 f}{\partial y^2}}\right) \Big/ \frac{\partial^2 f}{\partial x^2} \tag{7.62}$$

For this to give real solutions for α/β, we thus require

$$\left(\frac{\partial^2 f}{\partial x \partial y}\right)^2 \geq \frac{\partial^2 f}{\partial x^2}\frac{\partial^2 f}{\partial y^2} \tag{7.63}$$

Otherwise there are no (real) solutions for α/β, there are no directions for which δf is zero up to second order terms, and hence for all directions it must be always positive or always negative; this is then not a saddle point.

Thus our conditions for a saddle point are

$$\text{Saddle point} \ : \frac{\partial f}{\partial x} = \frac{\partial f}{\partial y} = 0; \quad \left(\frac{\partial^2 f}{\partial x \partial y}\right)^2 \geq \frac{\partial^2 f}{\partial x^2}\frac{\partial^2 f}{\partial y^2} \tag{7.64}$$

In contrast, if we want a maximum or a minimum, from conditions (7.60) there must be no directions given by α/β for which eqn (7.61) is satisfied. This will be so if

$$\left(\frac{\partial^2 f}{\partial x \partial y}\right)^2 < \left(\frac{\partial^2 f}{\partial x^2}\right)\left(\frac{\partial^2 f}{\partial y^2}\right) \tag{7.65}$$

This ensures that, for any possible value of α/β, the second order term will always be of the same sign.

We can also see from eqn (7.59) that the maximum will have $\partial^2 f/\partial x^2$ and $\partial^2 f/\partial y^2$ both negative, while for a minimum they are both positive. In view of the corresponding condition for the one-dimensional problem, these are not surprising.

In fact, given that a maximum or a minimum must satisfy (7.65), if one of the second derivatives $\partial^2 f/\partial x^2$ or $\partial^2 f/\partial y^2$ has the required sign, then automatically so will the other. This is because eqn (7.65) requires them to have the same sign, since their product is larger than the square of a real number.

Then we finally write our remaining conditions as

$$\text{Maximum}: \frac{\partial f}{\partial x} = \frac{\partial f}{\partial y} = 0; \quad \left(\frac{\partial^2 f}{\partial x \partial y}\right)^2 < \frac{\partial^2 f}{\partial x^2}\frac{\partial^2 f}{\partial y^2}; \quad \frac{\partial^2 f}{\partial x^2} < 0 \tag{7.66}$$

$$\text{Minimum}: \frac{\partial f}{\partial x} = \frac{\partial f}{\partial y} = 0; \quad \left(\frac{\partial^2 f}{\partial x \partial y}\right)^2 < \frac{\partial^2 f}{\partial x^2}\frac{\partial^2 f}{\partial y^2}; \quad \frac{\partial^2 f}{\partial x^2} > 0 \tag{7.67}$$

Eqns (7.64), (7.66) and (7.67) constitute the conditions we set out to determine.

It is important to realise that the first derivatives being equal to zero are the two simultaneous equations whose solutions determine the positions of the stationary values. We then use in turn each of these solutions for x and y to determine the numerical values of the the second derivatives, and then, by checking the conditions for them in eqns (7.64), (7.66) and (7.67), we determine the nature of each stationary point. This procedure corresponds in structure to the way we find and determine the nature of stationary values in one dimension.

In the next section, we apply the above to some simple functions of two variables.

7.7 Simple examples of two-dimensional stationary values

Here we will utilise a few simple functions, in order to illustrate the principles of the previous section. Just as in one dimension the simplest illustration of maxima and minima is with functions of the type ax^2, similarly here we will start by considering functions that are quadratic in x and y. Again for simplicity, we will initially deal with functions for which the stationary value has magnitude zero, and occurs at the origin; these restrictions are removed later.

7.7.1 Minima

The simplest quadratic function of two variables is

$$f = x^2 + y^2 \tag{7.68}$$

The best way to demonstrate the behaviour of the function is to draw a series of contour lines. Here we think of f as the height of a hill in the x–y plane; the contours show lines of constant height. They are thus given by

$$x^2 + y^2 = c \tag{7.69}$$

where c is one of a set of constants, giving the value of the contour (i.e. the height of the hill) along each line. From eqn (7.69), we see that they are all circles of radius $\sqrt{c}$; some of these are shown in fig. 7.9(a). The fact that they get closer and closer as we move away from the origin is because the hill is becoming steeper there. Conversely the absence of an accumulation of contours in the neighbourhood of the origin is because the function is approximately constant, since there is a stationary value at the origin. Since the function rises as we go away from the origin, the stationary value is a minimum.

Now we shall check these observations from our conditions (7.64), (7.66) and (7.67). First we need the derivatives of f. Thus

$$\frac{\partial f}{\partial x} = 2x \quad \frac{\partial f}{\partial y} = 2y \tag{7.70}$$

and

$$\frac{\partial^2 f}{\partial x^2} = 2 \quad \frac{\partial^2 f}{\partial y^2} = 2 \quad \frac{\partial^2 f}{\partial x \partial y} = 0 \tag{7.71}$$

On setting the first derivatives (7.70) equal to zero, we find that the only stationary value is at (0,0). We next should substitute ($x = 0, y = 0$) into eqns (7.71), in order to evaluate the second derivatives at the stationary value(s); in this particular case, this is in fact unnecessary, since the second derivatives are constants. Now since $(\partial^2 f/\partial x \partial y)^2$ is less than the product of $\partial^2 f/\partial x^2$ and $\partial^2 f/\partial y^2$, we are dealing with a maximum or a minimum. Finally, because $\partial^2 f/\partial x^2$ is positive, the stationary value is a minimum.

Thus we see that the conditions on the first and second derivatives give us the result we expected for this simple function.

Now we can try replacing the function (7.68) by something a bit more interesting. Thus if

$$f = a'x^2 + c'y^2 \quad (a' > 0,\ c' > 0) \tag{7.72}$$

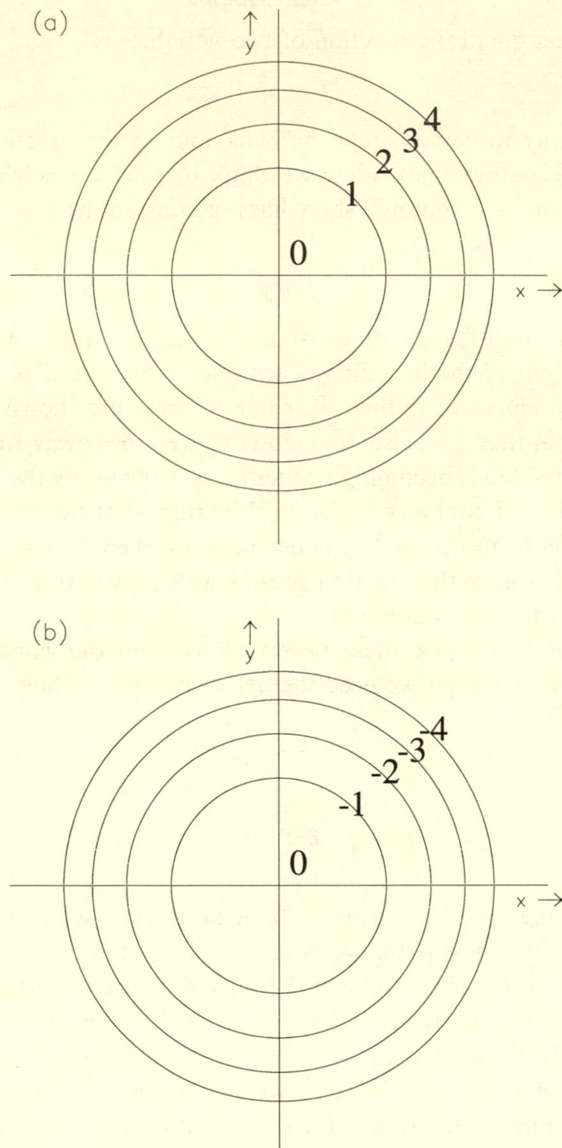

Fig. 7.9 Contour lines in the neighbourhood of (a) a minimum; and (b) a maximum. The stationary points are both at the origin, and have $f = 0$ there.

the minimum stays at the origin, but the contour lines become ellipses. We can also simulate the effect of rotating the axes, so that the function has the form

$$f = ax^2 + 2bxy + cy^2 \quad (a > 0,\ c > 0) \tag{7.73}$$

Now since the second derivatives are

$$\frac{\partial^2 f}{\partial x^2} = 2a \quad \frac{\partial^2 f}{\partial y^2} = 2c \quad \frac{\partial^2 f}{\partial x \partial y} = 2b \tag{7.74}$$

our function will have a minimum at the origin provided that

$$b^2 < ac \tag{7.75}$$

We can easily verify that if we start with the function (7.72) and rotate the axes, the coefficients of eqn (7.73) will satisfy the condition (7.75) (see Problem 7.3(i)).

Our final modification of the simple quadratic form is to have the minimum of height f_0 and at the position (x_0, y_0). Such a function is

$$f = f_0 + a(x - x_0)^2 + 2b(x - x_0)(y - y_0) + c(y - y_0)^2 \tag{7.76}$$

again with a and c both positive, and $b^2 < ac$. Application of the conditions (7.67) verifies the behaviour as expected.

7.7.2 *Maxima*

The change from minima to maxima is trivially achieved by writing a minus sign in front of each function. Thus the contours for

$$f = -(x^2 + y^2) \tag{7.77}$$

are as shown in fig. 7.9(b). Application of conditions (7.66) confirm that (7.77) has a maximum at the origin, as expected.

7.7.3 *Saddle points*

The contour lines of the function

$$f = xy \tag{7.78}$$

satisfy

$$xy = c \tag{7.79}$$

and hence are sets of hyperbolae as shown in fig. 7.10(a).

The function (7.79) is such that

$$\frac{\partial f}{\partial x} = y \quad \frac{\partial f}{\partial y} = x \tag{7.80}$$

and hence both derivatives are zero at the origin; this is our condition for a stationary value. Indeed we can see that along the axes $x = 0$ and $y = 0$, the function stays perfectly constant. Similarly if we move to a point $(\delta x, \delta y)$ a short distance away in an arbitrary direction, then f has no terms that are linear in δx and δy, and hence to first order the function stays constant. (This is equivalent to the fact that both first derivatives are zero.) Thus we see that the function indeed has a stationary value at the origin.

Now when we look at the behaviour of this function more carefully, we see that it increases away from the origin in both directions along the line AOA′, while it decreases in both directions along BOB′ (see fig. 7.10(a)). Thus the origin is a minimum with respect to (one-dimensional) variations along the line AOA′, but a maximum with respect to BOB′. Given that it is a stationary point, this implies that the origin is a saddle point.

In the terminology of the description of Section 7.6, we can imagine that the line AA′ represents the ridge of a mountain, with the height being positive or zero. If we want to cross this range, starting from B′ and ending up at B, it will be easiest to do so on a path that crosses the origin (e.g. along the line B′OB), which is always at a height which is negative or zero. As we travel along this route, we first go fairly steeply upwards, then up on a more gentle gradient until we reach the highest point of the track at the origin, after which we start to descend. However, the *highest* point on this track corresponds to the *lowest* point on the mountain ridge; this is the essential nature of this type of saddle point.

We have already checked that the first derivatives are zero, as required. The second derivatives are

$$\frac{\partial^2 f}{\partial x^2} = 0 \quad \frac{\partial^2 f}{\partial y^2} = 0 \quad \frac{\partial^2 f}{\partial x \partial y} = 1 \tag{7.81}$$

which as expected satisfy the saddle point condition (7.64). As with the function $x^2 + y^2$ which we chose to illustrate a minimum, similarly here the function (7.78) is so simple that the second derivatives are constants; in general they are functions of x and y, and we have to insert the values of x and y at each stationary point in order to determine whether (7.64) is satisfied.

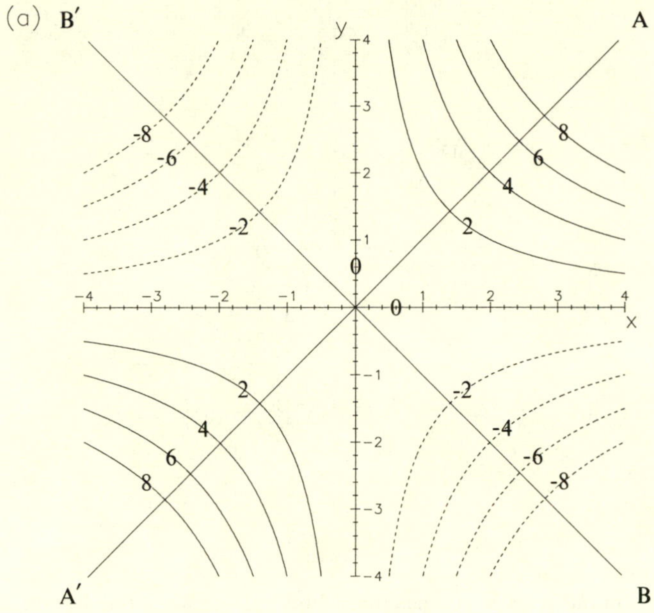

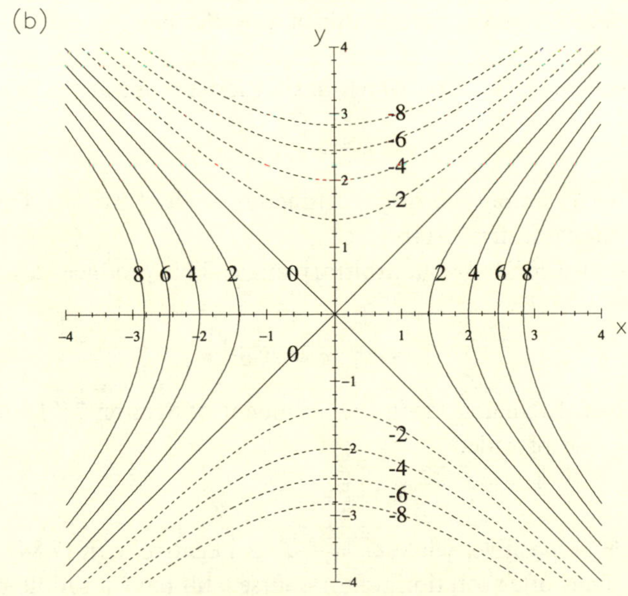

Fig. 7.10 For legend see next page.

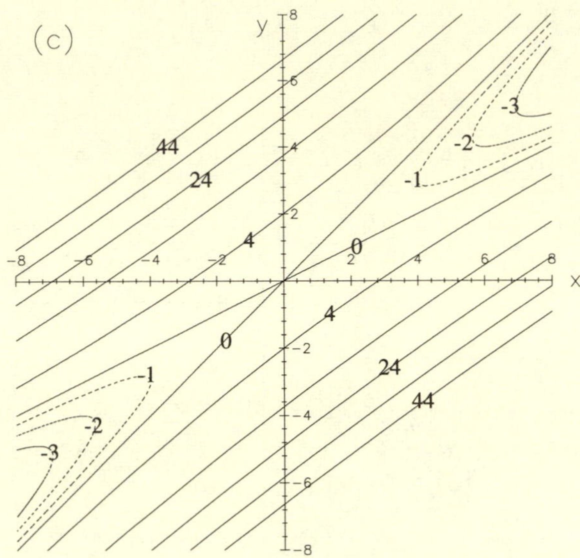

Fig. 7.10 Contour lines in the neighbourhood of saddle points for the functions (a) xy; (b) $x^2 - y^2$; and (c) $(x-y)(x/2-y)$. Contours at positive or zero height are drawn as solid lines; those at negative height are dashed. In all cases, there is a cross-shaped contour characteristic of a saddle point.

Another simple function which has a saddle point at the origin is

$$f = x^2 - y^2 \tag{7.82}$$

This in fact is simply the original function rotated by 45°. The contour lines are shown in fig. 7.10(b).

We can also rotate by an arbitrary angle. This produces a function of the form

$$f = ax^2 + 2bxy + cy^2, \tag{7.83}$$

with $c = -a$. In analogy with the argument of Section 7.7.1, this will be a saddle point provided

$$b^2 \geq ac \tag{7.84}$$

Because b^2 is positive, while $ac = -a^2$ is negative, eqn (7.84) is always satisfied. Thus after rotation we, of course, still have a saddle point.

We can, however do something a little more interesting. We saw in fig. 7.10(a) that the function (7.78) produces contour lines of height zero which are simply the x and y axes; this is because the condition for this contour line (see eqn (7.79)) is

$$xy = 0 \tag{7.85}$$

which implies

$$\left.\begin{aligned} x &= 0 \\ \text{or } y &= 0 \end{aligned}\right\} \tag{7.86}$$

These zero height contour lines divide the 'up' regions from the 'down' ones around the saddle point.

We can produce a distorted version of fig. 7.10(a), with the zero height contours having arbitrary gradients m_1 and m_2, by changing (7.78) to

$$f = (m_1x - y)(m_2x - y) \tag{7.87}$$

The contour lines are shown in fig. 7.10(c), for the choice $m_1 = 1$, $m_2 = 0.5$. Again (7.87) is of the form (7.83), and satisfies (7.84) (since $b = -\frac{1}{2}(m_1 + m_2)$, $a = m_1m_2$, and $c = 1$), as we expect for a saddle point.

With m_1 and m_2 in eqn (7.87) having the same sign, an important point emerges. As the contours of fig. 7.10(c) illustrate, if we examine the behaviour of the function f along the x axis, we see that it displays a minimum at the origin; and similarly there is a minimum at the origin for variations along the y axis. These two properties may fool us into thinking that the function $f(x, y)$ has a minimum at the origin. This is incorrect; the stationary value is a saddle point. The fact that, viewed separately in two orthogonal directions, a function has a minimum at the same point does not guarantee that it has a minimum in two dimensions there. What is required is that it satisfies conditions (7.67). What we have found is that, for eqn (7.87) with $m_1m_2 > 0$, at the origin

$$\frac{\partial f}{\partial x} = 0 \quad \frac{\partial f}{\partial y} = 0 \quad \frac{\partial^2 f}{\partial x^2} > 0 \quad \frac{\partial^2 f}{\partial y^2} > 0 \tag{7.88}$$

but this does not ensure that

$$\frac{\partial^2 f}{\partial x^2}\frac{\partial^2 f}{\partial y^2} > \left(\frac{\partial^2 f}{\partial x \partial y}\right)^2 \tag{7.89}$$

The function $f = (x - y)(x/2 - y)$ provides such an example.

It seems a little curious that we can produce distorted versions of the saddle point contours (e.g. fig. 7.10(c)), whereas in all the examples of maxima and minima, the relevant directions of the elliptical contours are always orthogonal. The resolution of this mini-paradox lies in the fact that we are comparing different quantities. In the maxima or minima problem, it is the directions of maximum and minimum curvature, which correspond to the axes of the ellipses of the contour lines in the neighbourhood of the stationary value, which are orthogonal. In fig. 7.10(c), whereas the contour lines through the saddle point are not orthogonal, the directions of the maximum and minimum curvature are; they are along the internal and external bisectors of these contour lines.

We shall return to the question of determining these directions for functions like (7.73), or similar ones for a larger number of independent variables, when we consider eigenvalues and eigenvectors in Chapter 16 of Volume 2.

Fig. 7.10 shows the characteristic form of contour lines in the neighbourhood of a saddle point. They form a cross passing through the saddle point, with the function increasing in one pair of regions, and decreasing in the other. The contour line cross through the stationary value is telling us that along these directions the function stays constant at its value at the stationary point. To the extent that the third order terms in the Taylor series can be neglected near the saddle point, the contour lines of the cross will remain straight. The functions depicted in fig. 7.10 have no third order terms (or higher), and hence the lines of the cross are straight for ever.

If we are trying to draw contour lines in the neighbourhood of a saddle point, it is most helpful if we know the orientations of the arms of the cross through the saddle point. These are easily determined since the solutions of eqn (7.61) for α/β are simply the reciprocals of their gradients. These solutions are given explicitly in eqn (7.62), in terms of the second derivatives of f at the saddle point (and which are hence known constants). Eqn (7.62) is not worth remembering, since (7.61) is derived very simply from the two-dimensional Taylor series, and everyone knows how to write down the solutions of a quadratic equation. Obtaining the gradients of the contour lines thus involves no more than the same equations and ideas than are involved in deducing the condition (7.63) for the existence of a saddle point. It is thus surprising that many students find the question of the gradients 'difficult', even though the majority have no problem with condition (7.63).

As an example for which we know the answer, we use the function (7.87). The equation (7.61) for the directions in which the second order contribution vanishes becomes:

$$m_1 m_2(\alpha/\beta)^2 - (m_1 + m_2)\alpha/\beta + 1 = 0 \tag{7.90}$$

whose solutions for the reciprocals of the gradients are

$$\alpha/\beta = 1/m_1 \text{ or } 1/m_2 \tag{7.91}$$

This agrees with the fact that we constructed the function (7.87) to have the constant value of zero along lines of gradients m_1 and m_2 passing through the origin.

It is worth mentioning that saddle points do not always have to look exactly as the classic form of fig. 7.10. This is because we have considered functions where the *inequality* of eqn (7.63) applies. If instead it is an equality, the stationary value is still called a saddle point. Provided the second order derivatives do not all vanish, this implies that the two arms of the contour line cross have closed up to coincide with each other; there is now only a single direction in which the function stays constant. Alternatively all the second order derivatives could be zero (e.g. as for $f = x^2y$ at the origin), and the behaviour would be determined by the first non-vanishing order of derivatives. In general, you can expect to meet only the simple form of saddle points, with two distinct directions along which the function is constant.

Finally a couple of fairly trivial points. Contour lines are usually drawn at fixed constant intervals. These might be such that we do not draw the contour which passes through the saddle point. This would happen, for example, for the function

$$f = xy + 1/2 \tag{7.92}$$

with the contours drawn at integral values of f. In that case, the cross of fig. 7.10 is missing, but the remaining structure of two regions of larger contours separated from two of lower values will be maintained.

The other remark is that, as stated at the beginning of this section, we have considered here only quadratic functions. This simplified the discussion, and enabled us to understand immediately the behaviour of the function, and to know what solutions we expected. Another consequence was that each of the functions investigated had only one discrete stationary value† and the pattern of contour lines around the stationary value extended out to infinity in much the same way. More complicated functions need not have these very ideal properties. We illustrate this with an example in the next section.

† We ignore functions like $(mx + y)^2, -10$ or y, which have respectively a line or a plane of stationary values, or none at all.

7.8 A more realistic example

We now turn to a more realistic example. We assume that we are asked to find the positions and the nature of the stationary values of the function

$$f = (x^2 - y^2)e^{-(x^2+y^2)} \tag{7.93}$$

and to sketch its contour lines.

First it is a good idea to see what symmetries the given function possesses. As far as its x dependence is concerned the function contains only x^2 terms, and hence is an even function of x. This implies that f stays unchanged when x is replaced by $-x$, which, in turn, means that if we find a stationary value at (x_0, y_0), there must be another of the same type at $(-x_0, y_0)$, (unless, of course, $x_0 = 0$) Similarly, the function is even in y, and so any stationary value at (x_1, y_1) is accompanied by another of the same type at $(x_1, -y_1)$.

Another symmetry we may notice is that if x and y are interchanged, the function becomes

$$(y^2 - x^2)e^{-(x^2+y^2)}$$

which is just the original function with the sign changed. Thus a reflection about the line $x = y$ simply changes the sign. This, in turn, implies that any stationary value at (x_2, y_2) requires another at (y_2, x_2) (unless $x_2 = y_2$). Furthermore, because of the sign change, these will be either both saddle points, or one maximum and one minimum.

The final simple property is that, because $(x^2 - y^2)$ is a factor, f is zero when $x = +$y or when $x = -y$ Thus these two lines with gradients ± 1 and passing through the origin are contour lines of height zero. This implies that we have a saddle point at the origin.

Of course, all the above discussion is useful, especially when we come to drawing the contour lines of f, but it is not absolutely necessary to discover all these features. If we go through the following procedure correctly, the relevant symmetry requirements will be automatically satisfied. We will use the above discussion to check that our answers are sensible.

We now employ the standard procedure of eqns (7.57) for finding the positions of the stationary points. We thus evaluate

$$\left.\begin{aligned} \frac{\partial f}{\partial x} &= 2x(1 - x^2 + y^2)E \\ \text{and } \frac{\partial f}{\partial y} &= -2y(x^2 - y^2 + 1)E \end{aligned}\right\} \tag{7.94}$$

where

$$E = e^{-x^2-y^2}$$

To find when these both equal zero as required, we note that the factor E is positive for any finite x and y. Thus we need

$$\left.\begin{aligned} x(1-x^2+y^2)=0 \\ \text{and } y(x^2-y^2+1)=0 \end{aligned}\right\} \qquad (7.95)$$

Hence

$$\left.\begin{aligned} x=0 \text{ or } y^2=x^2-1 \\ \text{and } y=0 \text{ or } x^2=y^2-1 \end{aligned}\right\} \qquad (7.96)$$

At this point, it is worth emphasising that the little words 'and' and 'or' are most important. Thus a combination like $y = 0$ and $x^2 = y^2-1$ would not ensure that we are at a stationary value. Appropriate combinations are

$$\left.\begin{aligned} x=0 \text{ and } y=0 \\ \text{or } x=0 \text{ and } x^2=y^2-1 \\ \text{or } y=0 \text{ and } y^2=x^2-1 \\ \text{or } y^2=x^2-1 \text{ and } x^2=y^2-1 \end{aligned}\right\} \qquad (7.97)$$

Of these, we can disregard the final combination, since the two conditions are mutually inconsistent. We are thus left with

$$\left.\begin{aligned} x=0,\ y=0 \\ x=0,\ y=\pm 1 \\ y=0,\ x=\pm 1 \end{aligned}\right\} \qquad (7.97')$$

as giving us the coordinates of the five stationary points.

We note that the above points satisfy the symmetry requirements mentioned earlier. Thus the symmetry in x results in $(+1,0)$ being paired with $(-1,0)$; $(0,+1)$ is accompanied by $(0,-1)$ because of the y symmetry; and the stationary values at $(0,\pm 1)$ require the ones at $(\pm 1,0)$ because of the $x \leftrightarrow y$ symmetry. (In fact, we would have expected even more accompanying stationary points had there not been so many zeroes in eqns (7.97′). Thus these symmetries would have implied that, for a stationary point at (a,b) with $a \neq 0$, $b \neq 0$ and $a \neq b$, there would have been others at $(a,-b)$ $(-a,b)$ $(-a,-b)$ (b,a) $(b,-a)$ $(-b,a)$ and $(-b,-a)$.)

Now we need to find the nature of these stationary points. We thus have to evaluate the second derivatives, and then check the conditions (7.64),

(7.66) and (7.67) for each of the five points of (7.97′). By differentiating eqns (7.94), we find

$$\left.\begin{aligned} \frac{\partial^2 f}{\partial x^2} &= 2E[(1-x^2+y^2)(1-2x^2)-2x^2] \\ \frac{\partial^2 f}{\partial y^2} &= 2E[(1-y^2+x^2)(1-2y^2)-2y^2] \\ \frac{\partial^2 f}{\partial x \partial y} &= 4Exy(x^2-y^2) \end{aligned}\right\} \qquad (7.98)$$

Before evaluating these explicitly, we note that E is a positive constant that occurs as a multiplicative factor in each of the three derivatives, and so can be factored out without affecting whether the relevant conditions are satisfied or not. Then we have:

$$\left.\begin{aligned} &\text{At } (0,0) \ : \frac{\partial^2 f}{\partial x^2} = 2E, \quad \frac{\partial^2 f}{\partial y^2} = -2E, \quad \frac{\partial^2 f}{\partial x \partial y} = 0 \\ &\text{At } (0,\pm 1) : \frac{\partial^2 f}{\partial x^2} = 4E, \quad \frac{\partial^2 f}{\partial y^2} = 4E, \quad \frac{\partial^2 f}{\partial x \partial y} = 0 \\ &\text{At } (\pm 1,0) : \frac{\partial^2 f}{\partial x^2} = -4E, \frac{\partial^2 f}{\partial y^2} = -4E, \quad \frac{\partial^2 f}{\partial x \partial y} = 0 \end{aligned}\right\} \qquad (7.99)$$

Thus

$$\left.\begin{aligned} &\text{At } (0,0) \ : \left(\frac{\partial^2 f}{\partial x \partial y}\right)^2 > \left(\frac{\partial^2 f}{\partial x^2}\right)\left(\frac{\partial^2 f}{\partial y^2}\right) && \text{and (7.64) is satisfied} \\ &\text{At } (0,\pm 1) : \left(\frac{\partial^2 f}{\partial x \partial y}\right)^2 < \left(\frac{\partial^2 f}{\partial x^2}\right)\left(\frac{\partial^2 f}{\partial y^2}\right), \frac{\partial^2 f}{\partial x^2} > 0 && \text{and (7.67) is satisfied} \\ &\text{At } (\pm 1,0) : \left(\frac{\partial^2 f}{\partial x \partial y}\right)^2 < \left(\frac{\partial^2 f}{\partial x^2}\right)\left(\frac{\partial^2 f}{\partial y^2}\right), \frac{\partial^2 f}{\partial x^2} < 0 && \text{and (7.66) is satisfied} \end{aligned}\right\}$$

We finally conclude that the stationary value at the origin is a saddle point; those at $(0,\pm 1)$ are minima; and there are maxima at $(\pm 1,0)$.

If we again check with our symmetry conclusions, we find that, as expected, the stationary points at $(0,1)$ and $(0,-1)$ are of the same type, as are those at $(1,0)$ and $(-1,0)$; while those at $(0,1)$ and $(1,0)$ are a minimum and a maximum. The motivation for doing this check is not because we suspect that there may be some impossibly peculiar behaviour of the function, but merely to ensure that our algebra has not given us a stupid result.

We now have to sketch the contour lines. Here the symmetries are very helpful, because it means that we need only investigate how the contours look in a limited region of the x–y plane, and then the symmetry

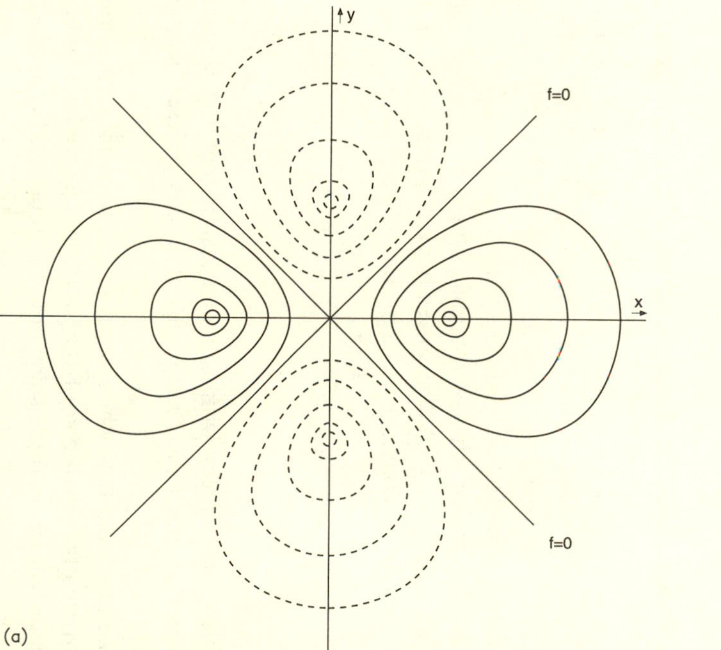

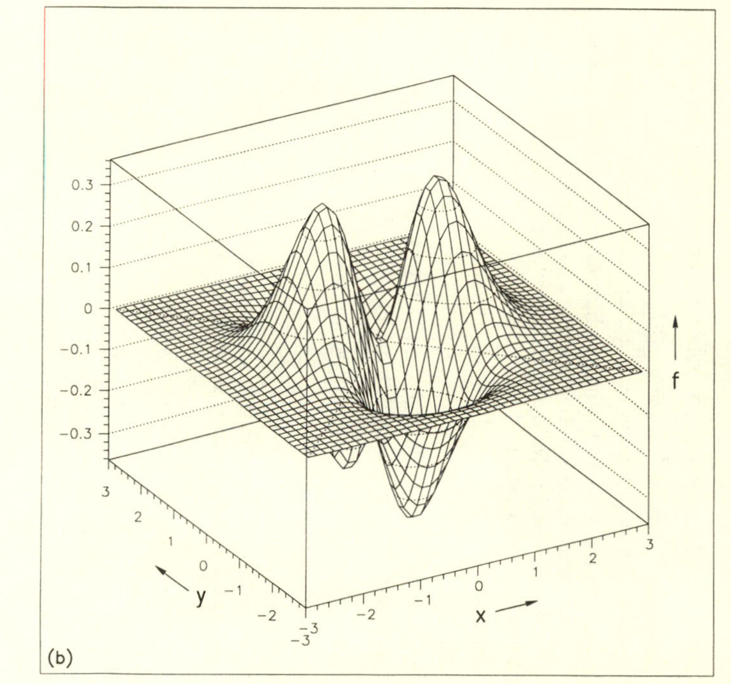

Fig. 7.11 (a) An approximation to some of the contour lines of the function of eqn (7.93), taking into account the specific considerations discussed in the text. The solid contours have positive f, and the dashed ones are for f negative. (b) A three-dimensional display of the function. The maxima at $(\pm 1,\ 0)$ are clearly visible, as is the minimum at $(0,\ -1)$; the second minimum at $(0,\ 1)$ is partially obscured by one of the maxima. At the origin is a saddle point; near this stationary point, the function increases in directions around the $\pm x$ axes, but decreases in other directions (close to the $\pm y$-axes). These different regions of behaviour are separated by the crossed contour lines at $y = \pm x$, along which the function stays constant.

Table 7.1. *Summary of procedure for finding stationary values and sketching contours*

(1) Look for symmetries (i.e. $f \to \pm f$), when x changes sign; or y changes sign, or x and y both change sign; or x and y are interchanged.
(2) Set $\partial f/\partial x = \partial f/\partial y = 0$, and solve for all possible (x, y) combinations.
(3) Evaluate the second derivatives, and check at each point from (2) whether
$$\left(\frac{\partial^2 f}{\partial x \partial y}\right)^2 \geq \left(\frac{\partial^2 f}{\partial x^2}\right)\left(\frac{\partial^2 f}{\partial y^2}\right)$$
(3′) If yes, this is a saddle point. Then use eqn (7.62) to find the gradients (β/α) of the contours through the saddle points.
(3″) If no, then check whether $\dfrac{\partial^2 f}{\partial x^2} > 0$ (minimum) or < 0 (maximum).
(4) Draw the contour lines by:
  (i) Using symmetries.
  (ii) Drawing closed loops round maxima or minima.
  (iii) Drawing crossed contour lines of appropriate gradients through saddle points.
  (iv) Ensuring contour lines do not cross anywhere else, and do not have too violent bends.
  (v) Deducing contours for simple values of f (e.g. $f = 0$).
  (vi) Finding how f behaves as x or $y \to \pm\infty$.

considerations allow us to continue them over the whole plane. Another very important aid is the location and nature of stationary values. Around maxima and minima, we have to draw closed curves that, at least in the vicinity of the stationary point, must look like ellipses or circles. Similarly at a saddle point, the contour lines cross, and the gradients of these lines are determined from eqn (7.62).

For our particular problem, the only saddle point is at the origin, and if we use the values of the second derivatives there as given in (7.99), we find the gradients g of the contours through the saddle point as

$$\begin{aligned} g = \left(\frac{\alpha}{\beta}\right)^{-1} &= [(-0 \pm \sqrt{0^2 - (2E)(-2E)})/2E]^{-1} \\ &= \pm 1 \end{aligned} \tag{7.100}$$

Very satisfactorily, this agrees with our conclusion that the contours of height zero pass through the origin with gradients ± 1 (see near the beginning of this section).

In passing we also note that when we draw contour lines, they must not cross anywhere other than one of the saddle points we have found.

Another useful feature is to consider how the function behaves as x

and/or y tend to infinity. In our case, the exponential factor ensures that the function tends to zero.

It may be possible to draw specific contour lines (e.g. for $f = 0$). This is useful in our problem since f factorises, and as we already saw at the beginning of this section, we could immediately deduce the existence of a saddle point at the origin.

We are now ready to sketch the contours. When we put together all the above considerations, we are virtually forced to produce a diagram like that of fig. 7.11(a). For comparison, a three-dimensional display of the function is shown in fig. 7.11(b).

A summary of this procedure is given in Table 7.1.

Problems

7.1 Evaluate e^x for $x = 0.9$ by using (i) a Taylor series, starting from $x = 1$; and (ii) the Maclaurin series, starting from $x = 0$. Use enough terms to achieve an accuracy of 1% in your answers. (The value of e is 2.7183.)

7.2 Find the maximum value of the function

$$f(\cos\theta) = (\cos^2\theta - 0.25)^2$$

Evaluate $f(0)$ and $f(0.9)$. Comment on your answers.

7.3 (i) Verify that, if the coordinate axes are rotated by an angle θ, then the quadratic expression (7.72) is transformed into a new function of the form (7.73), satisfying condition (7.75), irrespective of the magnitude of θ.

(ii) Use the conditions on the first and second derivatives to determine the position and nature of the stationary value for the function (7.76).

7.4 Find the location and the types of stationary values of the function

$$f(x, y) = xy(x - 2)(y + 4) + 8$$

Sketch contour lines for this function.

7.5 The function

$$z = x^3 + y^3 + 9(x^2 + y^2) + 12xy$$

has stationary values at points with (x, y) coordinates (0, 0), (−10, −10), (−4, 2) and (2, −4). Determine which of these are saddle points, and find the gradients (in the x–y plane) of the contour lines which pass through the saddle points.

8
Lagrangian multipliers

'Do not turn to the right or to the left; but go along the path...'
(Deuteronomy v 29,30)

8.1 What problem are we solving?

Mont Blanc is the highest mountain in Europe. Extremely expert climbers can reach the summit, at which point they are 4807 m above sea level. The less intrepid can alternatively enjoy an excellent view by going across the Mont Blanc range by telepherique. This does not reach the very top, but at its highest point it is 3842 m above sea level (see fig. 8.1).

The above situation illustrates two different types of maximisation (or more generally, stationary value) problems. In the first, we simply want to find the maximum value of a function. In the Mont Blanc case, we assume that someone has provided us with a function $h(x, y)$, giving the height at each point (x, y) on the mountain, where x and y are the distances east and north of Chamonix respectively. Then the summit is given by the values of x and y that maximise h (see Section 7.6).

The second type of problem is the *constrained* maximum: we wish to find the maximum value of our function, subject to some constraint condition among the variables. In this particular case, x and y could not both vary independently, but are restricted by the requirement that we remain in the telepherique. Presumably there is an equation of the form

$$c(x, y) = k$$

where k is a constant, which gives the trajectory of the telepherique, and it is this which provides the relationship between x and y.

In the unconstrained case, we know that the stationary values would

Fig. 8.1 View of the Alps near Chamonix. The highest point (unconstrained maximum) is at the peak of Mont Blanc. If we stay in the telepherique, the constrained maximum is at the point Aiguille du Midi.

satisfy

$$\frac{\partial h}{\partial x} = \frac{\partial h}{\partial y} = 0 \tag{8.1}$$

Since the constrained problem, in general, will have a different solution, we had better find a different set of equations to solve. In Section 8.2, we shall see that this is so.

Most maximisation problems that we encounter, either consciously or otherwise, in our daily lives probably involve constraints. For example, the combination of food items we buy each week can be regarded as being determined as an attempt to obtain the 'best' dietary mixture, subject to the constraint of spending a fixed amount of money. Similarly a headmaster of a school may arrange the schedule of his teachers (i.e. the subjects and classes that they teach) in order to maximise the educational benefit to the school as a whole, given that each of his staff teaches for only a specified number of hours each week.

Constraints among the variables can be of two types. In the first, they are equalities (e.g. $x^2 + y^2 = c^2$), and this is the situation we are considering in this chapter. If the constraints are inequalities (e.g. $x^2 + y^2 \leq c^2$), then there are two distinct possibilities. Thus the largest value may occur on the boundary of the region (i.e. $x^2 + y^2 = c^2$), which returns us to the subject of this chapter. Alternatively it may occur inside the region defined by the constraint, in which case it is a standard maximisation problem.

Inequality constraints can apply to a single variable (e.g. $|x| \leq 1$); if the largest value of the function occurs on the boundary, we may not have a true maximum there (see fig. 8.2). Of course, equality constraints can similarly be in terms of a single variable (e.g. $y = 1$), but it is hardly worth using the method of Lagrangian multipliers in such a trivial situation.

Some simple problems that are suitable for solving by the Lagrangian multiplier method are:

(i) The maximum area that a farmer can enclose in a rectangular field, using a fixed length of fencing.

(ii) Finding the axes of an ellipse which is skewed with respect to the x–y axes. This we do by determining the points in the x–y plane whose distance from the centre of the ellipse is a maximum or a minimum, subject to the constraint that the point is on the ellipse.

(iii) The minimum distance between two non-parallel lines in three dimensions. Here we minimise the distance between two points, subject to the constraints that one point is on each line.

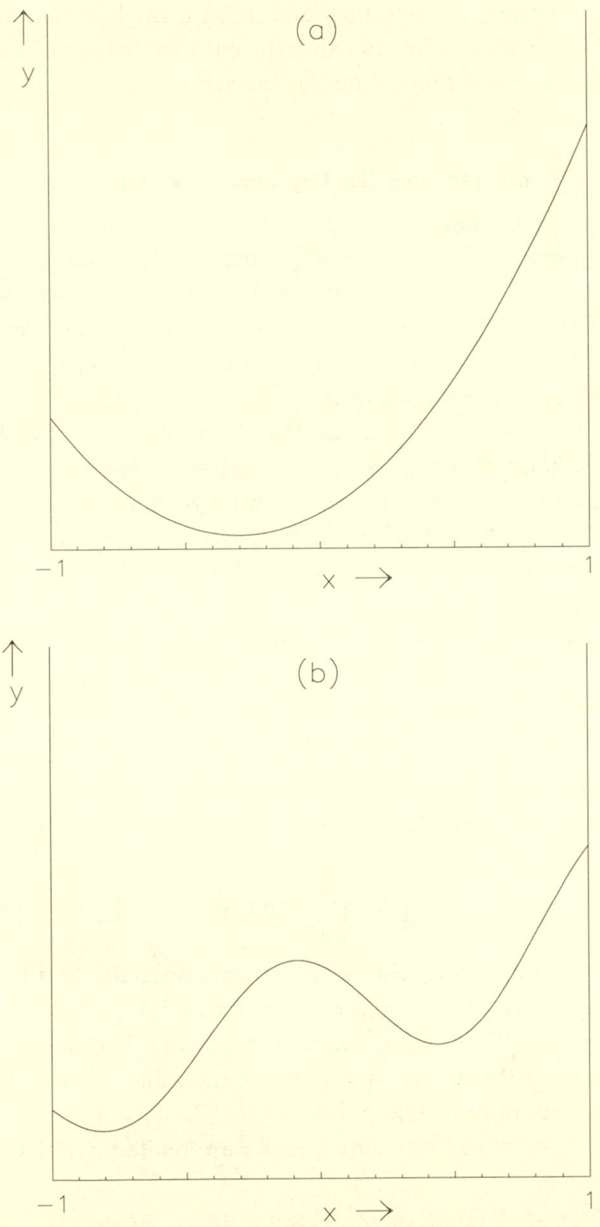

Fig. 8.2 The largest value of a one-dimensional function in a restricted range need not be a 'maximum'. In both (a) and (b), the largest value occurs at $x = +1$. In (a), the stationary value is a minimum, while in (b) there is a maximum, but it is not as large as the value at $x = 1$.

(iv) Demonstrating that when two bodies at different temperatures are placed in thermal contact, they end up at a single common temperature. This is done by maximising the entropy change of the system, subject to the constraint of energy conservation.

8.2 Deriving the Lagrange conditions

The best way of dealing with very simple constraints is to use them to eliminate some of the variables from the problem, and then to treat the remaining ones as independent. In this section, and indeed for the remainder of this chapter, we are going to analyse these problems by the more powerful technique of Lagrangian multipliers, which can then be used for more complicated situations.

In order to derive the conditions for a constrained stationary value problem, it is helpful first to remind ourselves of the essentials of the corresponding unconstrained situation. In particular, we will pay attention to how the constraints affect the problem, and how they necessarily change the equations we have to solve. For simplicity we consider a function of two variables, $f(x, y)$.

As we saw in Section 7.6, we start by writing down the Taylor series expansion

$$\delta f = \left(\frac{\partial f}{\partial x}\delta x + \frac{\partial f}{\partial y}\delta y\right) + \frac{1}{2}\left(\frac{\partial^2 f}{\partial x^2}\delta x^2 + 2\frac{\partial^2 f}{\partial x \partial y}\delta x \delta y + \frac{\partial^2 f}{\partial y^2}\delta y^2\right) + \ldots \tag{8.2}$$

Then we noted that if we are at a stationary point, the first order change in f must vanish, i.e.

$$\frac{\partial f}{\partial x}\delta x + \frac{\partial f}{\partial y}\delta y = 0 \tag{8.3}$$

Since x and y were *independent variables*, and since this condition must apply for any choice of the small changes δx and δy, $\partial f/\partial x$ and $\partial f/\partial y$ must *both* be zero. The higher order terms in the Taylor series merely serve to determine the nature of the stationary value, i.e. whether it is a maximum, minimum or saddle point.

Now in the constrained problem, x and y are related by the constraint, which we write as

$$c(x, y) = k \tag{8.4}$$

where k is a constant. For example, the constraint could be the line

$$y - 2x = 3$$

along which we are trying to find the stationary value of the function

$$f(x, y) = x^2 + 4y^2$$

It is almost a tautology to say that δx and δy are not independent. In fact

$$\frac{\partial c}{\partial x}\delta x + \frac{\partial c}{\partial y}\delta y = 0 \tag{8.5}$$

provides the relationship between δx and δy at any point. Thus although eqn (8.3) is still true, we cannot go on to deduce that the partial derivatives $\partial f/\partial x$ and $\partial f/\partial y$ are both zero, since this relied on the independence of δx and δy.

In order to deal with this new situation, the trick used is to introduce a Lagrangian multiplier whose value is at present completely arbitrary. We multiply eqn (8.5) by λ, and add it to (8.3) to obtain

$$\left(\frac{\partial f}{\partial x} + \lambda\frac{\partial c}{\partial x}\right)\delta x + \left(\frac{\partial f}{\partial y} + \lambda\frac{\partial c}{\partial y}\right)\delta y = 0 \tag{8.6}$$

We can now choose λ so that the first bracket in eqn (8.6) is zero. The equation then reduces to

$$\left(\frac{\partial f}{\partial y} + \lambda\frac{\partial c}{\partial y}\right)\delta y = 0 \tag{8.6$'$}$$

from which we deduce that the second bracket in eqn (8.6) is also zero.

Thus the equations that we must solve in order to find the constrained stationary value are

$$\left.\begin{aligned} \frac{\partial f}{\partial x} + \lambda\frac{\partial c}{\partial x} &= 0 \\ \frac{\partial f}{\partial y} + \lambda\frac{\partial c}{\partial y} &= 0 \\ \text{and } c(x, y) &= k \end{aligned}\right\} \tag{8.7}$$

This provides us with three equations for our three unknowns x, y and λ. In general we are interested in finding x and y; the value of λ itself is not too interesting, and it is merely introduced in order to help us solve the problem.

The reason that it is necessary to reintroduce the constraint $c(x, y) = k$ explicitly in eqns (8.7) is that up till that point in our derivation of the conditions, we have merely used the fact that $c(x, y)$ is a constant. The particular value of this constant determines which of the given family of relations between x and y applies in this specific case. Thus for example (i) of the previous section, the length L of fencing imposes the constraint

$2x + 2y = L$; a different constant on the right-hand side of this equation would imply a different length of fencing, and a different solution to the problem.

It is worth remarking that our method of deducing conditions (8.7) from eqns (8.3) and (8.5) deals with the variables in a rather asymmetric manner, in that λ was chosen to make the first bracket zero. The conditions (8.7) are, however, symmetric in x and y. In fact, as we may suspect, there are other ways of proceeding from (8.3) and (8.5) which are more symmetric in the way they treat the variables.

We also notice that our conditions apply only to the first derivatives of f (and c). This implies that we have merely found the stationary values, without determining their nature. The latter is beyond the scope of this book. In simple cases, the type of stationary value may be obvious from the specification of the problem and/or from the mathematical properties of the function and the constraints; or it may be possible to investigate it numerically.

Conditions (8.7) apply to a function of two variables, which are related by a single constraint. Clearly we could have more complicated problems. Thus the stationary values of a function $f(x, y, z, t)$ of four variables, subject to two constraints $c_1(x, y, z, t) = k_1$ and $c_2(x, y, z, t) = k_2$, are determined by

$$\left.\begin{aligned} \frac{\partial f}{\partial x} + \lambda_1 \frac{\partial c_1}{\partial x} + \lambda_2 \frac{\partial c_2}{\partial x} &= 0 \\ \frac{\partial f}{\partial y} + \lambda_1 \frac{\partial c_1}{\partial y} + \lambda_2 \frac{\partial c_2}{\partial y} &= 0 \\ \frac{\partial f}{\partial z} + \lambda_1 \frac{\partial c_1}{\partial z} + \lambda_2 \frac{\partial c_2}{\partial z} &= 0 \\ \frac{\partial f}{\partial t} + \lambda_1 \frac{\partial c_1}{\partial t} + \lambda_2 \frac{\partial c_2}{\partial t} &= 0 \\ c_1 &= k_1 \\ c_2 &= k_2 \end{aligned}\right\} \tag{8.8}$$

where we have now introduced two Langrangian multipliers λ_1 and λ_2, one for each constraint. A problem with six variables and four constraints is provided by example (iii) at the end of Section 8.1.

The general problem would involve a function $f(x_1, x_2, \ldots, x_i, \ldots, x_n)$ of n variables, among which there are m constraints of the form $c_j(x_1, x_2 \ldots x_n) = k_j$, where j is an index that runs from 1 up to m.

Then the conditions are

$$\left.\begin{aligned} \frac{\partial f}{\partial x_i} + \sum_{j=1}^{m} \lambda_j \frac{\partial c_j}{\partial x_i} = 0 \\ \text{and } c_j = k_j \end{aligned}\right\} \tag{8.9}$$

of which there are n equations of the first type and m of the second. Together these provide the $n + m$ equations for the $n + m$ unknowns (i.e. the n different x_i and the m Lagrangian multipliers). Thus as usual we have just enough equations for solving for the unknowns.

8.3 Some simple problems

In order to demonstrate how the method works, we here apply the Lagrangian technique to some very simple problems, which can mostly be solved more readily by other methods. It will, however, give us confidence that we know how to handle problems of this type.

8.3.1 Fencing around a field

We reverse the example (i) of Section 8.1, and assume that the farmer wishes to enclose a rectangular area of 100 m^2 with the minimum length L of fencing, and wants to know what should be the lengths of the sides x and y of the rectangle in order to achieve this. Thus the function we are minimising is

$$L = 2(x + y) \tag{8.10}$$

while the constraint is

$$c = xy = 100. \tag{8.11}$$

The required conditions (8.7) reduce to

$$\left.\begin{aligned} 2x + \lambda y = 0 \\ 2y + \lambda x = 0 \\ \text{and } xy = 100 \end{aligned}\right\} \tag{8.12}$$

From the first two we obtain

$$-\frac{\lambda}{2} = \frac{x}{y} = \frac{y}{x} \tag{8.13}$$

whence $x = \pm y$. Since the area xy is positive, we select the positive sign, and deduce that the optimum rectangular shape is in fact a square.

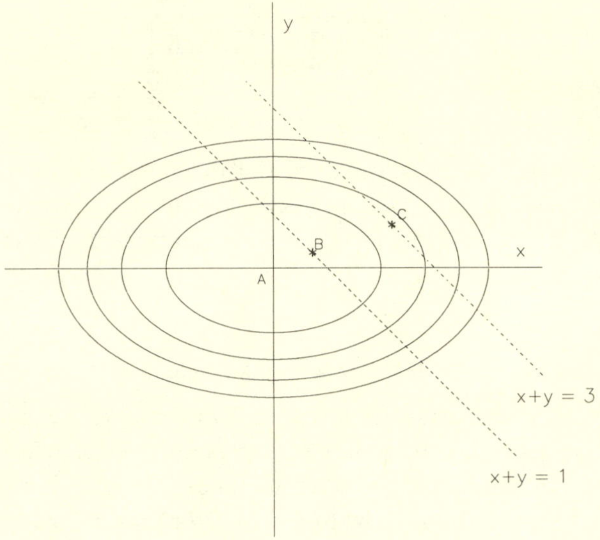

Fig. 8.3 The ellipses are the contour lines of the function (8.14), which has a maximum at the origin A. The dashed line is the constraint $x + y = 1$, which represents a railway line passing over the hill. The maximum height on this line is at the point B. An alternative track given by the dash-dotted line has its maximum at C.

Finally the last equation in (8.12) tells us that $x = y = 10$ m, and hence that the minimum length is 40 m.

8.3.2 The train on a hill

We shall assume that the height z of a certain hill which peaks at the origin can be approximated by the equation

$$z = A\ e^{-b(x^2+4y^2)} \tag{8.14}$$

and that a train travels on a straight path

$$c = x + y = 1 \tag{8.15}$$

across the side of the hill. This is illustrated in fig 8.3.

To find the maximum height of the train as it passes over the hill, we apply eqns (8.7) which yield

$$\left.\begin{aligned} -2Abx\, e^{-b(x^2+4y^2)} + \lambda = 0 \\ -8Aby\, e^{-b(x^2+4y^2)} + \lambda = 0 \\ x + y = 1 \end{aligned}\right\} \tag{8.16}$$

Eliminating the uninteresting Lagrangian multiplier from the first two equations, we find that

$$x = 4y$$

which on substitution into the constraint equation (i.e. the third one of (8.16)) yields

$$x = \frac{4}{5}, y = \frac{1}{5}$$

This is the position of the constrained maximum. It is to be contrasted with the position of the maximum height of the hill (i.e. the unconstrained maximum), which is at the origin.

We see that it is necessary to include the constraint in the conditions (8.7) or (8.16), in order to be able to solve for x and y (and λ). Had the railway line been given by $x + y = 3$, the first pair of equations in (8.16) would have been identical, and we would still have found that $x = 4y$, but the new constraint would have given us a different solution with coordinates (2.4, 0.6).

8.3.3 Axes of an ellipse

As mentioned in example (ii) in Section 8.1, the axes of an ellipse can be obtained by finding the maximum and minimum distances of a point on the ellipse to its centre.

In the trivial case of an ellipse given by

$$\frac{x^2}{a^2} + \frac{y^2}{b^2} = 1 \tag{8.17}$$

we find the stationary values of

$$D^2 = x^2 + y^2 \tag{8.18}$$

subject to the constraint (8.17). Then

$$\left.\begin{aligned} 2x + 2\lambda x/a^2 = 0 \\ \text{and } 2y + 2\lambda y/b^2 = 0 \end{aligned}\right\} \tag{8.19}$$

These imply

$$\left.\begin{aligned} x = 0 \text{ or } \lambda = -a^2 \\ \text{and } y = 0 \text{ or } \lambda = -b^2 \end{aligned}\right\} \tag{8.20}$$

We also remember that we still have to satisfy the constraint (8.18). The combination $x = y = 0$ fails to do this, so we are left with

$$\left.\begin{aligned} x = 0, \quad \lambda = -b^2, \quad y = \pm b \\ \text{or } y = 0, \quad \lambda = -a^2, \quad x = \pm a \end{aligned}\right\} \tag{8.21}$$

where the values of the Lagrange multipliers are related to the lengths of the axes, and the final part of each line of (8.21) is deduced from the constraint equation. Thus the ends of the axes are at $(0, \pm b)$ and $(\pm a, 0)$, and their lengths are $2b$ and $2a$.

The more interesting case is where the ellipse is rotated with respect to the axes, i.e.

$$Ax^2 + 2Bxy + Cy^2 = 1 \tag{8.22}$$

with B non-zero (see Problem 8.2). The constrained maxima and minima of D^2 are now given by

$$\left.\begin{aligned} x + \lambda(Ax + By) = 0 \\ y + \lambda(Bx + Cy) = 0 \end{aligned}\right\} \tag{8.23}$$

which can be rewritten as

$$\left.\begin{aligned} (A - v)x + By = 0 \\ Bx + (c - v)y = 0 \end{aligned}\right\} \tag{8.23$'$}$$

where $v = -1/\lambda$. (We make this somewhat curious choice for changing the variable for the Lagrange multiplier, in order to make it look suggestive of a matrix eigenvalue problem. We shall return to this point in Chapter 16 of Volume 2)

In order for these equations to be consistent (see Chapter 1), we require

$$-\frac{y}{x} = \frac{A - v}{B} = \frac{B}{c - v} \tag{8.24}$$

Thus

$$(A - v)(C - v) = B^2 \tag{8.24$'$}$$

which is a quadratic equation for v (or equivalently λ). There are two solutions, which provide us with two values of y/x. These are the gradients g of the axes. The individual values of x and y are obtained

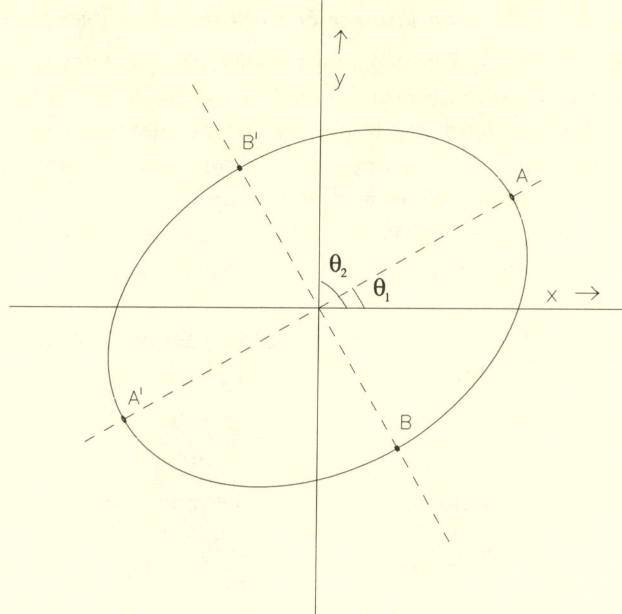

Fig. 8.4 An ellipse, given by equation (8.22). The end points of the axes (A and A′ for the major axis, and B and B′ for the minor one) are given by those points on the ellipse whose distances D from the origin are stationary values. The Lagrange condition results in first v and then the angles θ of the axes being given by eqns (8.24) ($g = \tan\theta = y/x$). Finally the coordinates of the points A, A′, B and B′ are determined from the equation of the ellipse.

by inserting each gradient g into the constraint eqn (8.22), which then becomes

$$(A + 2Bg + Cg^2)x^2 = 1$$

The two solutions of x for each value of g correspond to the two ends of each axis on the ellipse (see fig. 8.4).

For a three-dimensional ellipsoid, there are three axes. The distances from the origin to the ends of these axes are a maximum, a saddle point and a minimum.

8.4 More examples

Now that we are confident that we know how to apply the Lagrange method in simple situations, we go on to consider more interesting examples.

8.4.1 Shortest distance between two skew lines

In three-dimensional space, two non-parallel straight lines do not have to cross each other (see Section 2.5.3). For example, a railway line and the flight path of an aeroplane hopefully do not intersect. Of all possible pairs of points, one on each line, there is one combination which has the minimum separation. One way of finding this distance is by vector methods (see Section 3.5.1). We here use Lagrange multipliers to show that the vector joining the closest points is perpendicular to the directions of each of the lines.

The equation of a line (see Section 2.5.1) passing through the point (a_1, b_1, e_1) and in the direction of a vector (l_1, m_1, n_1) is given by

$$\frac{x_1 - a_1}{l_1} = \frac{y_1 - b_1}{m_1} = \frac{z_1 - e_1}{n_1} \tag{8.25}$$

To have these as standard constraints, we rewrite the equations as

$$\left.\begin{aligned} c_1 &= \frac{x_1 - a_1}{l_1} - \frac{y_1 - b_1}{m_1} = 0 \\ \text{and } c_1' &= \frac{x_1 - a_1}{l_1} - \frac{z_1 - e_1}{n_1} = 0 \end{aligned}\right\} \tag{8.26}$$

We also have a second line, which gives us constraints c_2 and c_2', which are obtained from (8.26) by changing all the subscript 1s to 2s.

The square of the distance D between (x_1, y_1, z_1) on the first line and (x_2, y_2, z_2) on the second is given by

$$D^2 = (x_1 - x_2)^2 + (y_1 - y_2)^2 + (z_1 - z_2)^2 \tag{8.27}$$

We minimise this with respect to the six variables $x_1, y_1, z_1, x_2, y_2, z_2$, subject to the four constraint equations $c_1 = c_1' = c_2 = c_2' = 0$.

Then

$$\frac{1}{2}\frac{\partial D^2}{\partial x_1} + \lambda_1 \frac{\partial c_1}{\partial x_1} + \mu_1 \frac{\partial c_1'}{\partial x_1} + \lambda_2 \frac{\partial c_2}{\partial x_1} + \mu_2 \frac{\partial c_2'}{\partial x_1}$$

$$= (x_1 - x_2) + \lambda_1/l_1 + \mu_1/l_1 = 0 \tag{8.28}$$

where we have made use of the fact that c_2 and c_2' are independent of x_1. Similarly, on differentiating with respect to y_1 and with respect to z_1 we find

$$(y_1 - y_2) - \lambda_1/m_1 = 0 \tag{8.29}$$

and

$$(z_1 - z_2) - \mu_1/n_1 = 0 \tag{8.30}$$

Eliminating the Lagrange multipliers λ_1 and μ_1 from the last three equations, we obtain

$$l_1(x_1 - x_2) + m_1(y_1 - y_2) + n_1(z_1 - z_2) = 0 \qquad (8.31)$$

which we recognise as the condition that the vector between the closest points on the two lines is perpendicular to the direction of the first line.

Not surprisingly, differentiating with respect to x_2, y_2 and z_2 would yield the similar condition that the difference vector must also be perpendicular to the second line. We have thus proved what we set out to do.

The two perpendicularity conditions together with the four constraint equations are then sufficient for us to find our six unknown coordinates. The Lagrange multiplier method is not the neatest one to use here, but it provides another example of its application to constrained problems.

8.4.2 Tossing a penny

At the risk of appearing to use an unnecessarily roundabout approach, we consider a very simple problem by the Lagrange method, simply in order to develop techniques that we are going to use in a more interesting example (see Section 8.4.3 below).

We ask the question of what is the most probable number of heads if we toss an unbiassed penny N times, where N is a large number. Our answer had better turn out to be $N/2$.

If the numbers of heads and tails are h and t respectively, the number of different orderings W in which these can occur is

$$W = \frac{N!}{h!\,t!} \qquad (8.32)$$

We are going to regard both h and t as variables, which are related by the constraint

$$c = h + t = N \qquad (8.33)$$

Then the most probable values of h and t are those that maximise W, subject to the constraint (8.33).

Because N and hence presumably h and t are large, we are going to ignore the fact that they are restricted to integral values, and thus not worry about differentiating with respect to them, as if they were continuous variables. In order to facilitate the differentiation, we shall change the variable from W to

$$\ln W = \ln N! - \ln h! - \ln t! \tag{8.32$'$}$$

and since N, h and t are all large, we shall use Stirling's approximation

$$\ln x! \approx x\ln x - x \tag{8.34}$$

for large x†.

Then eqn (8.32′) becomes

$$\ln W \approx N\ln N - N - h\ln h + h - t\ln t + t \tag{8.32$''$}$$

Differentiating this with respect to h and t, and incorporating the constraint (8.33) via a Lagrange multiplier λ, we obtain

$$\frac{\partial \ln W}{\partial h} + \lambda \frac{\partial c}{\partial h} = -\ln h + \lambda = 0 \tag{8.35}$$

and

$$-\ln t + \lambda = 0$$

Thus

$$\ln h = \ln t$$

and then using the constraint (8.33), we finally obtain

$$h = t = N/2 \tag{8.36}$$

We certainly do not claim that our method for deducing $h = N/2$ is the best one. Indeed our derivation is only an approximation valid for large N, whereas the result is true for all values (except of course for N odd, $h = (N \pm 1)/2$ are equally the most probable). We emphasise that we used it only in order to facilitate understanding the approach to the following physics problem.

8.4.3 A statistical mechanics problem

We are now going to use the method of Lagrangian multipliers, together with some fairly innocuous physics assumptions, in order to deduce the form of the distribution of speeds of molecules in a gas. Our approach will consist of the following steps:

† We can see that this is not an unreasonable approximation since

$$x! = x(x-1)(x-2)(x-3)\cdot\ldots\cdot 3\cdot 2\cdot 1$$

which is somewhat smaller than $x\cdot x\cdot x\cdot x\cdot\ldots\cdot x\cdot x\cdot x = x^x$

Thus $\ln x!$ is a bit smaller than $x\ln x$.

(i) Write down the number of ways W that the molecules can have any specific speed distribution.
(ii) Use Stirling's approximation for the logarithms of factorials of large numbers.
(iii) Maximise $\ln W$, subject to suitable constraints on the number of gas molecules with each speed. Thus we assume that the distribution of speeds assumed by the gas molecules will be the most likely one.

We assume we have a box containing N gas molecules at a fixed temperature. We imagine we divide up the speed scale into small regions of size δv, and that there are n_i molecules in the speed region centred on v_i. It is possible that not all the speed regions are equivalent, and so let g_i be the number of possible speed states within the ith region.

The number of possible ways of arranging the molecules as just described is†

$$W = \frac{N!}{\Pi n_i!} \Pi g_i^{n_i} \tag{8.37}$$

As in the previous problem of tossing a penny, we work with $\ln W$. Our use of Stirling's approximation is likely to be very well justified since the density of gas molecules at room temperature and pressure is $\sim 2 \times 10^{19}$ cm^{-3}. Thus under any normal circumstances N is very large, and for reasonable size speed bins, so will be the significant n_i. Then

$$\ln W \sim N \ln N - \sum (n_i \ln n_i - n_i) + \sum n_i \ln g_i \tag{8.37'}$$

This we want to maximise with respect to the n_i regarded as variables.

If all the n_i were independent, we could make W arbitrarily large simply by increasing all the n_i. However, to be realistic, we should impose two conditions. First the total number of molecules we are considering in our closed box should be equal to N, a constant. Second, as we imagine redistributing the molecules among the different speed regions of the distribution, we must do so in such a way that the total energy of all the molecules does not change, since this is determined by the temperature of the gas, which we regard as fixed. We write these constraints as

$$c_1 = \sum n_i = N \tag{8.38}$$

and

$$c_2 = \sum n_i \frac{1}{2} m v_i^2 = E \tag{8.39}$$

† The symbol Π means that we are required to take the product of the following variable, for all possible values of i.

Now our Lagrange approach will give us

$$\frac{\partial \ln W}{\partial n_i} + \lambda \frac{\partial c_1}{\partial n_i} + \nu \frac{\partial c_2}{\partial n_i} = -\ln n_i + \ln g_i + \lambda + \nu \frac{1}{2} m v_i^2 = 0$$

This must apply, for the same Langrange multipliers λ and ν, for all n_i and so

$$n_i = g_i e^{\lambda} e^{\nu \frac{1}{2} m v_i^2} \tag{8.40}$$

As usual, we must then use our constraints to determine the solution fully by eliminating λ and ν. We return to this question shortly.

One way of appreciating the need for the factor g_i is to consider what happens if we transform our number distribution (8.40) from having speed v_i as the dependent variable to energy ϵ_i instead. We obtain

$$\begin{aligned} n'_i &= g_i \frac{\partial v}{\partial \epsilon} e^{\lambda} e^{\nu \epsilon_i} \\ &= f_i e^{\lambda} e^{\nu \epsilon_i} \end{aligned} \tag{8.40$'$}$$

where n'_i is the number of molecules with energy in a small range $\delta\epsilon$ around ϵ_i. The factor $g_i \partial v / \partial \epsilon$ describes any intrinsic non-uniformity of the expected number of available states in our energy distribution. Since the energy of non-relativistic gas molecules of mass m is given by $\epsilon = \frac{1}{2} m v^2$, then $\partial v / \partial \epsilon = 1/mv$, and so it is impossible for both g_i and f_i to be simply constants.

Our starting point for deducing the functional form of g_i is that the distribution of the number of possible states s is uniform in velocity space, i.e.

$$\frac{\partial^3 s}{\partial v_x \partial v_y \partial v_z} = \frac{\partial^3 s}{v^2 \sin^2\theta \partial\theta \partial\phi \partial v} = A \tag{8.41}$$

where A is a constant, and θ and ϕ specify the direction of the molecule's velocity vector. The second form of the derivative for s is obtained from the first one simply by changing the variables from rectangular coordinates (v_x, v_y, v_z) in velocity space, to polar ones (v, θ, ϕ).

For the reasonable assumption that the speed distribution is independent of the direction of motion, we can integrate over θ and ϕ to obtain

$$g_i = \frac{\partial s}{\partial v} = 4\pi A v^2 \tag{8.42}$$

This is the exceptionally important physics result that the density of available states rises quadratically with the speed v. The underlying reason for this is that large speeds are intrinsically more likely than small ones since the latter require the unlikely combination of all three components of the velocity vector to be small. (This is equivalent to the fact that, if we imagine dividing up the earth into a large number of small cubic regions, there will be many more at large radii than at small ones.) An even stronger justification for this argument that g is proportional to v^2 can be derived within the context of quantum mechanics.

Thus eqn (8.40) becomes

$$n_i = 4\pi A v_i^2 e^{\lambda} e^{\nu \frac{1}{2} m v_i^2} \tag{8.43}$$

Presumably ν is negative, since otherwise n_i would diverge at large v_i. From here on, we use $\mu = -\nu$. We also drop the subscript i from n and v, since we now regard v as a continuous variable, rather than having a discrete set of values.

We now return to the overall constraints on the number N and energy E of the gas molecules. Given the distribution (8.43) for n_i, the average energy

$$\bar{\epsilon} = \frac{\int \epsilon n dv}{\int n dv} = \frac{\frac{1}{2} m \int v^4 e^{-\frac{1}{2}\mu m v^2} dv}{\int v^2 e^{-\frac{1}{2}\mu m v^2} dv} \tag{8.44}$$

where the ranges of integration extend from zero to infinity. The ratio of the integrals can readily be obtained to give

$$\bar{\epsilon} = \frac{3}{2\mu}$$

But from elementary kinetic theory, we know that the average kinetic energy of a molecule at temperature T is $\frac{3}{2}kT$, where k is Boltzmann's constant, so we deduce

$$\mu = -\nu = \frac{1}{kT} \tag{8.45}$$

The remaining Lagrange multiplier occurs in eqn (8.43) as the overall normalisation constant e^{λ} for the whole distribution. Thus its value is determined by the constraint

$$\int n dv = N$$

which gives

$$N = e^{\lambda} 4\pi A \left(\frac{m}{2kT}\right)^{\frac{3}{2}} \frac{1}{2}\sqrt{\pi/2} \tag{8.46}$$

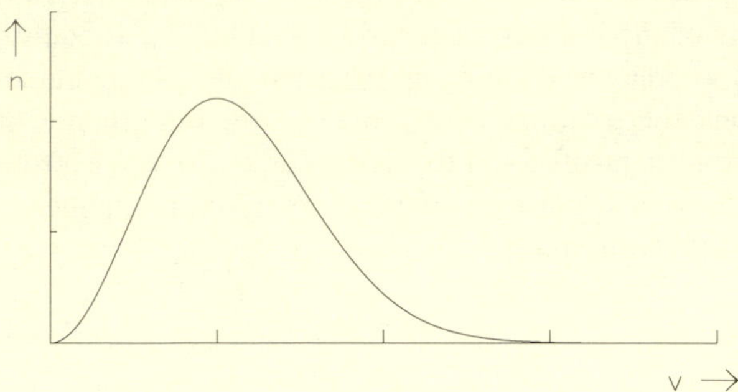

Fig. 8.5 The Maxwell–Boltzmann distribution (eqn 8.47) for the speeds v of molecules in a gas at a given temperature. This is derived by maximising the number of ways the speed distribution can be achieved, subject to the constraints of keeping the total number of molecules and their total energy fixed. The increase at small speeds is due to the v^2 factor, while the fall-off at large v is produced by the Boltzmann exponential factor.

which, in turn, can be substituted into (8.43) to give the final answer

$$n = N2\sqrt{2/\pi}\left(\frac{2kT}{m}\right)^{\frac{3}{2}} v^2 e^{-\frac{1}{2}mv^2/kT} \tag{8.47}$$

This then is the distribution of speeds of gas molecules in a container at constant temperature, which we deduced by finding the most likely distribution, subject to constraints on the overall number of molecules and on their total energy. Because constraints were involved, this provides a good example of the use of Lagrangian multipliers in a real physics problem.

Although eqn (8.47) has a forbidding look about it, it is simply the product of three factors whose origins are easy to understand. The last one $e^{-\frac{1}{2}mv^2/kT}$ arose from our constrained maximisation procedure; the v^2 term came from the 'density of available states' argument; and the rest is there simply to ensure that we have N molecules overall.

The speed distribution of eqn (8.47) is drawn in fig. 8.5. It is roughly bell-shaped. It goes to zero at small speeds because of the v^2. This density of states factor also occurs in such diverse fields as the black body radiation spectrum; the energy distribution of electrons emitted in beta decay; the speeds of free electrons in metallic conductors; etc. As explained earlier, it vanishes as v tends to zero because this requires all components of v to be small, which is relatively unlikely. It turns out that in an n-dimensional world, this results in a factor v^{n-1}; we live in three dimensions, so we have v^2.

At large v, it is the $e^{-\frac{1}{2}mv^2/kT}$ which is responsible for the fall-off. This is known as the Boltzmann factor, and is a specific example of the more general $e^{-\triangle E/kT}$, giving the relative probabilities of finding a system in two different states separated in energy by $\triangle E$, when the temperature is T. It is applicable over a very wide range of physical phenomena. If $\triangle E >> kT$, the occupancy of the higher state is likely to be very much smaller than that of the lower one. This sort of reasoning explains, for example, the reduction in pressure with height in an isothermal atmosphere; why electrons in atoms at ordinary temperatures are only very rarely to be found in excited states; the high resistance of a pure semi-conductor; etc.

Problems

The problems below should all be solved by the method of Lagrangian multipliers, even if simpler techniques are available.

8.1 (i) This is a two-dimensional problem. Find the point (x, y) on the line $2x = y - 1$ which is closest to the point $(1, -1)$. Confirm that the line joining (x, y) and $(1, -1)$ is perpendicular to the given line.

(ii) This is a three-dimensional problem. Find the point (x, y, z) on the plane $x + 2y - z = 5$, which is closest to the point (2,1,0). Confirm that the line joining (x, y, z) and (2,1,0) is normal to the plane.

8.2 Find the lengths of the axes of the ellipse $9x^2 - 4xy + 6y^2 = 1$, by determining the points in the x–y plane whose distances from the origin are either a maximum or a minimum, subject to the constraint that they lie on the ellipse.

8.3 A function $h(x, y)$ is given in terms of independent variables x and y. Show that the following problems have identical solutions.

(i) The maximum value of the third coordinate z, subject to the constraint $c = z - h(x, y) = 0$.

(ii) Maximisation of $z = h(x, y)$, as a function of the independent variables x and y.

8.4 A particle produced at the origin travels a short distance and then decays. The decay point is needed to determine how far it travelled, and hence how long it lived before it decayed. Its speed (which is required to convert the distance into a decay time) is determined from measurements on the observed decay products of the particle; these also give the direction along which the particle travelled as having direction cosine ratios (l, m, n). The estimated decay point is $(x_m \pm \sigma_x, y_m \pm \sigma_y, z_m \pm \sigma_z)$. Find the best estimate of the decay point (x_D, y_D, z_D) by minimising

$$\left(\frac{x_D - x_m}{\sigma_x}\right)^2 + \left(\frac{y_D - y_m}{\sigma_y}\right)^2 + \left(\frac{z_D - z_m}{\sigma_z}\right)^2$$

subject to the (two) constraints that (x_D, y_D, z_D) is in the known direction (l, m, n) from the origin. Show that the decay distance

$$D = \frac{\dfrac{lx_m}{\sigma_x^2} + \dfrac{my_m}{\sigma_y^2} + \dfrac{nz_m}{\sigma_z^2}}{\dfrac{l^2}{\sigma_x^2} + \dfrac{m^2}{\sigma_y^2} + \dfrac{n^2}{\sigma_z^2}}$$

NB1 From a mathematical viewpoint, this problem is simply a slightly generalised three-dimensional version of Problem 8.1(i).

NB2 This technique is actually used in high energy physics experiments in order to determine the decay times of short-lived particles, when the direct measurement of the decay position has relatively large errors (i.e. σ_x/x_m etc are not very small compared to unity). Then (x_D, y_D, z_D) is used as the decay position, rather than (x_m, y_m, z_m), and the decay distance is determined with a smaller error.

8.5 The discussion of Section 8.4.3 applied to a set of N gas molecules, which were assumed to behave classically and to be distinguishable. In fact, we ought to take into account that identical particles are indistinguishable, and according to the Pauli Exclusion Principle, if they are particles like electrons which have intrinsic spin of $\frac{1}{2}$ (measured in units of the reduced Planck constant $h/2\pi$), not more than one of them can occupy a given possible state. Then if we wish to distribute N such particles among a set of speed states, eqn (8.37) for the number of ways in which this can be done is replaced by

$$W = \Pi\left[\frac{g_i!}{(g_i - n_i)!n_i!}\right]$$

where n_i is the number of particles in the g_i possible states of speed within a bin of width δv centred at v_i.

Show that maximising this new W with respect to the n_i, subject to the constraints that the total number of particles N and their total energy E are constant, leads not to eqn (8.40) but instead to

$$n_i = \frac{g_i}{1 + e^{-\lambda} e^{-\nu \frac{1}{2} m v^2}}$$

where λ and ν are the arbitrary Lagrangian multipliers.

(This gives us the Fermi distribution, which governs, for example, the behaviour of a free electron gas in a metal. As in the text, $\nu = -1/kT$; λ is determined by the total number of particles present; and g_i is given by eqn (8.42), with the constant A determined from quantum mechanical considerations. The main feature of the above expresion is that $n_i \leq g_i$, as is expected since the Pauli principle excludes more than one particle from occupying each possible state.)

8.6 A moving billiard ball collides with a stationary one. The angles of the two billiard balls after the collision (with respect to the incident billiard ball direction) are measured as θ_1^m and θ_2^m, with accuracies σ_1 and σ_2 respectively. It is known that these angles should add up to $\pi/2$, but because of measurement inaccuracies they do not. Improved angles θ_1^f and θ_2^f are defined by finding the minimum (regarding θ_1^f and θ_2^f as variables) of the function

$$\frac{\left(\theta_1^f - \theta_1^m\right)^2}{\sigma_1^2} + \frac{\left(\theta_2^f - \theta_2^m\right)^2}{\sigma_2^2}$$

subject to the constraint

$$\theta_1^f + \theta_2^f = \pi/2$$

Use the method of Lagrangian multipliers to show that

$$\theta_1^f = \theta_1^m + \left(\pi/2 - \theta_1^m - \theta_2^m\right) / \left(1 + \sigma_2^2/\sigma_1^2\right)$$

(That is, an improved value of θ is obtained by correcting θ^m by a term which is proportional to the amount by which the sum $\theta_1^m + \theta_2^m$ differs from its expected value of $\pi/2$.)

Appendix A
Basic techniques

For a mirror mounted on a wall, the directions of up–down and left–right are equivalent. So why when you look into such a mirror do your right and left hands become interchanged, but your head does not become your feet?

Interesting conundrum

In this appendix, we collect a few basic ideas that are useful throughout the book.

A1 Systems of axes

In two dimensions, it is common to use either Cartesian (x, y) or polar (r, ϑ) coordinates to describe points in the plane. These are related by

$$\left.\begin{aligned} x &= r\cos\vartheta \\ y &= r\sin\vartheta \end{aligned}\right\} \quad \left.\begin{aligned} r^2 &= x^2 + y^2 \\ \tan\vartheta &= y/x \end{aligned}\right\} \tag{A1.1}$$

(see fig. A1). Which is more appropriate depends on the nature of the problem and its symmetries.

In three dimensions, Cartesian coordinates are (x, y, z). It is conventional to choose the direction of z so as to make a right-handed set of axes (see Section 2.3.1). An alternative is cylindrical polars, which involve r, ϕ and z. Here r and ϕ basically correspond to r and ϑ in the two-dimensional polar case, with z being added as the third coordinate (see fig. A2(a)). Then the relationships between the two sets of coordinates are

$$\left.\begin{aligned} x &= r\cos\phi \\ y &= r\sin\phi \\ z &= z \end{aligned}\right\} \quad \left.\begin{aligned} r^2 &= x^2 + y^2 \\ \tan\phi &= y/x \\ z &= z \end{aligned}\right\} \tag{A1.2}$$

It is common to choose ϕ as the symbol for the angular variable, so as to be consistent with the convention for spherical polars (see below); it is known as the azimuthal angle.

Another system is that of spherical polars, which utilises variables r, ϑ and ϕ, as shown in fig. A2(b). Here r is the distance of the space point from the origin, rather than that of the projection into the x–y plane as in cylindrical polars. Also

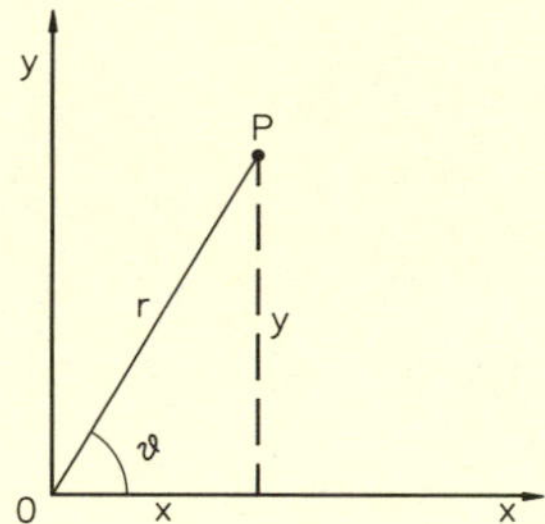

Fig. A1 The point P in two dimensions can be defined by Cartesian coordinates (x, y) or polar ones (r, ϑ).

ϑ is the polar angle between the z axis and the direction of the point. Thus r in spherical polars differs from r in cylindrical polars; and ϑ does not correspond to ϑ in two-dimensional polars.

Spherical polars are related to Cartesian coordinates by

$$\left.\begin{aligned} x &= r\sin\vartheta\cos\phi \\ y &= r\sin\vartheta\sin\phi \\ z &= r\cos\vartheta \end{aligned}\right\} \quad \left.\begin{aligned} r^2 &= x^2 + y^2 + z^2 \\ \tan\vartheta &= \sqrt{x^2 + y^2}/z \\ \tan\phi &= y/x \end{aligned}\right\} \tag{A1.3}$$

The whole of space corresponds to these variables covering the ranges

$$\left.\begin{aligned} 0 \le r \le \infty \\ 0 \le \vartheta \le \pi \\ 0 \le \phi \le 2\pi \end{aligned}\right\}$$

The relationships among integrals in three dimensions is

$$\begin{aligned} \int\int\int f(x, y, z)dxdydz &= \int\int\int f(r, \phi, z)rdrd\phi dz \\ &= \int\int\int f(r, \vartheta, \phi)r^2 \sin\vartheta drd\vartheta d\phi \end{aligned} \tag{A1.4}$$

Partial derivatives in the different sets of variables are related by the rule for changing variables as given in eqns (6.30). Thus for spherical polar coordinates

$$\begin{aligned} \frac{\partial f}{\partial \vartheta} &= \frac{\partial f}{\partial x}\frac{\partial x}{\partial \vartheta} + \frac{\partial f}{\partial y}\frac{\partial y}{\partial \vartheta} + \frac{\partial f}{\partial z}\frac{\partial z}{\partial \vartheta} \\ &= r\cos\vartheta\cos\phi\frac{\partial f}{\partial x} + r\cos\vartheta\sin\phi\frac{\partial f}{\partial y} - r\sin\vartheta\frac{\partial f}{\partial z} \end{aligned} \tag{A1.5}$$

A2 Transformations

An equation

$$f(x, y, z) = 0 \tag{A2.1}$$

in three dimensions represents a surface, such as a plane, an ellipsoid or something more complicated. In this section we are interested in the way the equation changes when either the surface or alternatively the axes are moved in some specified way. Two simple cases are translations or rotations.

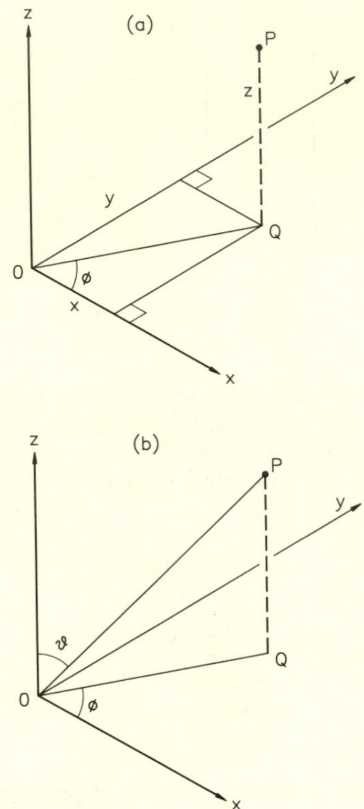

Fig. A2 The point P in three dimensions has Cartesian coordinates (x, y, z). Q is the foot of the perpendicular from P into the xy plane. Alternative descriptions of P are provided by (a) cylindrical polars (r, ϕ, z) and (b) spherical polars (r, ϑ, ϕ). In both cases ϕ is the azimuthal angle between OQ and the x axis. In (a), r is the length of OQ, while in (b) it is the length of OP.

A2.1 Translations

In a translation, the whole surface is moved without rotation, such that the point that was at the origin is now at (x_c, y_c, z_c) (see Fig. A3(a) and (b)). The recipe is that we replace x, y and z in eqn (A2.1) as follows:

$$\left.\begin{aligned} x &\to x - x_c \\ y &\to y - y_c \\ z &\to z - z_c \end{aligned}\right\} \tag{A2.2}$$

For example the sphere

$$x^2 + y^2 + z^2 = R^2$$

initially centred at the origin becomes

$$(x - x_c)^2 + (y - y_c)^2 + (z - z_c)^2 = R^2$$

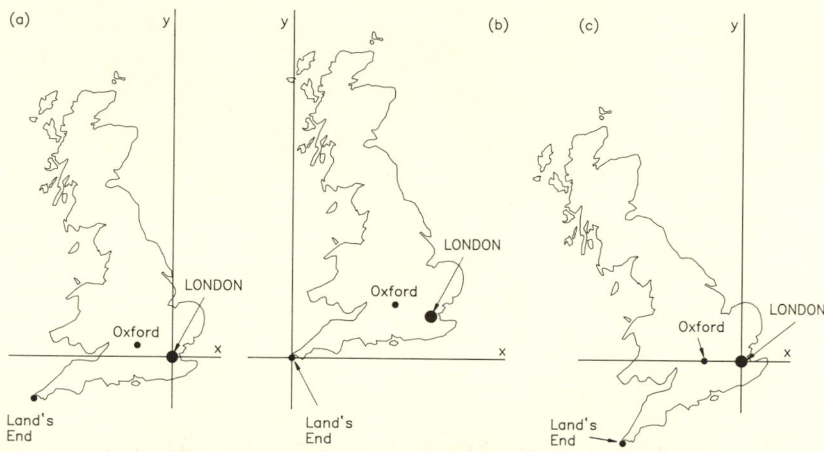

Fig. A3 (a) A map of England, Scotland and Wales, with the origin at London. (b) The same map, which has been translated so that Land's End is now at the origin. (c) It has now been rotated anticlockwise by 20°, so that both London and Oxford have $y = 0$.

In vector notation, (A2.2) is

$$\mathbf{r} \rightarrow \mathbf{r} - \mathbf{r}_c$$

where

$$\mathbf{r} = (x, y, z)$$

The rule (A2.2) is equivalent to the equations

$$\left.\begin{aligned} x' &= x + x_c \\ y' &= y + y_c \\ z' &= z + z_c \end{aligned}\right\} \qquad \text{(A2.2}'\text{)}$$

where the new coordinates x', y', z' are here distinguished from the old ones by having primes.

The above procedure of moving the object is known as an active transformation. A passive one is where we instead move the axes while leaving the object alone. In order to achieve the same displacement of the surface with respect to the new axes, we need to move the origin to $(-x_c, -y_c, -z_c)$. Thus (A2.2) or (A2.2′) also apply to this change of origin.

Although the form of these equations is almost self-evident, we do need to be careful about the signs in front of the x_c, y_c and z_c. It is thus advisable always to check with a simple example to ensure that they are correct.

Two-dimensional translations are discussed briefly in Section 2.2.2.1.

A2.2 Rotations

We consider the same surface defined by eqn (A2.1), but instead of translating it, we now rotate it by an anticlockwise angle θ around the z axis (see fig. A3(c)).

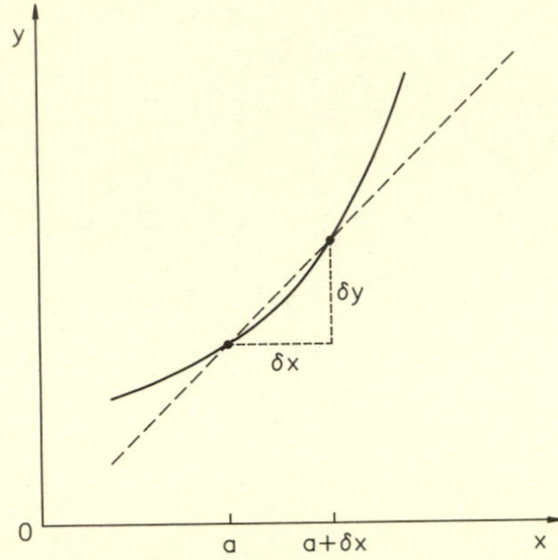

Fig. A4 The straight line passes through the neighbouring points at $x = a$ and $x = a + \delta x$ on the curve $y(x)$. As δx tends to zero, the ratio $\delta y/\delta x$ gets closer and closer to the gradient of the curve at $x = a$.

After some trivial but tedious trigonometry, we find that the coordinates change according to

$$\left.\begin{aligned} x &\to x\cos\theta + y\sin\theta \\ y &\to -x\sin\theta + y\cos\theta \\ z &\to z \end{aligned}\right\} \tag{A2.3}$$

Alternatively, if we denote the new coordinates by x', y' and z', then

$$\left.\begin{aligned} x' &= x\cos\theta - y\sin\theta \\ y' &= x\sin\theta + y\cos\theta \\ z' &= z \end{aligned}\right\} \tag{A2.3$'$}$$

The corresponding relationships for two dimensions (when z is absent) are given in eqns (2.11), where the logic of the $\sin\theta$ and $\cos\theta$ terms, and of the signs, is discussed. Again we notice sign changes between (A2.3) and (A2.3′), in a similar fashion to those of (A2.2) and (A2.2′).

Also we can achieve the same result by rotating the axes by an angle θ clockwise, so eqns (A2.3) and (A2.3′) apply for that situation as well. As with translations, care with signs is required.

The equations for rotations about the x or about the y axis are given in the text in eqns (2.55) and (2.56).

A3 Differentiation

The derivative of the function $y(x)$ at $x = a$ is the gradient of the curve of $y(x)$ at that particular value of x. It is defined by

$$y' = \frac{dy}{dx} = \lim\left[\frac{y(a + \delta x) - y(a)}{\delta x}\right] \tag{A3.1}$$

where 'lim' means the limit as δx tends to zero (see fig. A4).

Table A1. *List of derivatives.*[a]

$y(x)$	$y'(x)$
constant	0
x^n	nx^{n-1}
$\sin x$	$\cos x$
$\cos x$	$-\sin x$
$\tan x$	$\sec^2 x$
$\tan(x/2)$	$1/(1+\cos x)$
e^x	e^x
$\sinh x$	$\cosh x$
$\cosh x$	$\sinh x$
$ln x$	$1/x$
$f(x)g(x)$	$f'(x)g(x)+g'(x)f(x)$
$f(x)/g(x)$	$(g(x)f'(x)-f(x)g'(x))/(g(x))^2$
$\sin^{-1} x$	$\pm 1/\sqrt{1-x^2}$
$\cos^{-1} x$	$\pm 1/\sqrt{1-x^2}$
$\tan^{-1} x$	$\frac{1}{1+x^2}$
$\int_a^x y(t)dt$	$y(x)$
$u(x)w(x)-\int w(x)\frac{du}{dx}dx$	$u(x)v(x)^b$, where $w(x)=\int v(x)dx$

NB All the trigonometric functions must have the variable in radians.
[a] You should make your own additions to this list.
[b] Formula for integration by parts.

Differentiation is straight-forward. There are simple rules telling us how to differentiate a specified function $y(x)$. Thus Table A1 gives the derivatives of functions like x^n, $\sin x$, e^x and $f(x)g(x)$; very straight-forward extensions then give the derivatives of x^{-n}, $\sin^3 x$, e^{ax} and $e^{\sin x}$.

It is always possible to check that a derivative is correct by differentiating using first principles. Thus

$$\begin{aligned}\frac{d}{dx}(x^n) &\sim \frac{(x+\delta x)^n - x^n}{\delta x} \\ &= x^n\left[\left(1+\frac{\delta x}{x}\right)^n - 1\right]\Big/\delta x \\ &\sim [nx^{n-1}\delta x + \frac{n(n-1)}{2}x^{n-2}\delta x^2 + \ldots]/\delta x \\ &\sim nx^{n-1}\end{aligned} \qquad \text{(A3.2)}$$

where in the third line we have used the binomial theorem, and in the last line we have neglected terms which involve extra powers of the infinitesimal quantity δx. This procedure, of course, takes longer than remembering the answer.

Another way of checking a derivative is to substitute suitable numerical values, using a calculator. Thus at $x = 1$, the derivative of the function $y = x^3$ is estimated as

$$\frac{y(1.01)-y(1)}{1.01-1.00} = \frac{(1.01)^3-1}{0.01} \sim \frac{1.0303-1}{0.01} \sim 3.03 \qquad \text{(A3.3)}$$

This compares with the analytic answer of $3x^2$, which implies 3. Some care is needed in choosing a suitable value of δx. Since the derivative is defined in terms of a limit as δx tends to zero, too large a value may simply result in an estimate that differs from the correct gradient; while too small a value may cause problems because of rounding on a calculator. A final word of caution is that one numerical check can never prove that an analytic answer is correct.

Inverse functions can be differentiated simply by rewriting them. Thus for $y = \sin^{-1} x$, we write

$$x = \sin y$$

so that

$$\frac{dx}{dy} = \cos y = \pm\sqrt{1-x^2} = 1 \bigg/ \frac{dy}{dx} \tag{A3.4}$$

A little thought about the graph of $\sin^{-1} x$ confirms that the $\pm$ sign is required, as there are two possible gradients at any $|x| < 1$; and that values of $|x| > 1$ are unphysical.

Any inverse trigonometric function can be dealt with in this way. So can $y = \ln x$, which implies $x = e^y$.

Functions of functions are best dealt with by substitutions. Thus for $y = \sin x^2$, we define $z = x^2$, and then obtain

$$\frac{dy}{dx} = \frac{dy}{dz}\frac{dz}{dx} = (\cos z)(2x) = 2x\cos x^2 \tag{A3.5}$$

Similarly

$$y = \ln(\sin^{-1}\sqrt{1-2x^2})$$

yields to a series of substitutions.

If the function involves x raised to a power involving x, taking the logarithm of both sides can help. For example, if

$$y = x^x$$

then

$$\ln y = x \ln x$$

so that

$$\frac{1}{y}\frac{dy}{dx} = 1 + \ln x \tag{A3.6}$$

A4 Integration

Integration is the inverse of differentiation. Thus if

$$y(x) = \int^x z(x)dx \tag{A4.1}$$

then

$$\frac{dy}{dx} = z(x) \tag{A4.2}$$

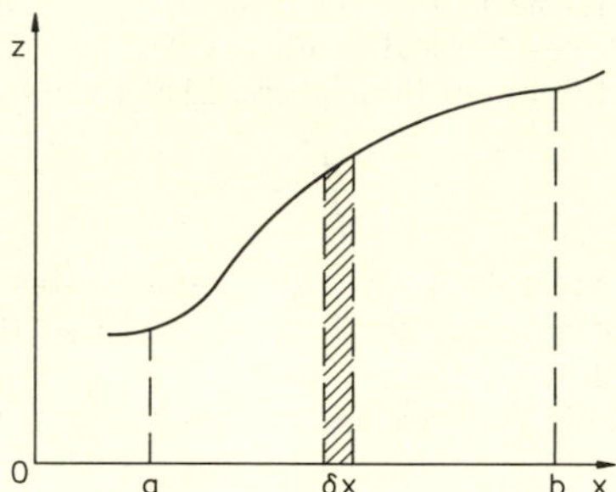

Fig. A5 The area of a very narrow shaded strip of width δx and of height $z(x)$ is approximately $z(x)\delta x$. The total area between $x = a$ and $x = b$ is found by adding up all such strips between these limits. As δx for each strip tends to zero, this is given by the integral $\int_a^b z(x)dx$.

This means that if we want to integrate $z(x)$, we first look at Table A1 to see if the function appears in the right-hand column under $y'(x)$. If so, its integral is simply $y(x)$.

In terms of the graph of $z(x)$ in fig. A5, $z(x)\delta x$ is approximately the area of the shaded thin strip. Then $\int_a^b z(x)dx$ is the result of adding up all the similar narrow strips between a and b, and then going to the limit of very small δx for every strip. This then gives the area under the curve, between a and b.

If we consider a small increase of δx beyond b, the change in the integral is just the extra little area added on, i.e. $z(b)\delta x$. Thus from eqn (A3.1), the derivative of the integral is

$$\frac{d}{dx}\left[\int_a^x z(x)dx\right] = \frac{z(b)\delta x}{\delta x} - = z(b) \tag{A4.3}$$

This confirms eqn (A4.2), in that the derivative of an integral is just the original function, and hence differentiation and integration are inverse processes.†

A4.1 Definite and indefinite integrals

An integral with specific limits, like

$$\int_{\pi/4}^{\pi/2} \sin x dx$$

is the area under a $\sin x$ curve, for x between $\pi/4$ and $\pi/2$. From Table A1, the integral of $\sin x$ is $-\cos x$, and so the area is

$$-\cos\frac{\pi}{2} + \cos\frac{\pi}{4} = 1/\sqrt{2} \quad §$$

† Except that integration can involve an arbitrary constant, so if a function is differentiated and then integrated, the result can differ from the original function by a constant.

§ This is not implausible since the height of the curve is less than 1, and the x range is $\pi/4$, so we expect an answer somewhat less than $\pi/4$.

Note that the actual variable in the integral disappears from the final answer; we would have obtained the identical result if x had been replaced by y, z, θ, etc.

An integral in which the upper limit is specified by a variable symbol, like

$$y(z) = \int^{z} \sin x dx$$

is called an indefinite integral. They are common when we are dealing with questions of just integrating a specific function, without the values of the limits of integration being defined.

The same entry in Table A1 tells us that integrating $\sin x$ yields $-\cos x$, but now the answer is

$$-\cos z + k$$

where k is a constant. This arises because we are now not specifying the lower limit of integration, so there is an essential ambiguity in the answer due to the arbitrariness of this choice. Alternatively, since the derivative of a constant is zero, we can always incorporate a constant in our answer when we integrate a function. This arbitrary constant does not appear in the definite integrals considered earlier.

This constant is not merely a boring formality; omitting it can simply make the answer wrong. Also it may turn out that the integral may be equal to some function of the variables (see, for example, Section 5.8). Thus perhaps

$$-\cos z + k = y \cot z$$

Then the solution is

$$y = k \tan z - \sin z$$

Leaving out the uninteresting-looking constant then results in our missing a significant z-dependent part of the solution.

As with definite integrals, the exact variable appearing inside the integral sign (i.e. x in the example above) is irrelevant and does not appear in the answer, which is a function of the upper limit variable z.

This distinction between these variables is, however, beclouded when the same symbol is employed for the two separate usages, e.g.

$$\int^{x} \sin x dx$$

Even more so, sometimes the upper limit is not explicitly specified, and is simply taken as the variable used in the integration, i.e. $\int \sin x dx$ is assumed to mean

$$\int^{x} \sin x dx = \int^{x} \sin \beta d\beta$$

A4.2 Methods of integration

As already mentioned, the simplest method is simply to find the function we need to integrate in the right-hand column of the table. However, not all possible functions appear there. Indeed some simple functions do not have explicit analytic integrals. An example is $\int_0^x e^{-x^2} dx$, which is important for the theory of experimental errors, the speed distribution of molecules in a gas, etc. It can, of course, be tabulated as a function of x, by performing the integration numerically; and the definite integral over the range 0 to infinity can be evaluated as $\sqrt{\pi}/2$, by the neat trick of changing from Cartesian to polar coordinates. However, apart from defining a new function as being the integral of e^{-x^2}, this has no analytic integral.

A4.2.1 Substitution

We can try using the substitution $x = x(y)$ to rewrite an integral

$$\int f(x)dx = \int f[x(y)]\frac{dx}{dy}dy \tag{A4.4}$$

It is essential to remember to replace the dx in the left-hand integral not merely by dy, but by $(dx/dy)dy$.

For example, in order to integrate $\tan x$, we can try the substitution $y = \cos x$. Then

$$\frac{dy}{dx} = -\sin x,$$

so the integral becomes

$$\begin{aligned}\int \tan x dx &= \int \left(\frac{\sin x}{y}\right)\left(-\frac{1}{\sin x}\right) dy \\ &= -\ln y + k \\ &= \ln(\sec x) + k\end{aligned} \tag{A4.5}$$

Choosing a suitable substitution requires experience and/or insight.

For definite integrals, there is no need to perform the last step of changing back to our original variable. It is, however, necessary to change the limits of integration to those appropriate for the new variable. Thus

$$\begin{aligned}\int_0^{\pi/4} \tan x dx &= [-\ln y]_1^{1/\sqrt{2}}, \\ &= \frac{1}{2}\ln 2\end{aligned} \tag{A4.6}$$

A4.2.2 Integration by parts

The 'integrating by parts' procedure is readily derived from that for differentiating a product. Thus since

$$\frac{d}{dx}[f(x)g(x)] = f'(x)g(x) + g'(x)f(x) \tag{A4.7}$$

on integration we obtain (without having to perform any serious integration)

$$f(x)g(x) = \int f'(x)g(x)dx + \int g'(x)f(x)dx \tag{A4.8}$$

We then simply rewrite this as

$$\int h(x)g(x)dx = H(x)g(x) - \int H(x)g'(x)dx \tag{A4.8$'$}$$

having defined

$$\left.\begin{aligned} h(x) &= f'(x) \\ \text{and} \quad H(x) &= \textstyle\int h(x)dx \end{aligned}\right\} \tag{A4.9}$$

This is an important rule, and is worth remembering. This is most easily achieved in terms of some form of words. For example, 'To integrate a product, write down one factor and integrate the other, and subtract the integral of what you have just obtained by integration times the differential of the other'.

If we are performing a definite integral, it is necessary to apply the limits to both the separate terms on the right-hand side.

This procedure is useful for products, where one of the factors behaves particularly simply on differentation. Thus

$$\begin{aligned}\int_0^x xe^{-x}dx &= \left[-xe^{-x}\right]_0^x + \int_0^x e^{-x}dx \\ &= -xe^{-x} + 1 - e^{-x}\end{aligned} \tag{A4.10}$$

Another integral which yields to this technique is

$$I_n = \int_0^x \sin^n x dx \tag{A4.11}$$

On writing the integrand as $(\sin x)(\sin^{n-1} x)$, a recurrence relation is obtained between I_n and I_{n-2}.

A final example is $\int \ln x dx$, which does not appear to contain a product. However, we consider it as

$$\begin{aligned}\int (\ln x)(1)dx &= [(\ln x)(x)] - \int (x)(1/x)dx \\ &= x\ln x - x + k\end{aligned} \tag{A4.12}$$

In definite integrals, some product functions are such that, on integration by parts, the first term vanishes. This is the case for

$$I_n = \int_0^\infty x^n e^{-x}dx \tag{A4.13}$$

The resulting expression is then somewhat simpler. In this case, a recurrence relation

$$I_n = nI_{n-1} \tag{A4.14}$$

is established.

A4.2.3 Numerical methods

Definite integrals like $\int_a^b f(x)dx$ can be estimated by any of the methods for finding the area under a known curve between specific limits. Of these the simplest conceptually (but one of the worst for numerical accuracy) is as follows: n equispaced x values are chosen to span the range a to b, and then the integral is estimated as

$$\frac{b-a}{n}\sum_i^n f(x_i) \tag{A4.15}$$

(see fig. A6).

Numerical integration thus usually involves more computation than numerical differentiation, but is a more robust process. Thus for a function $f(x)$ which oscillates very rapidly between two close values f_1 and f_2, the derivative may not even by defined, but the integral is between $(b-a)f_1$ and $(b-a)f_2$.

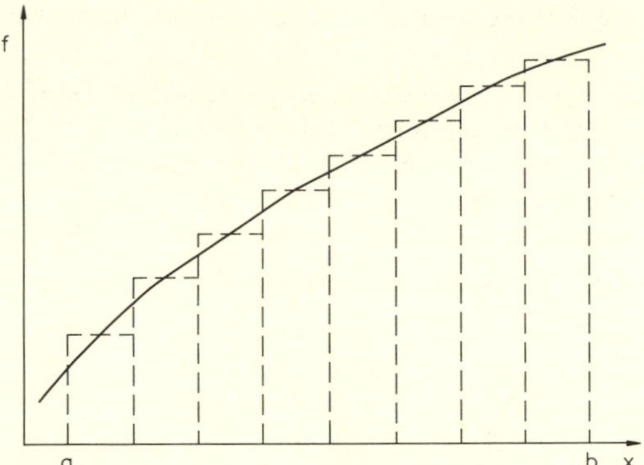

Fig. A6 The integral of the curve of $f(x)$ between $x = a$ and $x = b$ can be estimated by dividing the range into n equal steps (here $n = 8$), each of width $(b - a)/n$, and then adding the areas of each of the dashed rectangles, whose heights are simply the value of f at the middle of each step. As n is made larger, the numerical estimate of the integral improves.

A4.2.4 Simple checks

It is worth checking that an answer for a definite integral is not unreasonable. Thus, for $\int_a^b h(x)dx$ where $h(x)$ is positive throughout the range a to b (and b is greater than a), the answer must be positive, and be such that it increases as a decreases or as b increases.

If $f(x)$ is an odd function and is integrated between symmetric limits, the integral must be identically zero (see Appendix A8). This can circumvent a large amount of effort, as it is not necessary to perform the integration explicitly, and avoids the danger of algebraic slips. In contrast, for an even function $g(x)$,

$$\int_{-L}^{+L} g(x)dx = 2\int_0^L g(x)dx \tag{A4.16}$$

A5 Proof by induction

The sum of the n terms of an arithmetic progression

$$t_1 + t_2 + t_3 + \ldots + t_n$$

is given by

$$\sum_{i=1}^{n} t_i = \frac{n}{2}[2t_1 + (n-1)g] \tag{A5.1}$$

where g is the difference between neighbouring terms. This is an example of a formula that depends on the integer parameter n. Here n is the number of terms; it could alternatively be the number of times some operation is performed, etc.

In such cases, one method of verifying the given formula is that of induction. The steps involved are:

(1) Assume the formula is true for some particular value of n.

(2) Show that it then follows that the formula also applies for $n+1$.
(3) Demonstrate that the formula is true for some simple value of n (such as $n=1$).

Then since from step (3) it is true for $n=1$, it follows from step (2) that it is also valid for $n=2$. But we can repeat this argument to show that it is also true for $n=3$; and then for $n=4,5,6,\ldots$. Thus it follows that the formula applies for all positive integral values of n.

As an example, we return to the arithmetic series. The steps are

(1) We assume eqn (A5.1) is true for some n.
(2) If so

$$\begin{aligned}\sum_{i=1}^{n+1} t_i &= \left(\sum_{i=1}^{n} t_i\right) + t_{n+1} \\ &= \frac{n}{2}[2t_1 + (n-1)g] + (t_1 + ng) \\ &= (n+1)t_1 + \frac{n}{2}(n+1)g \\ &= \frac{n+1}{2}(2t_1 + ng) \end{aligned} \tag{A5.2}$$

This is just eqn (A5.1), with n replaced by $n+1$. Hence if (A5.1) is true for n, it is also true for $n+1$.

(3) We now test the formula with $n=1$. We obtain

$$\sum_{i=1}^{1} t_i = t_1 = \frac{1}{2}(2t_1)$$

which is clearly true.

Thus, by induction, we have shown that eqn (A5.1) for the sum of an arithmetic progression is true for any number of terms n.

Other examples that can be verified by the method of induction include the sum of a geometric progression; the sum of the cubes of integers; Leibnitz's theorem (see Appendix A6); and formulae involving integrals such as

$$I_{2n} = \int_{-\infty}^{+\infty} x^{2n} e^{-\alpha x^2}\, dx = \frac{(2n-1)!!}{2^n}\frac{\sqrt{2\pi}}{\alpha^{(2n+1)/2}} \tag{A5.3}$$

where the double factorial is given by the product

$$(2n-1)!! = (2n-1)(2n-3)(2n-5)\ldots 3\cdot 1$$

A6 Leibnitz's theorem

This gives the nth derivative of a product function as a series of terms

$$\begin{aligned} D^n(uv) = (D^n u)v + n(D^{n-1}u)(Dv) + \frac{n(n-1)}{1.2}(D^{n-2}u)(D^2 v) + \ldots \\ \ldots + n(Du)(D^{n-1}v) + u(D^n v) \end{aligned} \tag{A6.1}$$

where $D^m f$ means that the function f has to be differentiated m times.

Each term on the right-hand side contains n derivatives in total, some being on u and the others on v. The numerical factor in front of each term is simply a binomial coefficient ${}^nC_m$. Thus eqn (A6.1) can be rewritten as

$$D^n(uv) = \sum_{m=0}^{n} {}^nC_m(D^{n-m}u)(D^m v) \tag{A6.1$'$}$$

Eqns (A6.1) and (A6.1′) are of course symmetric in the two functions u and v.

If we set $n = 1$, we obtain the well known result

$$D(uv) = (Du)v + u(Dv) \tag{A6.2}$$

Differentiating (A6.2) again yields

$$D^2(uv) = (D^2u)v + (Du)(Dv) + (Du)(Dv) + uD^2v \tag{A6.3}$$

which again is consistent with (A6.1). Repeated differentiation then makes the Leibnitz result plausible. It can in fact be verified by the method of induction (see Appendix A5).

A6.1 Uses of Leibnitz's theorem

The evaluation of high order derivatives is particularly simple if one of the functions u or v is a low power of x. For example

$$\begin{aligned} D^{50}(x^2 \sin x) &= (D^{50} \sin x)x^2 + 50(D^{49} \sin x)(Dx^2) \\ &\quad + \frac{50 \times 49}{1 \times 2}(D^{48} \sin x)(D^2x^2) + \ldots \\ &= -x^2 \sin x + 100x \cos x + 1225 \sin x \end{aligned} \tag{A6.4}$$

All the terms represented by the dots in the first line are identically zero, because if x^2 is differentiated more than twice it vanishes.

Leibnitz's Theorem can be used to obtain recurrence relations between different orders of derivatives of a function, and this, in turn, can be employed to derive the function's Taylor or Maclaurin series (see Chapter 7). For example, the function

$$y = e^{a \sin^{-1} x} \tag{A6.5}$$

can readily be shown to satisfy the differential equation

$$(1 - x^2)\frac{d^2y}{dx^2} - x\frac{dy}{dx} - a^2y = 0 \tag{A6.6}$$

If we use Leibnitz's Theorem in order to differentiate $n - 2$ times, we obtain

$$(1 - x^2)y^{(n)}(x) - (2n - 3)xy^{(n-1)}(x) - (n^2 - 4n + 4 + a^2)y^{(n-2)}(x) = 0 \tag{A6.7}$$

where $y^{(n)}(x)$ is the nth derivative of y.† On now setting $x = 0$, this gives the recurrence relationship between different derivatives, evaluated at $x = a$:

$$y^{(n)}(0) = [(n - 2)^2 + a^2]y^{(n-2)}(0) \tag{A6.7$'$}$$

Now $y^{(0)}(0)$ and $y^{(1)}(0)$ are simply the function itself and its first derivative at $x = 0$, and can be evaluated directly from eqn (A6.5) as 1 and a respectively. Then we can use eqn (A6.7′) to evaluate all the derivatives $y^{(n)}(0)$ required for the Maclaurin series.

† It is worth checking that, for $n = 2$, eqn (A6.7) reproduces (A6.6).

Finally we obtain the power series expansion for y as

$$\begin{aligned} y(x) &= y(0) + y^{(1)}(0)x + \frac{1}{2!}y^{(2)}(0)x^2 + \frac{1}{3!}y^{(3)}(0)x^3 + \dots \\ &= 1 + ax + \frac{1}{2!}a^2x^2 + \frac{1}{3!}(1+a^2)ax^3 + \frac{1}{4!}(4+a^2)a^2x^4 + \dots \end{aligned} \tag{A6.8}$$

A7 Limits

We are sometimes required to find the limit of a function $f(x)$ as x approaches some particular value a (which is most often zero, sometimes infinity, and occasionally other fixed values such as ± 1, π, etc.). The point is that $f(x)$ is either a product

$$f(x) = g(x)h(x) \tag{A7.1}$$

where $g(a)$ or $h(a)$ is zero, and the other is infinity; or it is a ratio

$$f(x) = l(x)/m(x) \tag{A7.2}$$

where $l(a)$ and $m(a)$ are both zero. Then direct evaluation of $f(a)$ is impossible. The required answer is the limit that $f(x)$ reaches as x gets progressively closer and closer to a.

L'Hôpital's Rule can be used for ratios like (A7.2).† The rule says that the limit is given by

$$f(a) = \frac{D^n l}{D^n m} \tag{A7.3}$$

where $D^n l$ is the nth derivative of $l(x)$, evaluated at $x = a$. We choose n large enough so that at least one of $D^n l$ and $D^n m$ does not vanish, while all lower derivatives of l and m at $x = a$ do vanish. Then $f(a)$ is either zero, infinity or some other constant, depending on whether one or both of $D^n l$ and $D^n m$ are non-zero.

Thus for the limit as x tends to zero of

$$f(x) = \frac{x - \sin x}{x^3} \tag{A7.4}$$

the various steps of this procedure are as outlined in Table A2(a), and give an answer of 1/6. A second example,

$$f(x) = \frac{\tan x - \cot x}{(x - \pi/4)^2} \tag{A7.5}$$

as x tends to $\pi/4$, is set out in Table A2(b); the resulting limit is infinity.

This is all very satisfactory. However, although differentiation is not difficult in principle, in practice it can be laborious. Especially if we have to differentiate many times, it is all too easy to make an algebraic error.

In general it is better to perform a series expansion for $l(x)$ and $m(x)$ around the point $x = a$. Then the limit is zero, a constant or infinity depending on whether the first term of $l(x)$ is of a higher, the same or a lower power in $(x - a)$ than that for $m(x)$.

† It can also be used for products like (A7.1) by rewriting them as a ratio:

$$f(x) = g(x)/[1/h(x)]$$

Table A2. *L'Hôpital's rule in action.*
(a) Limit as x tends to zero of $(x - \sin x)/x^3$.

	Numerator $l(x)$		Denominator $m(x)$	
	Analytic form	Value at $x = 0$	Analytic form	Value at $x = 0$
Function	$x - \sin x$	0	x^3	0
First deriv	$1 - \cos x$	0	$3x^2$	0
Second deriv	$\sin x$	0	$6x$	0
Third deriv	$\cos x$	1	6	6

Therefore $f(x) \to 1/6$

(b) Limit as x tends to $\pi/4$ of $(\tan x - \cot x)/(x - \pi/4)^2$.

	Numerator $l(x)$		Denominator $m(x)$	
	Analytic form	Value at $x = \pi/4$	Analytic form	Value at $x = \pi/4$
Function	$\tan x - \cot x$	0	$(x - \pi/4)^2$	0
First deriv	$\sec^2 x + \operatorname{cosec}^2 x$	1	$2(x - \pi/4)$	0

Therefore $f(x) \to \infty$

Thus for the function $f(x)$ of (A7.4), we can write the numerator for small x as

$$x - (x - \frac{x^3}{3!} + \ldots) = \frac{x^3}{6} + \text{ higher powers of } x \tag{A7.6}$$

When we divide by x^3, we obtain

$$f(x) \sim \frac{1}{6} + \text{ a power series in } x \tag{A7.7}$$

Thus the limit is $1/6$ as before, but with less working.

It takes a little experience to become efficient at using the power series method. In particular, we need to develop a feeling for how far we have to expand our functions in order to keep the leading term in the numerator and in the denominator. Thus in (A7.6) it was not necessary to expand beyond its $x^3/3!$ term. On the other hand, if we had written down only its first term (i.e. $\sin x \sim x$), we would incorrectly have obtained a value zero for the numerator, and perhaps also for the answer. If in doubt, it is clearly better to write down too many terms and waste a bit of time, than to use too few and obtain a wrong result.

In this day and age of pocket calculators, it is a crime not to check a limit numerically. All that is required is to evaluate $f(x)$ for a few choices near $x = a$, and to see experimentally what the trend of the function is. Of course, it is necessary to have x near enough to a so that we are indeed investigating the limit; but not so close that the main digits on the calculator cancel, and we are

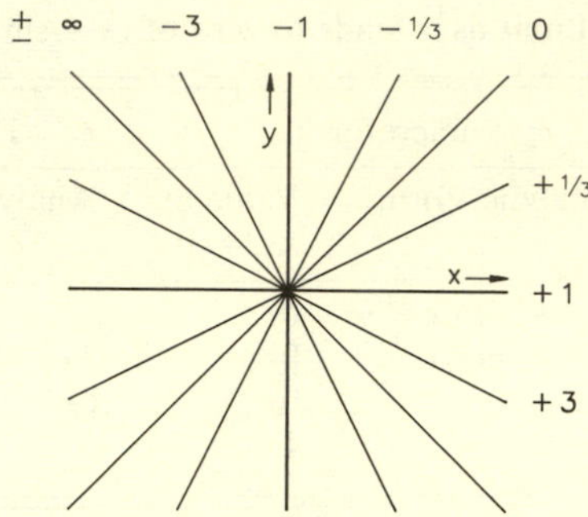

Fig. A7 Contour lines of the function $f(x, y) = (x - y)/(x + y)$. As the origin is approached along a straight line, f remains constant but its value depends on the gradient of the line. The limit of f as x and y tend to zero thus depends on the direction in which the origin is approached.

left with some rounding errors which result in a fairly random estimate. Thus with my calculator, I obtain for the function of (A7.4):

$$\left.\begin{array}{ll} x = 1 & f(x) = 0.84\ldots \\ x = 0.5 & f(x) = 0.1646\ldots \\ x = 0.1 & f(x) = 0.16658\ldots \\ x = 0.01 & f(x) = 0.1666653\ldots \\ x = 10^{-3} & f(x) = 0.1666 \\ x = 10^{-4} & f(x) = 0.16 \\ x = 5 \times 10^{-5} & f(x) = 0.144 \\ x = 10^{-5} & f(x) = 0 \end{array}\right\} \tag{A7.8}$$

Thus there is a reasonable range of x for deducing that the limit is around 1/6.

There are also problems about limits in more than one dimension, and these need a little more care. For the one-dimensional case there are at most two possibilities for a particular limit, depending on whether we approach $x = a$ from above or from below (and for a function $f(x)$ which is continuous at $x = a$, these will coincide). For two-dimensional functions, however, the situation can be more interesting.

Thus for the deceptively simple-looking function

$$f(x, y) = \frac{x - y}{x + y} \tag{A7.9}$$

we can consider the limit at the origin. 'Clearly' the answer is $+1$, since for $y = 0$

$$f(x, 0) = \frac{x}{x} = 1 \tag{A7.10}$$

independent of x. However,

$$f(0, y) = \frac{-y}{y} = -1 \tag{A7.11}$$

so the limit as y decreases to zero now appears to be -1. What has gone wrong?

The answer, of course, is that nothing has gone wrong. The contour lines of the function (A7.9) are shown in fig. A7. If we move along a straight line of constant gradient g in the x–y plane, the function is a constant, with value $(1-g)/(1+g)$. Since the function is constant along this line, this also is the limit as we approach the origin *along the line.* The only point is that the limit is different for each direction. Thus the limit as we approach the origin does not have a unique value. If we tried to guess the answer by choosing random points closer and closer to the origin, we may have difficulty in recognising this behaviour from our numerical answers.

A8 Parity

There is a kindergarten joke, which consists of drawing a squiggle and asking someone what it is. When they give up, you draw an identical one and tell them 'I don't know either, but here is another'. More or less the same basic idea is used in the concept of parity.

A8.1 Odd and even functions

If we are given a function $f(x)$, the fundamental question here is what happens at $f(-x)$. There are two special cases

(i) $f(-x) = f(x)$ for all x (A8.1)
(ii) $f(-x) = -f(x)$ for all x (A8.2)

In the former case the function is said to be 'even' and in the latter it is 'odd'. Examples of even functions are $\cos x$, $\ln(|x|+3)$ and x^n where n is an even integer (which explains why these functors are called even). In contrast $\sin x$, $\tan x$ and x^n (for n odd) are odd functions. Some even and odd functions are drawn in Fig. A8.

Differentiation converts an odd function into an even one, and vice versa. Thus the derivative of the even function $\cos x$ is the odd one $-\sin x$. Another useful property is that if we represent an odd function by -1 and an even one by $+1$, then the product of several factors is odd or even, as given by the product of the relevant factors of ± 1. Thus $x^3 \sin^2 x$ is odd because $\sin^2 x$ is even and x^3 is odd.

Of course not every function is even or odd. An example is e^x, since for example $e^1 = 2.718$ while $e^{-1} = 0.368$ and is neither $\pm e^1$. Nevertheless any arbitrary function can be expressed as the sum of an even and an odd part. Thus

$$\begin{aligned} e^x &= \frac{1}{2}(e^x + e^{-x}) + \frac{1}{2}(e^x - e^{-x}) \\ &= \cosh x + \sinh x \end{aligned} \tag{A8.3}$$

where $\cosh x$ and $\sinh x$ are respectively even and odd.

The most useful property of odd functions $f(x)$ is that their integral between symmetric limits is identically zero, i.e.

$$\int_{-L}^{+L} f(x)dx = 0 \tag{A8.4}$$

This is simply because, for every contribution to the integral at positive x, there is a corresponding one of equal magnitude but opposite sign at $-x$. Thus the value of the integral from $-L$ to 0 exactly cancels that from 0 to L. This is an exceptionally useful property, since it enables us to write down the value of the

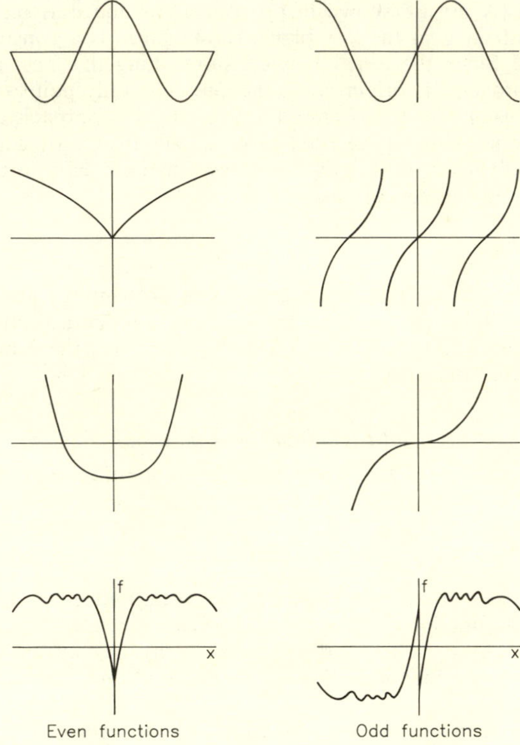

Fig. A8 Examples of even and of odd functions.

integral by inspection, without having to do any explicit integration. Indeed this result is true even if it is impossible to integrate $f(x)$ analytically.

Conversely, if we find that the answer to an integral is zero, we may be able to look at it and see whether this is obviously true. However, if it emerges that the integrand is always positive, we should realise that we must have made a mistake.

The above ideas find a ready extension to more than one dimension. Thus the function $g(x, y, z)$ of the three Cartesian coordinates x, y and z is even or odd if

$$g(x, y, z) = \pm g(-x, -y, -z)$$

That is, there is a simple relationship between the values of the function at any pair of points which are reflections of each other through the coordinate origin. Alternatively, if spherical polar coordinates are used (see Section A1), the reflection through the origin requires that

$$g(r, \vartheta, \phi) = \pm g(r, \pi - \vartheta, \pi + \phi)$$

As in the one-dimensional case, if g is odd, its integral over a suitably symmetric region about the origin will be identically zero.

A8.2 Applications to physics

The above ideas find wide application in physics, where the subject is known as 'parity'. Thus $f(x)$ is said to have odd or even parity depending on whether it is an odd or an even function.

Consider, for example, an atom with an electron in an excited state, and which decays to the ground state by emitting a photon i.e. $A^* \rightarrow A + \gamma$. According to quantum mechanics, the rate at which such a process occurs depends on $|I|^2$, the square of an integral. I has the form

$$I = \int_{-\infty}^{+\infty} f(x)dx$$

where $f(x)$ involves the product of three functions of x:

(i) the quantum mechanical wave function describing the electron in its excited state;
(ii) a similar wave function for the ground state; and
(iii) a term describing the electromagnetic interaction responsible for the decay process.

In order for the rate to be non-zero, I must not vanish, and this means that $f(x)$ must not be an odd function of x. This, in turn, imposes restrictions on the relative parities of the wave functions for the two states involved in the transition; if these two states have the wrong relative parities, this particular type of transition is forbidden. 'Selection rules' governing such decay processes are simply statements about the relative properties (e.g. parity and angular momentum) of the states involved in order for the transition to be possible. The rule for the Bohr atom stating that the orbital angular momentum of the electron must change by one unit is just such an example.

Similar rules apply to other transition processes, such as nuclear alpha, beta and gamma decays.

For a long time it was believed that parity was conserved in all physical processes. This is equivalent to the statement that it makes no basic difference whether right-handed or left-handed axes are used to describe a physical system. Thus if we were shown a filmed recording of some physical process, we would be unable to tell whether the pictures had been taken directly or via a mirror.

An equivalent way of describing this is to consider imaginary beings on some very distant planet. Remarkably enough, they have managed to learn English, and their only remaining uncertainty is to know whether they have got right and left the correct way round. If parity is conserved, there is no way in which you would send them a message telling them to perform a physics experiment that would resolve this problem. (Suggesting that they look at specific star formations or put their hand on their hearts is regarded as cheating.)

However, in 1957, it was suggested by Lee and Yang and shortly afterwards confirmed by Wu that nuclear beta decay processes violate parity. This means that left and right are not equivalent, and we can instruct the distant beings to perform a beta decay experiment that will enable them to distinguish their right hand from their left one.

Thus we see that very simple ideas about the oddness or evenness of functions leads us to some surprises about the very nature of the space of the Universe we inhabit.

Appendix B
Useful formulae

This appendix contains a brief summary of the crucial equations from each of the chapters.

Chapter 1 Simultaneous equations

Solutions of

$$\left.\begin{aligned} a_{11}x + a_{12}y + a_{13}z &= c_1 \\ a_{21}x + a_{22}y + a_{23}z &= c_2 \\ a_{31}x + a_{32}y + a_{33}z &= c_3 \end{aligned}\right\} \tag{1.10}$$

given by

$$x = \frac{\begin{vmatrix} c_1 & a_{12} & a_{13} \\ c_2 & a_{22} & a_{23} \\ c_3 & a_{32} & a_{33} \end{vmatrix}}{D}, \text{ etc} \tag{1.11}$$

where

$$D = \begin{vmatrix} a_{11} & a_{12} & a_{13} \\ a_{21} & a_{22} & a_{23} \\ a_{31} & a_{32} & a_{33} \end{vmatrix} \tag{1.12}$$

For

$$\left.\begin{aligned} a_{11}x + a_{12}y + a_{13}z &= 0 \\ a_{21}x + a_{22}y + a_{23}z &= 0 \\ a_{31}x + a_{32}y + a_{33}z &= 0 \end{aligned}\right\} \tag{1.20}$$

solution other than

$$x = y = z = 0 \tag{1.21}$$

requires $D = 0$. Solution is then line or plane through origin.

Chapter 2 Three-dimensional geometry

Angle β between two directions:

$$\cos\beta = \frac{a_1a_2 + b_1b_2 + c_1c_2}{\sqrt{(a_1^2 + b_1^2 + c_1^2)(a_2^2 + b_2^2 + c_2^2)}} \tag{2.19}$$

Plane:

$$ax + by + cz = d \tag{2.21}$$

or

$$a(x - x_0) + b(y - y_0) + c(z - z_0) = 0 \tag{2.26}$$

Line:

$$\frac{x - x_0}{l} = \frac{y - y_0}{m} = \frac{z - z_0}{n} \tag{2.32}$$

or simultaneous equations for two non-parallel planes.

Shortest distance from (x_0, y_0, z_0) to plane $ax + by + cz = d$ (see Problem 2.6)

$$p = \frac{d - ax_0 - by_0 - cz_0}{\sqrt{a^2 + b^2 + c^2}} \tag{2.24}$$

Shortest distance between two skew lines is perpendicular to both. Lines intersect if

$$\begin{vmatrix} l_1 & l_2 & x_2 - x_1 \\ m_1 & m_2 & y_2 - y_1 \\ n_1 & n_2 & z_2 - z_1 \end{vmatrix} = 0 \tag{2.51}$$

Sphere:

$$(x - x_c)^2 + (y - y_c)^2 + (z - z_c)^2 = R^2 \tag{2.53}$$

Translation, moving the object:

$$\left.\begin{aligned} x &\to x - x_c \\ y &\to y - y_c \end{aligned}\right\} \tag{2.7}$$

Rotation of object by θ anticlockwise about z-axis:

$$\left.\begin{aligned} x' &= x\cos\theta - y\sin\theta \\ y' &= x\sin\theta + y\cos\theta \end{aligned}\right\} \tag{2.11}$$

Chapter 3 Vectors

Scalar product:

$$\mathbf{a}\cdot\mathbf{b} = ab\cos\theta \tag{3.13}$$

$$= a_x b_x + a_y b_y + a_z b_z \tag{3.18}$$

Vector product:

$$\mathbf{a}\wedge\mathbf{b} = ab\sin\theta\,\hat{\mathbf{n}} \tag{3.14}$$

$$= \begin{vmatrix} \mathbf{i} & \mathbf{j} & \mathbf{k} \\ a_x & a_y & a_z \\ b_x & b_y & b_z \end{vmatrix} \tag{3.22}$$

Triple scalar product:

$$\mathbf{a}\cdot(\mathbf{b}\wedge\mathbf{c}) = (\mathbf{a}\wedge\mathbf{b})\cdot\mathbf{c} = \begin{vmatrix} a_x & a_y & a_z \\ b_x & b_y & b_z \\ c_x & c_y & c_z \end{vmatrix} \qquad (3.30'),\ (3.31)$$

$$\{\mathbf{a}\,\mathbf{b}\,\mathbf{c}\} = -\{\mathbf{a}\,\mathbf{c}\,\mathbf{b}\} \tag{3.32'}$$

Triple vector product:

$$(\mathbf{a} \wedge \mathbf{b}) \wedge \mathbf{c} = (\mathbf{a} \cdot \mathbf{c})\mathbf{b} - (\mathbf{b} \cdot \mathbf{c})\mathbf{a} \tag{3.34}$$

Plane

$$\mathbf{n} \cdot \mathbf{r} = d \tag{3.39}$$

Line

$$\mathbf{r} = \mathbf{r}_0 + \lambda \mathbf{v} \tag{3.42}$$

Distance between skew lines:

$$d = \frac{(\mathbf{s} - \mathbf{t}) \cdot (\mathbf{v} \wedge \mathbf{w})}{|\mathbf{v} \wedge \mathbf{w}|} \tag{3.44}$$

Reciprocal vector set:

$$\mathbf{a}' = \mathbf{b} \wedge \mathbf{c}/\{\mathbf{a}\,\mathbf{b}\,\mathbf{c}\}, \text{etc} \tag{3.54}$$

$$\mathbf{f} = (\mathbf{f} \cdot \mathbf{a}')\mathbf{a} + (\mathbf{f} \cdot \mathbf{b}')\mathbf{b} + (\mathbf{f} \cdot \mathbf{c}')\mathbf{c} \tag{3.53''}$$

$$\frac{d}{dt}(\mathbf{a} \cdot \mathbf{b}) = \mathbf{a} \cdot \frac{d\mathbf{b}}{dt} + \mathbf{b} \cdot \frac{d\mathbf{a}}{dt} \tag{3.67}$$

$$\frac{d}{dt}(\mathbf{a} \wedge \mathbf{b}) = \mathbf{a} \wedge \frac{d\mathbf{b}}{dt} + \frac{d\mathbf{a}}{dt} \wedge \mathbf{b} \tag{3.68}$$

Acceleration in rotating system of axes:

$$\mathbf{a} = \mathbf{a}_r + 2\boldsymbol{\omega} \wedge \mathbf{v}_r + \boldsymbol{\omega} \wedge (\boldsymbol{\omega} \wedge \mathbf{r}) + \frac{d\boldsymbol{\omega}}{dt} \wedge \mathbf{r} \tag{3.74''}$$

Chapter 4 Complex numbers

Modulus of product is product of moduli.
Argument of product is sum of arguments.

de Moivre's Theorem:

$$(\cos\theta + i\sin\theta)^n = \cos n\theta + i\sin n\theta \tag{4.36}$$

$$L\frac{d^2q}{dt^2} + R\frac{dq}{dt} + \frac{q}{C} = V_o\cos\omega t \tag{4.56'}$$

has particular integral

$$\begin{aligned} \mathscr{I} &= j\omega\mathscr{Q} \\ &= \frac{j\omega\mathscr{V}}{(1/C - \omega^2 L) + j\omega R} \end{aligned} \tag{4.73}$$

Resonant frequency:

$$\omega_R = 1/\sqrt{LC} \tag{4.90}$$

Quality factor:

$$Q = \frac{\omega_R}{2\delta\omega} \tag{4.98}$$

$$Q = \frac{\omega_R L}{R} = \frac{1}{R}\sqrt{\frac{L}{C}} = \frac{1}{\omega_R C R} \tag{4.99}$$

Phase variation:

$$\tan\phi = -\frac{\omega L - 1/\omega C}{R}$$
$$= Q\left[\frac{\omega_R}{\omega} - \frac{\omega}{\omega_R}\right] \qquad (4.101)$$

'Uncertainty Principle':

$$T\delta\omega = 1 \qquad (4.105)$$

Chapter 5 Ordinary differential equations

Important terms: order, degree, linear, homogeneous, complementary function, particular integral.

Methods of solving differential equations: see Table 5.3

Functional forms of particular integrals of second order differential equations: see Table 5.1

Rules and manipulations for D operators: see Table 5.2

Graphical techniques, series solutions, change of variables, numerical approximations.

Chapter 6 Partial derivatives

$$\frac{df}{dt} = \frac{\partial f}{\partial x}\frac{dx}{dt} + \frac{\partial f}{\partial y}\frac{dy}{dt} \qquad (6.24)$$

$$\frac{df}{dx} = \left(\frac{\partial f}{\partial x}\right)_y + \left(\frac{\partial f}{\partial y}\right)_x \frac{dy}{dx} \qquad (6.25)$$

$$\left(\frac{\partial f}{\partial x}\right)_u = \left(\frac{\partial f}{\partial x}\right)_y + \left(\frac{\partial f}{\partial y}\right)_x \left(\frac{\partial y}{\partial x}\right)_u \qquad (6.29)$$

$$\left.\begin{aligned} \frac{\partial f}{\partial t} &= \frac{\partial f}{\partial x}\frac{\partial x}{\partial t} + \frac{\partial f}{\partial y}\frac{\partial y}{\partial t} \\ \text{and } \frac{\partial f}{\partial u} &= \frac{\partial f}{\partial x}\frac{\partial x}{\partial u} + \frac{\partial f}{\partial y}\frac{\partial y}{\partial u} \end{aligned}\right\} \qquad (6.38)$$

$$\left(\frac{\partial f}{\partial x}\right)_y = -\left(\frac{\partial f}{\partial y}\right)_x \left(\frac{\partial y}{\partial x}\right)_f \qquad (6.41')$$

$$\text{and} \quad \left(\frac{\partial f}{\partial x}\right)_t = +\left(\frac{\partial f}{\partial y}\right)_t \left(\frac{\partial y}{\partial x}\right)_t \qquad (6.42')$$

Jacobian:

$$J = \frac{\partial(x, y)}{\partial(r, \theta)} = \begin{vmatrix} \dfrac{\partial x}{\partial r} & \dfrac{\partial x}{\partial \theta} \\ \dfrac{\partial y}{\partial r} & \dfrac{\partial y}{\partial \theta} \end{vmatrix} \qquad (6.67)$$

$$J' = \frac{\partial(r, \theta)}{\partial(x, y)} = 1/J \qquad (6.80)$$

$$\frac{\partial(s,t)}{\partial(x,y)} = \frac{\partial(s,t)}{\partial(u,v)}\frac{\partial(u,v)}{\partial(x,y)} \qquad \text{(Problem 6.2)}$$

Role of J in integrals with change of variables

Chapter 7 Taylor series

One-dimensional Taylor series:

$$f(a+\delta x) = f(a)+f'(a)\delta x+\frac{1}{2!}f''(a)(\delta x)^2+\frac{1}{3!}f'''(a)(\delta x)^3+\ldots+\frac{1}{n!}f^n(a)(\delta x)^n+\ldots \tag{7.8}$$

Mean Value Theorem:

$$f(a+\delta x) = f(a) + f'\delta x \tag{7.27}$$

One-dimensional maximum:

$$f'(a) = 0, \quad f''(a) < 0 \tag{7.34}$$

Two dimensional Taylor series:

$$\begin{aligned} f(a+\delta x, b+\delta y) = f(a,b) & \\ & + (\frac{\partial f}{\partial x}\delta x + \frac{\partial f}{\partial y}\delta y) \\ & + \frac{1}{2!}\left(\frac{\partial^2 f}{\partial x^2}\delta x^2 + 2\frac{\partial^2 f}{\partial y \partial x}\delta x \delta y + \frac{\partial^2 f}{\partial y^2}\delta y^2\right) \\ & + \ldots \end{aligned} \tag{7.42}$$

$$\frac{\partial^2 f}{\partial x \partial y} = \frac{\partial^2 f}{\partial y \partial x} \tag{7.45}$$

$$\text{Saddle point}: \frac{\partial f}{\partial x} = \frac{\partial f}{\partial y} = 0; \quad \left(\frac{\partial^2 f}{\partial x \partial y}\right)^2 \geq \frac{\partial^2 f}{\partial x^2}\frac{\partial^2 f}{\partial y^2} \tag{7.64}$$

$$\text{Maximum}: \frac{\partial f}{\partial x} = \frac{\partial f}{\partial y} = 0; \quad \left(\frac{\partial^2 f}{\partial x \partial y}\right)^2 < \frac{\partial^2 f}{\partial x^2}\frac{\partial^2 f}{\partial y^2}; \quad \frac{\partial^2 f}{\partial x^2} < 0 \tag{7.66}$$

$$\text{Minimum}: \frac{\partial f}{\partial x} = \frac{\partial f}{\partial y} = 0; \quad \left(\frac{\partial^2 f}{\partial x \partial y}\right)^2 < \frac{\partial^2 f}{\partial x^2}\frac{\partial^2 f}{\partial y^2}; \quad \frac{\partial^2 f}{\partial x^2} > 0 \tag{7.67}$$

Chapter 8 Lagrangian multipliers

$$\left.\begin{aligned} \frac{\partial f}{\partial x} + \lambda\frac{\partial c}{\partial x} = 0 \\ \frac{\partial f}{\partial y} + \lambda\frac{\partial c}{\partial y} = 0 \\ \text{and } c(x,y) = k \end{aligned}\right\} \tag{8.7}$$

Index